VARIORUM COLLECTED STUDIES SERIES

Before and After Darwin

M.J.S. Hodge

Before and After Darwin

Origins, Species, Cosmogonies, and Ontologies

Routledge
Taylor & Francis Group

LONDON AND NEW YORK

First published 2008 by Ashgate Publishing

2 Park Square, Milton Park, Abingdon, Oxfordshire OX14 4RN
711 Third Avenue, New York, NY 10017

Routledge is an imprint of the Taylor & Francis Group, an informa business

First issued in paperback 2018

ISBN 978-0-7546-5938-9 (hbk)
ISBN 978-1-138-37519-2 (pbk)

British Library Cataloguing in Publication Data
Hodge, M. J. S. (Michael Jonathan Sessions), 1940–
    Before and after Darwin : origins, species, cosmogonies, and ontologies.
    – (Variorum collected studies series ; no. 897)
    1. Darwin, Charles, 1809–1882 2. Evolution (Biology) 3. Cosmogony 4. Ontology
    I. Title
    576.8'09

Library of Congress Cataloging-in-Publication Data
Hodge, M.J.S. (Michael Jonathan Sessions), 1940–
    Before and after Darwin : origins, species, cosmogonies, and ontologies / by
    M.J.S. Hodge.
        p. cm.
    Includes bibliographical references.
    ISBN 978-0-7546-5938-9 (alk. paper)
    1. Evolution (Biology) 2. Cosmogony. 3. Ontology. 4. Darwin, Charles,
    1809–1882.
    I. Title.

    QH360.H62 2008
    576.8–dc22   2007052266

VARIORUM COLLECTED STUDIES SERIES CS897

CONTENTS

This volume contains xx + 340 pages

INTRODUCTION

This volume is one of a pair. The other volume will appear later this year and will include only articles on Charles Darwin's life and work. This volume collects articles concerning pre-Darwinan and post-Darwinian topics. The articles are here arranged in three groups. The two articles in the first group address big-picture, long-run interpretative issues, especially concerning theorising about origins and species from ancient times to the present day. The five in the second group concentrate on four ambitious pre-Darwinian theorists of modern times, theorists who were theorising comprehensively about the earth, life and man: Georges Buffon, Jean-Baptiste Lamarck, William Whewell and Robert Chambers. The principal writings of these four were spread over roughly a hundred years from the mid-eighteenth to the mid-nineteenth century. The articles in the third group treat of the theory of natural selection, as it was argued for by Charles Darwin in his 1859 book *On The Origin of Species* and as it has been developed and deployed since. Here the discussion often moves from the history of past science and past philosophy to the current philosophy of current science; so there is much in this third group of articles that will interest philosophers as much as historians.

The articles, identified by Roman numerals, are, with one exception, reproduced here just as they first appeared and with their original pagination. The exception is article (V) which is presented here in the English version that I wrote rather than in the translation given it for its printed publication. For some articles, the bibliographies have had to be extracted from composite bibliographies for the collaborative volumes in which the articles were published. But these extractions involve no additional editorial changes. Following this introduction, I have provided notes about the original publications of the articles.

A word may be useful here about the title given to this volume. The conjunction of the terms origins and species will seem familiar enough for obvious reasons. When 'evolutionists' debate 'creationists' today both sides are agreed that animal and plant species have had origins; where they disagree is over how species originate, over what origins species have: origins in gradual natural modifications of earlier species or origins in sudden, supernatural acts of God? The big divides have not always run on just those lines. Among the Greek cosmologists, Plato taught that species were those eternal, immaterial forms – recipes and standards with no origins – looked to by a craftsman God in working chaotic matter into an ordered world; while Democritus found origins for species in fortuitous concatenations of mindlessly-moving material atoms. To favour Platonic forms over Democritean atoms was to favour one account over another of the order of the cosmos and, at the same time, to favour one account of being, of what is, one ontology, over another. From the Greeks there descended then diverse traditions that divided over issues than are at once cosmological (concerning the nature of the world's orderliness) cosmogonical (concerning the origins of that orderliness) and ontological (concerning the being of the origins and constituents of that order). As an aside, I might add that when Geoffrey Cantor and I collaborated on writing a historical overview of ether concepts to introduce a volume of papers on that subject, we saw there too the indispensability of all those old commonplaces, for such they are, about the legacy of diverse Greek integrations of cosmologies with ontologies.

The rest of this introduction will be largely devoted to discussing in turn the three groups of articles. The first group can be read as arguing that in the history of the life sciences we still need critical historians' histories of science, alternatives to what may be called partisan scientists' histories. Partisan, participants' histories are familiar enough in political history or military history, written by the ministers or the generals who took part. And one reason we have a profession of history is to ensure that there are alternatives to those histories; and likewise for the profession of the history of science: we need alternatives to the histories the scientists write. For some periods and topics that has already been achieved and the kind of campaigning I call for is no longer needed. Writers on the new natural philosophies of the seventeenth century are no longer under the spell of Ernst Mach and other physicists – often positivistic ones – as that prominent young historian of science and Newton scholar Bernard Cohen was only six decades ago. The shift to historians' history of science has been made in some places; but in others it has still to be made. To be convinced of this all one needs to do is to look at recent standard texts by authoritative historians of science, texts whose titles signal that they chart, to use conventional phrasing, the history of evolutionary biology. For if one then compares those texts with the essays and books with similar titles written over a century ago by Darwin's defender T.H. Huxley and his associates, one finds manifest continuities in how the history is interpreted and written. Indeed, in some cases, one will find explicit recognition that the historian of science of our time is knowingly perpetuating Huxleyan historiograhical precedents.

In an article written in 2005 for a special issue of the *Journal for the History of Biology* devoted to reassessing the so-called Darwinian Revolution (an article which will appear in the partner volume to this), I was tactless. I suggested that for historians of science in our time to have made no deliberate, sustained break with that Huxleyan tradition of scientists' history of science is remarkable not to say bizarre. In 'Origins and species before and after Darwin' (I) I am not so tactless but the aim is the same: to provoke a long-overdue rejection of that tradition and a quest for alternatives. In that piece, I offer one possible alternative but without wishing to imply that only this one is all we need. The tactless stance is more evident in the second article: 'Canguilhem and the history of biology' (II). I would justify the lapses of tact in two ways. First, for good reasons, Canguilhem has enjoyed during his life and since, great, merited international academic respect. Second, again for good reasons, those folk whom one's French friends used to call the Anglo-Saxon historians and philosophers of science have often looked to the Continent for critical insights into the limitations in Anglo-Saxon views. However, neither of these considerations seems to me to be adequate grounds for not raising critical questions about just how far Canguilhem can help us as historiographers of science, especially in our efforts to make those shifts away from scientists' historiographies, shifts still needed in major areas of our discipline. It did seem appropriate therefore to ask why Canguilhem's history of science seems so often to be written in ways that remind us too much, to put it crudely, of scientists' history of science. Is it despite his philosophical sophistication, or rather because of it? The answer has to be a more complex one than I could offer in one short article. But, I argue there, it has to do, among many other things, with Canguilhem being too much under the influence of that other positivist, one Mach often sought to vindicate: Auguste Comte. The connections between positivist influences and scientists' history of science form a vast and instructive topic not faced in either of my articles. But any enquiry, into the nineteenth-century sources of those connections, would confirm that the original ideological rationale of the positivism pioneered by Comte and upheld by Mach was to legitimate the claims professional scientists were then making for unlimited autonomous and hegemonic authority for science as the only modern progressive way to think and act in any domain. So a positivist history of science must reconstruct a progress from persistent religious and metaphysical dogmas and speculations to successive scientific corrections of them; such a history is also then the most fitting, reassuring, celebratory, edifying, triumphant history for any scientists to give of their science.

Turning now to the second group of articles, there is an issue about labels and phrasing that connects directly with what has just been said about scientists' history of science. In identifying Buffon and the others as 'pre-Darwinian' as I did earlier, and in using that phrase in the title of (I), I did of course do what we should not be doing: namely, talking of the people before Darwin as those who came before him and so falling into two lots: the unwittingly prescient preparers and anticipators whose efforts were later consummated by Darwin, and the preventers, resisters and losers whose role was to be the precursors of Darwin's own opponents. There is plainly a simple antidote to this kind of talk and to the fallacious presumptions it promotes: talk, think and write about Lamarck, say, not as someone who came before some one other with one view, but as someone who came after many others with many diverse views: Plato, the priestly redactor of *Genesis*, Lucretius, Descartes, Newton, Linnaeus and Buffon and so on and on. The advantage of working at postcursorial interpretations is not merely that successions can be read the right way round, as they happened , from earlier to later, but that the narrowing imposed by one-dimensional relatings to one future development can be replaced with much broader, plural relatings to the many past resources, influences, contexts and so on being responded to consciously or otherwise, intentionally or not so, with embracings or with repudiations and so on and on. My hope then is that the articles on Lamarck and the other three can not only exemplify such familiar dicta, but do that with enough evident, specific, detailed advantages as to make those dicta look like more than just pious and truistic homilies.

The period from Buffon to Chambers, from the mid-eighteenth to the mid-nineteenth century, spans of course the Enlightenment and its sequel the Counter-Enlightenment, including the romantic movement; and no less familiarly the French Revolution and the Industrial Revolution. Or at least that is what students used to be told. Now, quite properly, it is explained that some, if not all of these labels, should go in quotation marks to signal recent questionings. So in the case, say, of the 'Industrial Revolution' there is no denial that many of the transformations historians used to write about under that heading did indeed take place, but their timings, their pacings, their causes and effects are now understood in ways making inappropriate that old label with its connotations of suddenness and totality. Historians of science can only welcome such revisions, fitting well as they often do with their revisings of older views in their own fields. Buffon is rightly seen as an Enlightenment *philosophe* of the *ancien regime*, Lamarck as no less an Enlightenment *philosophe* but, from mid-career on, as a citizen professor in the reorganised national museum of the revolutionary regime. But attempts to relate either of them to trends in urbanisation or increases in machinofacturing or the rise of the middle classes seem less plausible. And here the historian of science can be encouraged by social and economic historians' newer emphases on the role in the maturing of European modernity of agrarian capitalism and imperial capitalism. Too narrow a focus on the sources and consequences of urban, bourgeois machinofacturing capitalism can hamper and misguide any efforts to relate the natural history and natural philosophy of these epochs to their proper social and economic contexts in other forms of capitalist life.

Before moving to discuss the articles in the third group, it may be useful to indicate how far those in the first two groups connect to the doctoral thesis I wrote in the late sixties, and which was published much more recently without revision but with a new extended preface in 1991. The ancient, medieval and early modern authors treated very hurriedly in the article 'Origins and species' (I), Plato, for example, and Aristotle, Augustine, Aquinas, Descartes, Burnet and Whiston, all got much more sustained, albeit pretty amateurish attention, in that thesis. Likewise, the relations of Darwin and Wallace to Lyell, relations relevant to the Whewell and Chambers articles printed here, are given extensive treatment in that thesis, more extensive, regarding Wallace's biogeography especially, than in anything I have published before or since. The new preface provided a chance to explain how the state of interpretation on various topics had moved on in the seventies and eighties. Obviously things have moved on further now. To take just two examples: anyone studying Lamarck today goes to the indispensable website on his life and work run by Pietro Corsi and colleagues and thereby engages many archival, explicational and contextual possibilities that were hitherto much less easily available, if at all; while no one takes up Chambers as a subject today without steeping themselves in Jim Secord's marvellous monograph: *Victorian Sensation: the extraordinary publication, reception, and secret authorship of* Vestiges of the Natural History of Creation (Chicago, 2000).

Turning now to that third group of articles, I can begin with a familiar pair of points about what has happened in evolutionary biology since Darwin. The first point in the pair emphasises that Darwin's theory about the main cause of evolutionary change – his theory of natural selection – is alive and well in current textbooks setting out today's consensual conclusions. Most professional evolutionary biologists are natural selectionists, as Darwin was. So there is continuity, agreement, with Darwin on that much. The second point is that, for today's biologists, natural selection works mostly with small hereditary variations just as it does in Darwin's account; but since Darwin the understanding of the sources of these hereditary variations have been transformed by three developments: the elaboration in the early twentieth century of Mendelian cytological genetics, in the mid-century of Mendelian mathematical and experimental population genetics; and, in the later twentieth century of molecular genetics. So, on what is now called the genetics of evolutionary change, major changes in theories and techniques have come in since Darwin.

The papers in this third group contribute to this question of continuity and discontinuity between Darwin and today. On the theory of natural selection as an element of continuity, they aim to contribute in two ways. For a start, the first four papers defend an analysis of what natural selection is for Darwin and of how his various arguments for it are integrated in the one long argument of the *Origin.* Put simply the main claim is that for Darwin natural selection was not a law or a principle but a cause, a causal process or agency, just as artificial selection was for him a causal process or agency. Accordingly, he argues that this cause exists in nature, that it is capable of causing to new species to be formed in the modification of older ones, and that it has been the main process or agency responsible for producing the species that have lived on earth. In sum, Darwin adduces evidence for the existence of this cause, adduces independent, separate evidence for its power to produce species and again separate, further, evidence that it has indeed been responsible for doing so. Now, what is insisted on throughout the last batch of papers in this volume is that, as for Darwin so today, the theory of natural selection is a causal theory appropriately evidenced in just such ways; and is a theory providing, as it did for Darwin, causal explanations for the production of new species from old and so causal explanations for all kinds of generalisations – ecological, embryological , geographical etc generalisations – about species. Today then as of yore it is a causal-explanatory theory.

But now what about all those changes in thinking since Darwin about heredity? There is a simple story one can tell here and it is not entirely misleading. It says that Darwin held assumptions about heredity that did not cohere at all well with his theory of natural selection. However, the story goes on, those assumptions about heredity were discredited in the early twentieth century; so that a trio of mathematical population geneticists, Haldane, Fisher and Wright, could show that with those errors eliminated natural selection works fine with hereditary variations as newly understood.The trouble with this story is that it overlooks too many instructive complications. In one article (XII) I have tried to show how this story is too simple-minded in what it says about the changes that took place in theories about reproduction or generation and about heredity between Darwin in the 1830s and the evolutionary genetics of the 1930s. Especially I urge that in order not to be leaving vital considerations out of the history one has to focus on changes in thinking about sexual and nonsexual reproduction and even life and death themselves. And once again, the conclusion is that only by doing so can we avoid the limitations and distortions inherent in the scientists' histories of these subjects over these decades. In another article (XIII) I join others in emphasising that, in the case of Sewall Wright and Ronald Fisher, there were along with their agreements – about how to integrate the new genetics with the old theory of natural selection – deep and instructive disagreements about how to understand evolution and indeed the cosmos as a whole. For Fisher, the cosmos divides into two worlds: first, the living world subject to all the entropic (disordering) causes and subject also to natural selection, the only counterentropic (ordering) cause in the cosmos, and, second, the rest of the world which is subject to entropic causation alone. For Wright, in keeping with his panpsychist metaphysics (mind is what is ultimately real in everything all the way down to the matter of elementary particles), there is no such divide. His science contrasts not entropic and counter-entropic causes but heterogenesing and homogenising causes; and he holds that all science whether physics or biology may make any process more intelligible by disclosing how that process involves both kinds of causal tendencies in shifting balances or moving equilibria, this last a phrase drawn from Herbert Spencer. Wright's shifting balance theory of evolution is, then, not only opposed to Fisher's mass-selectional alternative to it, it is integrated with cosmological assumptions quite at odds with Fisher's own. The controversy between Fisher and Wright, still continued today by their scientific descendants, is then yet another example of the light thrown even on much modern science by legacies from ancient Greek traditions.

Going back now to natural selection as a causal-explanatory theory: two articles, (IX) and (XIV), pursue this theme in ways that connect directly with current controversies in the philosophy of biology. In (IX) I argue that two recent, general, philosophical views about scientific theories – van Fraassen's and Kitcher's – have been used to throw welcome light on Darwin's theory in the *Origin*; but that both views fail to do justice to its causal-explanatory character. In (XIV) I contrast any correctly causal-explanatory view of the theory with any view such as the Machian positivist Pearson held early in the twentieth century: namely that Darwin's theory is properly understood as all physical theories are, not as causal-explanatory but as correlational-descriptive. Now, a view of natural selection, a view reasonably called neo-Pearsonian, is currently being defended by some philosophers of biology who think that the theory of natural selection when properly understood is not a causal theory but a statistical theory. In writing (XIV) and in arguing that Pearson's view was an aberrant and misleading minority view of the theory, I did not have these more recent neo-Pearsonians in mind, for they had not arrived on the scene yet. But anyone familiar with the lively debates they have provoked in the last few years will find in (XIV) resources of a historico–philosophical kind very adaptable to countering their stimulating but mistaken neo-Pearsonian take on the theory of natural selection. Indeed, I am working towards joining in myself in future publications. In doing so I will find myself once again taking issue with the historico-philosophical legacies of nineteenth-century positivism. But I should not let that remark give a false impression. All the articles here reprinted were not written by an author consciously bent on taking up one issue after another with those positivist legacies. It is just that countering the legacies from nineteenth-century scientists' histories of science can often develop so that it overlaps with any countering of legacies from that century's positivist philosophies of science.

In closing this brief introduction, I am delighted that there is now this chance to express my most grateful thanks to Greg Radick for first thinking of this republication project, for discussing it with Ashgate and for continuing encouragement since; to Annie Jamieson for contributing so expertly to the preparation of text copies, bibliographies and other material and for the securing of permissions from holders of copyrights; to Martin Rudwick for setting decisive precedents for these two volumes in his own earlier ones; and to John Smedley, our publisher, for welcoming and overseeing the project and taking it on to the final stages.

JONATHAN HODGE

Leeds
December 2007

NOTES ON THE ARTICLES

'Origins and species before and after Darwin' (I) was written for the *Companion to the History of Modern Science* (1990), a comprehensive volume edited by four of us then at the University of Leeds, with chapters from around five dozen contributors; a main aim of the volume being to promote critical and sophisticated historiographical interpretations. Though dated somewhat now it is still, we editors are glad to observe, a widely-valued volume. 'Canguilhem and the history of biology' (II) was a contribution to an issue of the *Revue d'Histoire des Sciences* devoted, posthumously, to the work of Georges Canguilhem. All the papers were very properly respectful; mine was also critical. 'Two cosmogonies (theory of the earth and theory of generation) and the unity of Buffon's thought' (III) was originally a paper for the Buffon conference proceedings that the volume published. The volume took in all of Buffon's career and writings. 'Lamarck's science of living bodies' (IV) was, in an earlier version, a doctoral thesis chapter. 'Lamarck's great change of mind' (V) was translated into French for its publication in a volume honouring posthumously the teaching and research of Jacques Roger. 'The history of the earth, life, and man: Whewell and palaetiological science' (VI) was a chapter in a collaborative volume on all aspects of William Whewell's polymathic life and labours. An earlier version of 'The universal gestation of nature: Chambers' *Vestiges* and *Explanations*' (VII) was a thesis chapter.

'The structure and strategy of Darwin's "long argument"' (VIII) was an essay review of Robert Stauffer's edition of the vast, incomplete, previously-unpublished book by Darwin of which his *Origin of Species* (1859) was a condensed version. 'Darwin's theory and Darwin's argument' (IX) was a tribute to David Hull and appeared in a Festschrift for him. The volume includes David Hull's magnanimous responses to our essays. 'Discussion: Darwin's argument in the *Origin*' (X) was an attempt to sort out what was right and wrong in various views then current as to how that argument related to the *vera causa* ideal.

'Knowing about evolution: Darwin and his theory of Natural selection' (XI) continued that campaign; it countered, along the way, the views of Michael Ruse about Darwin's debts to Whewell, views set out in another chapter in that same volume. The editors' introduction to the volume provides an excellent guide to the issues. 'Generation and the origin of species (1837–1937): an historiographical suggestion' (XII) is from a number of the *British Journal for the History of Science* devoted to the history of genetics and eugenics and commemorating the life and work of Bernard Norton. 'Biology and philosophy (including ideology): a study of Fisher and Wright' (XIII) grew from a paper given to a conference at Boston University on the foundings of evolutionary genetics, a conference that brought historians together with practitioners such as J.F. Crow and John Turner. 'Natural selection as a causal, empirical and probabilistic theory' (XIV) was originally a contribution to a major collaborative seminar project at Bielefeld University. In addition to the two volumes on *The Probabilistic Revolution* a further volume came out of that project: *The Empire of Chance: How Probability Changed Science and Everyday Life* (Cambridge: Cambridge University Press, 1989) edited by Gerd Gigerenzer and others.

BIBLIOGRAPHY

G.N. Cantor and M.J.S. Hodge, eds., 1981. *Conceptions of Ether: Studies in the History of Ether Theories, 1740–1900*. Cambridge: Cambridge University Press.

Hodge, M.J.S. 1971. 'Lamarck's science of living bodies.' *British Journal for the History of Science* 5: 323–352.

—.1972. 'The universal gestation of nature: Chamber's *Vestiges* and *Explanations.*' *Journal of the History of Biology* 5 (1): 127–151.

—. 1977. 'The structure and strategy of Darwin's "long argument".' *British Journal for the History of Science* 10: 237–245.

—. 1987. 'Natural selection as a causal, empirical and probabilistic theory.' In L. Krüger, et al., eds., *The Probabilistic Revolution* (2 vols.), vol. 2, pp. 233–270. Cambridge, MA: The MIT Press.

—. 1989. 'Darwin's theory and Darwin's argument.' In Michael Ruse, ed., *What The Philosophy of Biology Is: Essays Dedicated to David Hull*, pp. 163–182. Dordrecht: Kluwer.

—. 1989. 'Generation and the origin of species (1837–1937): a historiographical suggestion.' *British Journal for the History of Science* 22: 267–281.

—. 1990. 'Origins and species before and after Darwin.' In R.C. Olby, G.N. Cantor, J.R.R. Christie and M.J.S. Hodge, eds., *Companion to the History of Modern Science*, pp. 374–395. London: Routledge.

—. 1991. 'The history of the earth, life, and man: Whewell and palaetiological science.' In M. Fisch and S. Schaffer, eds., *William Whewell: A Composite Portrait*, pp. 255–288. Oxford: Clarendon Press.

—. 1991. *Origins and Species: A Study of the Historical Sources of Darwinism and the Contexts of Some Other Accounts of Organic Diversity from Plato and Aristotle on*. New York: Garland.

—. 1992. 'Two cosmogonies (theory of the earth and theory of generation) and the unity of Buffon's thought.' In Jean Gayon, et al., eds., *Buffon 88: Actes du colloque international pour le bicentenaire de la morte de Buffon*, pp. 241–254. Paris: Librarie Philosophique Vrin.

—. 1992. 'Biology and philosophy (including ideology): a study of Fisher and Wright'. In Sahotra Sarkar, ed., *The Founders of Evolutionary Genetics: A Centenary Reappraisal*, pp. 231–293. Dordrecht: Kluwer.

—. 1992. 'Discussion: Darwin's argument in the *Origin.' Philosophy of Science* 59: 461–464.

—. 1995. 'Lamarck: un grand changement de cadre conceptuel [Lamarck's great change of mind]'. In C. Blanckaert, J-L. Fischer and R. Rey, eds., *Nature, Histoire, Société: Essais en homage à Jacques Roger*, pp. 227–239. Paris: Klincksieck.

—. 2000. 'Canguilhem and the history of biology.' *Revue d'Histoire des Sciences* 53: 65–81.

—. 2000. 'Knowing about evolution: Darwin and his theory of natural selection.' In Richard Creath and Jane Maienschein, eds., *Biology and Epistemology*, pp. 27–47. Cambridge: Cambridge University Press.

—. 2005 'Against "revolution" and "evolution".' *Journal of the History of Biology* 38: 101–121.

Secord, James A. 2000. *Victorian Sensation: The Extraordinary Publication, Reception, and Secret Authorship of* Vestiges of the natural history of creation. Chicago, IL and London: University of Chicago Press.

ACKNOWLEDGEMENTS

Grateful acknowledgement is made to the following institutions and publishers for their kind permission to reproduce the articles included in this volume: Thomson Publishing Services, London (article I); Presses Universitaires de France, Paris (II); Librarie Philosophique Vrin, Paris (III); Cambridge University Press, Cambridge (IV, VIII, XI, XII); Klincksieck, Paris (V); Oxford University Press, Oxford (VI); Springer Science and Business Media, Berlin (VII, IX, XIII); University of Chicago Press, Chicago, IL (X); The MIT Press, Cambridge, MA (XIV).

PUBLISHER'S NOTE

I

ORIGINS AND SPECIES BEFORE AND AFTER DARWIN

1. AN ALTERNATIVE TO THE DOMINANT HISTORIOGRAPHIC TRADITION

Most languages distinguish different kinds of things and most peoples have theories about the origins of things, including the origins of animal and plant kinds, their diverse designs for living and diverse degrees of life, from the highest to the lowest. Diversity in designs and degrees has been an explicit preoccupation throughout the long run of Western thought. This preoccupation has always concerned the nature and fate of our own species, and so the understanding of the soul, morality and polity. The Jewish account of the Creation includes the forming and the falling of the first humans, a narrative later elaborated into the Christian doctrine of original sin. Among the Greek philosophers, Aristotle taught that any species is distinguished by the purposes embodied in its natural actions; and this teleology of specific natures provided foundations for his politics, his views on man as a political animal, no less than for his interpretation of plant reproduction.

Today, as for over a century now, such issues are always likely to be subsumed under two catchwords: evolution and Darwin. For even people who find neither congenial are conditioned in their belief and attitudes by both. Whenever a culture engages modern Western science, there to be confronted is evolution: the proposal that extant species, including man, have arisen by the modification of earlier ones, in an extremely long process that began with a few simple forms of life – a proposal first made scientifically respectable, and also more generally notorious, by Charles Darwin (1809–82) in his book *On the Origin of Species* (1859).

Ever since the great debates of the 1860s, and initially in efforts to win them for Darwin, people have been rewriting the history of Western thought as the history of *evolution*: the rise of *evolution*, as an idea or world-view or whatever, and its eventual triumph, with Darwin, over its natural contrary, *creation* or *stasis* or some combination of both, *static creationism*, say. Surprisingly, that historiographic tradition continues to dominate even current specialist writing. The present article will be more easily understood if it is explained straightaway how it seeks to show that at least one alternative to that tradition is possible.

The trouble with all *evolution* historiographies is that they project back into earlier centuries the issues and distinctions, the conflicts and the alliances, that were decisive for the great debates of the 1860s. One way to avoid this fault is to adopt an *origins* and *species* historiography instead. This has at least three advantages. Firstly, we are working with words and therefore issues that the people saw themselves as working with at the time, any time at least since the middle of the last millenium BC. Secondly, we can operate not with a binary scheme of contrary options, but with as many differences as we find reason to recognise among the positions that have been held on questions about origins and species. Thirdly, we can interpret any steps taken away from earlier positions as they were taken, that is as deliberate departures from what was previously held, rather than as so many unknowing steps towards the options, evolutionary or otherwise, that will later be contested in the wake of Darwin's work.

These three points are to be kept especially in mind as we turn to Medieval Christian attempts to integrate Greek and Hebrew intellectual legacies concerning questions of origins and species. Consideration of such attempts is indispensable for any understanding of how Western scientific thought on these questions has developed in the four centuries since the Renaissance.

Plato's theory of Forms or Ideas was a theory both of origins, *archai*, and of species, *eide, gene*. Moreover, it provided for objective moral and political values, as well as for the foundations for the mythic cosmogony of his dialogue, *Timaeus*. For not only does the Form Justice constitute a standard for judging acts in so far as they are just, the Forms of natural kinds – Fire, Horse, Man and so on – were the recipes and standards looked to by the Craftsman when he made an orderly cosmos by imposing on a chaos the order eternally available in the Forms. Plato's pupil Aristotle also took forms to be *archai*. But he rejected Plato's view that forms can exist separately from formed individuals; Horse and Man as forms are not, he held, separable from individual horses or men; and he denies that the cosmos ever had a beginning in time. The form of an individual has an origin in the generation of the individual from its parents, but man or horse as species are perpetuated, not originated thereby. There is then, for Aristotle, a cyclic perpetuation of forms, in species life cycles, among animals and plants, matching the endless, everlasting cycling of the divine, unchanging heavenly bodies, the planets and stars. Forms, as species, are *archai* of formed individuals, and are origins so original as to have, themselves, no origins.

By contrast, for the Greek atomists, as represented most fully in the Roman Lucretius's long poem *De Rerum Natura*, species are in no sense *archai*. For the atomists divided and distributed Being not into so many distinct Forms, as did Plato, but into myriad uncuttable chips of matter, the atoms. For that is all there is: material atoms moving mindlessly in the beingless void of space. Species must then be referred to atoms, as the *archai* for everything. Lucretius finds in the motions and shapes of atoms his explanation for why there are the species of animals and plants there are. When the earth was young and hotter than now, all sorts of permutations of animal and plant parts were formed on the earth's surface. Only some, however, could survive and reproduce to continue their lives in stable species, and it is those that have lived ever since.

Any Medieval Christian found many themes affirmed in the Biblical account of God's making and populating the heavens and the earth, the account set forth in the first thirty-five verses of the book of Genesis. God is supreme and worked without assistance; the world is a good work; plants and animals have been fitted to life on land and in the seas, and following their first appearance have perpetuated their species under God's command to increase and multiply; man is unique in being made in God's image; the Sabbath is a special day of rest; the work of making the heavens, earth, plants, animals and men was completed then and God has done no comparable work since.

As a representative writer on creation in the thirteenth century, Thomas Aquinas (1225–74) is typical in continuing the preference for the Platonic over the atomist tradition, but he is an innovator in being the first to give a sustained explication of the work of the six days as the creation of an Aristotelian cosmos. Confronting the issue of why a single, simple God could and would create many differing creatures, Aquinas upholds the Platonic tradition in arguing that God can do so, because the singularity and simplicity of his essence is imitable in creatures in many diverse ways existing as so many distinct Ideas in the divine mind; and in arguing that God would do so, because he creates so that his goodness shall be communicated to creatures and represented in them, and because that goodness can only be represented by a multitude of creatures distributed in the scale of perfections, from the lowliest of the corporeal to the highest of the angelic species. The distinction of creatures according to their species must be God's work and not left to any delegated agency, because, Aquinas insists, a craftsman who intends to produce an ordered whole intends all the elements of its order; and the ordered elements that together constitute the order of the universe are the specific natures forming the scale of natures.

For Aquinas, therefore, the doctrine of God as the *arche* without *arche*, *principium* of all else, is integrated with Plato's view of forms as exemplars of the cosmic Craftsman's ordering work and with Aristotle's view of forms as sempiternally embodied natures. The work of the six days is thus constitutional work for species. In introducing forms into matter, in working initially to instantiate these exemplars, the Divine Ideas, God is giving species their constitutions. Then, once complete with the creation of man, God's working has been administration, sustaining and governing species in accord with their constitutions given them during the six days. No introduction of new forms into matter can come within the administrative course of nature, because such introductions are completed in the constitutional work. Here, therefore, species, as the *naturae* of the *scala naturae*, are the ultimate constituents of the very order of the cosmos. Aquinas, as an Aristotelian, believes that for God to form a cosmos is for Him to distinguish in matter a central, terrestrial realm from a surrounding celestial realm replete with heavenly bodies and angelic souls, and for Him to complete the cosmos by adorning the earth with species of all the animal and plant natures in the full *scala*. Thus, the forming of the earth and its first stocking with life is constitutional work.

Christian writers in the Middle Ages made various selective uses of Greek traditions concerning origins and species, both in making sense of the general teaching in Genesis that God was the ultimate origin of everything and therefore of specific diversity, its degrees and designs; and in making sense of the difference between what was involved in the originating work of the six days and the way the world has run in the four or five thousand years since. Only by keeping these uses of these traditions in mind can we understand the implications, for Western thought about origins and species, of the fundamental shifts in cosmology and natural philosophy that took place in the seventeenth century.

2. ORIGINS AND SPECIES IN THE AGE OF ENLIGHTENMENT

To understand the implications of these shifts in thought, we should start with René Descartes (1596–1650). For, according to Descartes, the earth is not stationary and central in a finite spherical cosmos, but is a planet circling the sun in a boundless material universe whose workings are ultimately the mechanical transactions of a universal matter of a single common essence, transactions conforming to a few universal laws of motion. Descartes's cosmogony can therefore differ fundamentally from Aquinas's, for it can treat the earth's origin within a general theory of planet formation. Any planet has arisen from a swirling vortex of material particles. The individual case of the earth's formation, including the acquisition of its shape, atmosphere, mountains, soil and inhabitants, can be given, then, in a separate narrative. Because Descartes is a mechanical philosopher, the earth's formation need not belong to the original imposing of lawful motion on matter, but can come much later in the subsequent governing of matter in conformity with those laws. The Cartesian geogony was, explicitly, only a hypothetical narration, only a demonstration, as Descartes puts it prudently, of how God, had He so chosen, could have produced the earth and its inhabitants gradually by means of natural, secondary causes; rather than working, as Descartes assures us He really did, a miraculous creation in the Biblical six days.

Thomas Burnet (1635–1715) was the first deliberately to reject this contrast between hypothetical Cartesian narration and sacred Biblical history. The new astronomy, he says, allows answerable questions about terrestrial history to be separated from unanswerable mysteries about the earlier origin of the whole universe. Genesis concerns only this, our earthly world. By referring the events it records to lawful Cartesian matter in motion, a hypothesis can be had as to how a chaos became this habitable world. An early earth, enjoying a perpetual spring thanks to rotation at right angles to its orbit, would provide the warmth and nutrition of a womb. The sun could naturally initiate generations of the first animals and plants in this soil. New species never arise today, but they did once and all at once. Nature's laws are constant, but the conditions natural agencies operated in made them more productive then than nowadays.

Newton's protégé William Whiston (1667–1752) upheld the Newtonian rather than the Cartesian account of matter and motion. But, like Burnet, he assumes that Moses treats only of our planet and its origin from its own chaos. Moses does not say, Whiston notes, what the earth owes to nature and what to miracle. Among the items that Whiston decides are beyond nature's ability, and so due to God's direct, miraculous action, are Adam and Eve, their bodies and souls, and the production of all the animals and planets that were ever to live on the earth. Whiston holds that these were formed minutely encapsulated in the first members of each species, in accord with the new boxes-within-boxes *emboîtement* theory of pre-existent germs.

Whiston's official Newtonian replacement for Burnet's theory of the earth shows that the eighteenth century inherited from the seventeenth a host of new issues, at once historical and causal, concerning two cosmogonies: the macrocosmogony of the origin of the solar system in general and of the earth in particular, and the microcosmogony of any individual animal or plant as it appears to move from seminal chaos to adult organisation. Many ambitions motivated Georges Buffon (1707–88) to write what he did in the many volumes of the *Histoire Naturelle*, which began appearing in 1749 and was incomplete still at his death. But one ambition throughout was to uphold Newton's account of matter and motion – and force – while developing natural Newtonian macrocosmogonical and microcosmogonical alternatives to all Divine miracles, or Biblical chronologies, such as Whiston had included in his integration of Moses and *emboîtement*.

Buffon's alternative to pre-existent boxed germs was his theory of organic

I

molecules and organic moulds. The molecules are miniscule, universal components of living bodies. The diversity of living species is grounded, therefore, in the diversity of the ordinary, brute matter with which the organic molecules are associated. The distinctive combination of organic molecules and brute matter that forms the body of an organism acts as an internal mould. By internal mould, Buffon means a constellation of active, penetrating forces, analogous to gravitational force. Such an internal mould goes to work on incoming nutrients, which will include more organic molecules, and moulds are stable in that they make more of themselves.

Organic molecules can differ in their activity and their numerousness. The organisation of the highest animals is due to many active molecules and their moulds are the most stable. Within lower grades of life there is less stability, and so one finds families of species that have diversified from a common stock. The numerousness and activity of the organic molecules depend in turn upon heat. Gravity, which is attractive, and heat, which is repulsive in its action, are the ultimate agents in nature for Buffon. The history of nature's productions has then to be reconstructed as the history of the diverse effects of gravity and heat; and, since gravity is constant and never wanes, the changes in the living world must be referrable to changes in heat. Buffon is resolute in his thermal determinism. The same degrees of heat, he says, would, other conditions being the same, always produce the same grades and types of organisation. Eventually, after conducting experiments on cooling spheres, Buffon is ready to provide a chronology for the habitability of the earth, and indeed of other planets. His chronology – which has the earth first cool enough to sustain life tens, perhaps hundreds, of thousands of years ago – is a direct alternative to all Scriptural chronologies which it challenges throughout.

The early heat of the nascent earth formed the first organic molecules from unorganised matter, and these molecules organised themselves into the first species. Nature was then in her first force and elaborated organic matter with a greater heat; hence the early, giant, extinct quadrupeds and molluscs that we find today only as fossil remnants. As Buffon's treatise on the *Epoques de la Nature* (1779) proceeds, it explains how the first and hottest land was formed in the northern regions. In the Old World the highly organised products of that first land were able to migrate into the younger land formed to the South; hence the grandeur of African fauna. By contrast the mountains of Central America blocked migration in the New World. The South American fauna is less grand, because it was formed from organic molecules produced there, when nature was older and cooler. The llama differs from the giraffe because it is due to a sparser and more feeble stock of organic molecules. Thus did Buffon find, in the history of Newtonian heat and gravity, sources for the diversity of the species that have originated at different times and places on our planet.

The great innovators of the seventeenth century, most obviously Descartes

379

and Newton, to whom Buffon was heir, had seen themselves as replacing once and for all the Aristotelian cosmology. Moreover, in the theory of matter, Newton, especially, had drawn knowingly although not exclusively, on Greek atomism as transmitted by writers such as Gassendi. But it would be a mistake to think that we have with Buffon a simple substitution of the Lucretian heritage for the Aristotelian. The *Epoques de la Nature* is not a rewriting of *De Rerum Natura*. Enlightenment cosmogonists were not reduced to making new choices among old, classical options. There is, however, this element of accord between Buffon and Lucretius: for both, as contrasted with Aristotle, species are not origins in a way that precludes them having origins within the workings of nature.

Just as we can not identify Buffon simply as Lucretius updated, nor can we characterise his celebrated contemporary Carl Linné (Linnaeus in Latin: 1707–78) as a recycled Aristotle. Certainly, Linnaeus's main aim and role was to reform and consolidate a tradition in natural history that he knew went back to Andrea Cesalpino (1519–1603), a man who had, in turn, appointed himself to bring Aristotle's logic of classification to bear on the classifying of plants, on the understanding that, as Cesalpino saw it, in doing this one was displaying the constitution of an Aristotelian cosmos. Certainly, too, Linnaeus had no time for the theories of the earth in the cosmogonical sense of Burnet, Whiston and Buffon. But his cosmogonical reticence and his commitment to the Aristotelian taxonomy inaugurated by Cesalpino, did not mean that Linnaeus always took the species living today, and demanding classification into genera, families and so on, as merely given; as perpetuated by nature now, having been initiated by God as so many distinct kinds tracing to the first, Mosaic week of creation. By the end of his life, Linnaeus was proposing that there are many more species now than God made in that first week, because within each of the broad groupings – families and orders – later species have arisen from the hybrid crossings of earlier ones. Sexual generation, newly found to be as ubiquitous among plants as among animals, is a means not only for species to perpetuate themselves but also to join in making others. (See art. 19.)

Buffon and Linnaeus set very contrasting precedents, both for systematic taxonomy and for the understanding of origins and species. Jean Lamarck (1744–1829) had debts to both men in his early work, but in 1800 he began taking a position that is best understood as a new Newtonian alternative to Buffon's Newtonian account of life's place in nature.

For Lamarck's most general claim, eventually made quite explicit, is that, yes, nature – which is to say the collaboration of attractive and repulsive forces of heat and gravity – has indeed produced all bodies, living and lifeless, organised and otherwise, the animals and planets no less than the minerals. However, among living bodies, nature can only make the simplest directly, in direct spontaneous generations from lifeless matter; the more complex have therefore been produced indirectly through the gradual complexification of these over vast eons of time. Where Buffon has nature making organic molecules from brute matter and then these molecules suddenly making animals and plants, simple and complex, Lamarck has the simplest organisms formed as intermediaries and the rest formed gradually from them. Buffon although correctly Newtonian about the nature's agencies, is incorrect about their powers and their history, according to Lamarck. The earth's heat is constantly revived by the circulation of fire to and from an unfading sun; so nature now can do no more, and no less, than at any time in our planet's apparently unlimited past. Always obliged to begin the production of living bodies with the simplest, she has always to produce the higher forms of life by modifying the simplest in a progressive, successive production of the classes forming the two or more series of animal and of plant organisation. This successive production of the several class series involves a successive production of species that are unlimitedly mutable.

This unlimited mutability serves two complementary ends for Lamarck. Firstly, fossil remains of animals that match no known living species need not be interpreted, as they were by Buffon, as relics of species that have failed to survive changing circumstances and ended in extinction; but as relics of species that have survived by being changed by those circumstantial changes. Secondly, the serial rankings of the main classes of animals and plants can be interpreted as giving the order of their production over long periods of time, an order that is not determined by circumstantial change but by a tendency of the movements within the living bodies that constitute life itself, a tendency to produce increases in organisational complexity that are conserved and so accumulated in reproduction. Species have the successive origins they have in Lamarck, because the indirect production of all living bodies except the simplest has necessarily been successive.

Lamarck was exceptional in his generation in being a professional zoologist who responded to the full range of natural philosophy and natural history integrated in Buffon's macrocosmogonical and microcosmogonical hypotheses. The norm among those in the early nineteenth century who saw themselves contributing to zoology or botany or geology as branches of knowledge, as sciences, appropriate to professorial profession, was to pursue such sciences in ways that postponed or circumvented direct sustained engagement with the problems those hypotheses purported to solve.

Georges Cuvier (1769–1832), Lamarck's Parisian colleague, developed the Linnaean quest for a natural taxonomic system in bringing to it the comparative anatomy of internal organisational structures and functions. He extended the results to cover the fossil animals, insisting that they were indeed often remains of extinct species that had lived and died at different periods in the earth's long past. He was prepared to imply that the oldest extinct species had been replaced by newer, younger species; but he declined to say how such replacements took place, insisting that the origin of life and organisation lay beyond science, even inclining to the pre-existence of germs with its corollary that the origin of individual organisation lay there too. Alexander von Humboldt (1769–1859) likewise insisted that the science of the geographical distribution of plants and animals is concerned with where species have originated, but leaves as a mystery beyond science the causes for their originating when and where they have.

Lamarck did have allies, if not followers. Another Parisian colleague, Etienne Geoffroy Saint-Hilaire (1772–1844), accepted the indefinite modifiability of species and supported that thesis from his work on embryology. However, the precedent set by Buffon's integration of the theory of the earth and theory of generation was not an invitation most professional naturalists and natural philosophers were inclined to accept. Even those, many of whom were in Germany, who were pursuing new interests in comparative embryology on the assumption that pre-existent germs were discredited once and for all, did not move willingly and comprehensively from their theorising about generation to the theory of the earth. Developments in geology were, however, opening up new contexts of inquiry where integrative efforts might find a place. It was in such contexts that Darwin was to pursue his work on origins and species.

3. THE EMERGENCE OF DARWIN'S ZOONOMICAL PROGRAMME

Darwin has often been portrayed as an amateur, gentleman beetle collector who had learned little from his formal education, and who was innocent of intellectual ambition or ideological orientation. But such portrayals are now known to be highly misleading. A better sense of the man's early years can be had by noting the obvious biographical landmarks. His school years were followed by, successively, five student years (1825–31) at Edinburgh and then at Cambridge, five voyage years (1831–36) and then five London years (1837–42). It was during the London years that he arrived at almost all the main theoretical conclusions that were to dominate the rest of his life, especially the theory of natural selection – his theory of the origin of species – and, very probably too, pangenesis – his theory of reproduction, including heredity and variation. Even before going to Edinburgh, he had developed serious scientific interests, especially under the influence of his older brother Erasmus who went with him to study at the Scottish university. Once there, he engaged, through reading, lectures and field trips, a range of scientific issues, especially in general zoological theory. Likewise, at Cambridge the botanical lectures he attended provided far more than an introduction to taxonomy and field work. Again, the voyage years were as much years of reading (there were some four hundred volumes on the boat) and writing and reflecting as they were of observing and collecting. It is, then, simply a mistake to think that Darwin's biological theorising is confined

I

to later London years that followed an earlier phase of observation by a naïve voyaging naturalist.

It is also a mistake to make Darwin the young theoriser nothing but a precursor of the mature author. We should not reduce our study of his earlier years to a hunt for the germs of those ideas that are familiar from his later books. To do this would be to overlook much that mattered to him when he first began working on issues of general theoretical import.

We can confirm these biographical and historiographical themes by beginning our narrative with Darwin in mid-1834, when he was halfway through his voyage and about to leave eastern South America for the western coast. His main writing throughout the voyage was in two scientific diaries (still unpublished today): one on zoology, one on geology. A good way to grasp Darwin's permanent preoccupations as a biological theorist is to see what issues of theory were dominant in these two diaries in 1834.

One cluster of issues traces directly to what can be called his Grantian generational preoccupations. For at Edinburgh he had learned from Robert Grant (1793–1874) to compare and contrast various modes of generation or reproduction, sexual and otherwise, in invertebrate animals, and to reflect on the way a colony of more or less independent individuals often grows and acts in a co-ordinated way as if pervaded by a single principle of life or vital force. Thanks to this Grantian heritage, Darwin in mid-1834 was investigating whether the matter making up the central living mass of a polyp colony was growing continuously so as to connect the component individual polyps; and he was concluding that the gemmules, minute buds, whereby many invertebrate animals reproduced, often arise from a distinctive granular matter in the parent.[1]

As for geology, by mid-1834 Darwin was embracing the controversial theoretical teachings of Charles Lyell (1797–1875) as set out in his *Principles of Geology* (1830–33). All the causes of change are presumed, in Lyell's system, to persist undiminished into the present, the human period, and on into the future. Now, as at all times, habitable dry land is being destroyed by subsidence and erosion in some regions, while it is being produced by sediment consolidation, lava eruption and elevationary earthquake action in others. Equally, Lyell has the long succession of faunas and floras brought about by a continual, one-at-a-time extinction of species and their replacement by new ones. In defending this last thesis about species replacement, Lyell had explicitly countered the views of Humboldt and others that the origins of species were mysteries beyond all science. The geologist, Lyell insisted, has to consider whether species are now coming into being and what is determining the timing and placing of these species' origins. He had considered at length Lamarck's system and rejected it in favour of a separate creation of fixed species. How these creations occurred Lyell left unspecified, but he did lay down that adaptational constraints alone determine when and where species originate and so determine which genera and families of species originate in which areas of the world.

In mid-1834, Darwin had no disagreement with these Lyellian doctrines. By mid-1837 he had shifted far away, and knowingly so, from Lyell's account of the organic world, while remaining committed to Lyell's views on the inorganic, physical world. To understand Darwin's moves away from Lyell's views on the world of life, we have to see that they were being made by the same Darwin who was preoccupied with the issues deriving from his Grantian generational heritage. Nothing conditioned Darwin's entire life as a scientific theorist more than his lifelong habit of integrating and adjudicating between these two (Scottish) intellectual legacies: the Lyellian-geological (including biogeography and ecology) and the Grantian-generational.

A first break from Lyell came in 1835, when Darwin rejected Lyell's view that species become extinct when physical changes in an area lead to competitive imbalances and so competitive defeats, wherein some invasions of alien species succeed and some retreats of other species do not. Darwin favoured instead a generational theory of extinctions, whereby the succession of individuals in a species is propagating a limited, original duration of life, and so dies out eventually of old age just as a single animal does or a grafted succession of plant shoots had often appeared to.

A second break came, it seems, in mid-1836 shortly before the voyage ended. For it was then, there is reason to believe, that Darwin first inclined toward the view that new species arise by the modification of earlier ones. The rationale for this inclination was most likely a conviction that there were resemblances between very similar but distinct congeneric species that could not be explained, *à la* Lyell, as due to adaptation to similar conditions; for these species had evidently originated in areas with very different conditions, desert, rain forest and so on. If new species can arise in the modification, the transmutation, of earlier ones, then such resemblances can be explained as due to heredity rather than adaptation.[2]

The third and final break with Lyell came in the spring of 1837, a few months after Darwin's return to England. New judgements by London naturalists, especially on the birds and fossil mammals he had collected, seem to have convinced Darwin that his 1836 rationale for favouring transmutation was dramatically vindicated by several groups of biogeographical and paleontological facts. Accordingly, he decided to go all the way with transmutation and to construct a system of laws of life, a zoonomical system, as comprehensive as that system of Lamarck's which Lyell had rejected.

Darwin's zoonomical system, which is articulated in the first two dozen pages of his *Notebook B*, opened in July 1837, is a generational system in its two most general claims. Firstly, change in species and so adaptation, is credited ultimately to the distinctive powers of sexual generation, as contrasted with all asexual modes of generation. Secondly, all organisation, high and low, is traced to organisational progress beginning with the simplest organisms of all – monads, so called – whose eventual issue over eons of time is a branching propagation of species, analogous to the branching propagation of buds whereby a tree grows. Thanks to sex, a tree of life can arise from a monad.

From the start, Darwin was explicitly including in his system man and his mental life. Within a year he was also explicit in taking a materialist view of human mental, including moral, faculties. His metaphysics, at this time, was theistic as well as materialistic, however. The lawful causation whereby species are adapted and diversified is taken to be benevolently instituted originally by a Divine Creator.

His notebook programme set itself two tasks. The first was to understand how sexual generation in changing conditions allowed species to change into new species adapted to the new conditions. The second was to show how far otherwise inexplicable facts in biogeography, paleontology and morphology can be explained on the theory that many new species could have arisen in the branching and rebranching descent from a single ancestral stock. Here Darwin extended quite generally the form of argument first encountered in his use of transmutation in resolving biogeographical problems. Resemblances – for instance, in the forelimbs of bats, moles and monkeys – are often not explicable as adaptations to a common way of life, for those lives differ; they are explicable as a shared heritage from a common ancestor. Differences – among those fore-limbs, say – can be referred to adaptive divergences in distinct lines of descent from a common ancestral species.

The branching tree of descent arises from the reiteration of adaptive species formation. The most general quest is, therefore, for causes that are adequate for the formation of species and their unlimited, arboriform, adaptive diversification into distinct genera, families and so on.

4. THE CONSTRUCTION, PRESENTATION AND RECEPTION OF THE DARWINIAN THEORY

Darwin's initial presumption, in 1837, was that sexual generation would ensure the adaptiveness of change, because in changing conditions adaptive innovations of character would be impressed on the maturing organisation of the offspring from sexual generation, and only adaptive acquisitions would be transmitted to future generations.

The movements Darwin made away from this view in the next year, 1838 – movements which eventually brought him to his theory of natural selection – are much more complex than even many specialist writers on Darwin seem willing to recognise. Misled, still, by Darwin's later brief recollections, they often imply that Darwin read one book, Thomas Malthus's *Essay on the*

Principle of Population (the fifth edition, 1826), and on one day, September 28th, 1838, in a few notebook sentences, reached the theory then and there in all its essentials. The truth is, however, that Darwin's theory of natural selection was not arrived at until late November or early December that year, and that it was not, indeed could not have been, reached in the way that various versions of the myth of a single moment of Malthusian insight all imply.[3]

Certainly, the reflections on reading Malthus were decisive. Up to then, Darwin inclined to think that a whole species is able to adapt to changes in the conditions throughout its area, because local adaptations to local changes are eliminated, as are occasional congenital peculiarities, by crossing with unaffected individuals in the same species. Malthus impressed him that were it not for limited food, populations – in animals as well as man – would increase at very great rates. So, in the struggle to survive and reproduce, many lose out in each generation. Here, then, is a force making for adaptation, through the elimination in a few generations of maladaptive variations and the retention of adaptive variations.

This conclusion left in place, however, Darwin's confidence that the impressionable maturations and blending crossings, distinctive of sexual as contrasted with asexual generation, are what ultimately makes adaptive change possible. The Malthusian crush of population supplements this causation, but its inclusion in Darwin's theorising calls for no rethinking of the necessity and efficacy of sex. Such a rethinking happens two months later, when Darwin comes for the first time to think that species in the wild owe their adaptive formation to a process analogous to the selective breeding of hereditary variants practised by animal breeders, in producing distinctively adapted races of domesticated species: greyhounds and bulldogs, race horses and draught horses, for instance. With this selection analogy, Darwin sets aside the assumption that adaptation is ensured by the maturation and crossings distinctive of sex; for, rather, adaptation – whether to the needs of a wild species or to man's ends – now depends on the selection to which any variation is subject, even variation that may be arising by chance, and not as an adaptive response to changed conditions. Species are now, for Darwin, like the races of dogs and horses, in being adapted products of selected breeding; but they are more slowly and therefore more permanently adapted; and they are more discriminatingly, consistently and comprehensively selected and therefore more perfectly adapted.

It is this argument concerning chance variation, adaptation and selection that remains for Darwin, ever thereafter, as constituting his theory of the origin of species. Over the next twenty years, he refines it, extends it and supports it in many new and significant ways, while never quite getting around to publishing it.

He was prompted finally to prepare his book, *On the Origin of Species*, for publication in 1859, when, quite unexpectedly the year before, he received a

I

sketch of a theory of natural selection very like his own theory, sent by a younger English naturalist Alfred Russel Wallace (1823–1913), who was then working at natural history in the East Indies. Wallace's sketch and some manuscript pieces by Darwin were read as a joint paper at the Linnean Society and appeared in print in 1858. Wallace was like Darwin in following the line of historical, biogeographical enquiries opened up by Lyell, and he was like Darwin in taking a materialist view of man and mind. There was, however, nothing in Wallace corresponding to the generational preoccupations that can be traced in Darwin back to Edinburgh and medicine. Wallace was content to be a junior partner in the whole business, rather than claiming equal entitlement; and, although members of the scientific community were aware that the theory had been converged on independently by two men, they and the wider public associated it principally with Darwin.

The argument of the *Origin* makes three successive cases for natural selection: that it exists, that it is sufficient to produce, adapt and diversify species, and that it has probably been the main agent responsible for producing the species extant today and those extinct ones found as fossils. The existence of selection in nature is argued for on the grounds that it follows from the existence of hereditary variation, together with the struggle to survive and reproduce that is entailed by the powers of reproductive multiplication and limited resources, especially food. The adequacy of natural selection to produce species is defended by arguing that man, as a selective breeder, had produced distinctive races adapted to different uses within domesticated species; and by arguing that because natural selection is far more prolonged, precise and comprehensive as a selective breeding process, it is able to produce not only distinct races, but distinct species, starting from a single common ancestral species. The races produced by natural selection will meet the criteria for counting as species: inability or disinclination to interbreed; distinction by character gaps with no intermediate varieties; and true breeding, that is, perpetuation of characters over successive generations.

The third and final section of Darwin's argument presents a revised version of the view he had taken back in 1837. The theory that species have arisen in branching divergences due mainly to natural selection is more probable than the theory of separate creations of fixed species, because the branching divergence theory allows for more facts of various kinds – biogeographical, morphological and so on – to be explained by being referred to general laws rather than to particular Divine willings.

The reception of Darwin's proposals was also a much more complicated affair than is sometimes appreciated. Too often, a single, unrepresentative moment is taken to indicate the overall trend. That moment is the meeting at Oxford of the British Association for the Advancement of Science, where Darwin's supporter T. H. Huxley clashed with the Bishop of Oxford, Samuel

Wilberforce and, according to legend, won the day for Darwin against the Bishop's reactionary religious prejudices and flippant debating ploys. One way to move towards a less misleading account of that occasion in particular, and of the issues generally in play at that time, is to see that Darwin's argumentation was often raising questions of authority, especially the authority of physics on the one hand and of the churches on the other.

The authority of what was taken to be the best science of the day, namely the Newtonian celestial mechanics of the solar system, was such that reviewers of Darwin's book and appraisers of his theorising were expected to ask how fully Darwin was satisfying the standards set by this, the best science. They often concluded that his science was defective by those standards. A main difficulty was in supporting the theory by comparing its consequences with relevant, ascertainable facts. With Newton's theory of gravitation, it seemed possible to deduce precisely the consequences of the forces acting under specifiable conditions; to establish when those conditions have been met in the world, and to confirm the theory by showing that its consequences are matched by what has occurred under those conditions. By contrast, what the consequences of natural selection would be under any specifiable conditions was not easily decided from the theory itself. What is more, comparing those consequences with the changes species have undergone was not at all easy, for the obvious reason that there was no direct access (via a moving picture record of life's changes in the past) to those changes. A theory, it was implied by the critics, that was subject to these difficulties was not sufficiently like Newton's physics to count as good science. This criticism raised, therefore, the question of whether a scientist working on the problems Darwin had addressed should be expected to submit to the authority of physics.

The authority issue as it concerned the churches took a different form. It had long been agreed between some men of science, at least, and some men of the church that some possible conflicts of authority should be avoided by a suitable distinction of responsibilities. For instance, geologists had concerned themselves with the history of the earth and its animal and plant inhabitants before the arrival of man; leaving churchmen, including those scientists who were churchmen, responsible for the subject of man's life on earth under the moral government of God, a government that had no place in the pre–human epochs. The Darwinian camp's willingness, signalled already in the *Origin*, to subsume our species under the general theory of descent obviously indicated a deliberate unwillingness to negotiate any comparable divisions of responsibility in constructing and presenting that theory. Darwin's general account of species, independently of the case of man, made trouble for various versions of the argument from design: the argument that only an intelligent designer, identifiable with God, could have fitted structure to function as we find it fitted in a bird's wing or rabbit's eye. Even if one insisted that natural selection was God's chosen means of

I

bringing species naturally into being, there remained the difficulty that it seemed a very unreliable, wasteful and cruel way for the Divine intentions to be carried out. Darwin did, then, seem to take the goodness out of nature and to deprive Adam and Eve of literal life. A God who would choose natural selection was not a God who would send his Son as saviour; an earth that had never been trodden by the first Adam was not an earth that needs that saviour as the second Adam.

Historians of science have been moving away in recent decades from the familiar thesis, now a century old, that science and religion have been forever destined to conflict by virtue of their permanent and opposed natures. In place of this conflict thesis they have often proposed another that it is surely time to insist is equally indefensible, namely that Western science was a legitimate child of Christianity and so normally and properly, if not invariably, has lived in natural accord with Christianity to the benefit of both parent and offspring. One trouble with both the conflict thesis and this revisionist alternative to it is that they appeal equally to assumptions of a natural relationship between two abstract entities, a relationship whose naturalness traces to the characters of the entities that are presumed invariant. Whatever else we can learn from considering the long run of theories about origins and species, including the issues arising in the wake of Darwin's work, we must see that any such assumptions are highly problematic. (See art. 50.)

5. WIDER MOVEMENTS AND DEEPER SIGNIFICANCE

Ever since the 1860s people writing about Darwinian science have sought the ultimate implications of what Darwin was doing, and in seeking these they have sought to place his science in some larger context of major trends in history, including social and economic as well as intellectual history. The scope and variety of these proposals illustrates once again that there is much more to the history of science than might at first appear.

Writers of philosophical bent have often wished to fit Darwinian science into some comprehensive generalisations about the way fundamental conceptions of the universe have altered through the ages. For example, it has been suggested that for the Greeks the universe was a permanently adult organism, while for the seventeenth-century metaphysicians it was a lifeless machine; and that, thanks to Darwin and others, there was a shift in the nineteenth century to a universe conceived of as a growing organism.

There are always two sorts of difficulties with any such proposals. First, even when put in more qualified and sophisticated forms, they often seem open to the obvious objection that, at any one time, there is serious disagreement over what the science of the day presupposes or implies as to the ultimate nature of the universe. Second, there is uncertainty as to how we are to understand such a proposal in relation to particular individuals. True, Newton did take a stand on the question of whether nature was ultimately mechanical – although it was, contrary to legend, a stand against that view; but Darwin nowhere commits himself concerning the universe as a whole, and among those who have taken his science most seriously there is little consensus.

However, even if we find these two difficulties insuperable, we can still consider what insights can be had concerning Darwinian science, by seeing what range of different metaphysical positions it has been enlisted to support. For example, several writers, notably John Dewey early in this century, and Ernst Mayr more recently, have argued that Darwinism is fatal to any version of Platonic essentialism, any version, that is, of the view that the contents of the world fall into so many types – like types of figures in geometry: triangle, square and the rest – distinguished by essential differences specifiable in appropriate definitions. For, argue anti-essentialists, any theory of gradual change brought about by selective accumulation of variation within species cannot accept that species are delimitable and definable types, with the differences within species understood as defects due to imperfections in the individual embodiments of the type. So, they also argue, the typological conception of species that supports racialist doctrines, of more and less perfectly human races within mankind, is discredited by Darwinian theory. Whatever we make of this line of argument, we can certainly benefit, as historians, from being prompted by it to consider how far the emergence of Darwinian science involved departures from the metaphysics and ideology of the original sources of Platonism, Plato himself and his many followers.

The search for a social history of Darwinian science has been much developed in recent years. One point of departure for such quests has been the Marxist tradition of relating the emergence of Darwinian science to the changes historians identify as the Industrial Revolution. On the face of it the existence of some such relation would seem an obvious conclusion. The timing seems right: England in the 1830s had just become the world's first industrial nation. Darwin's theory has ideas of progress and competition integral to it. The conclusion would seem to be, therefore, that the theory represents one of the ways whereby the new economic order of industrial capitalism was expressed intellectually. (See arts. 6 and 66.)

Sometimes proposals of this kind are met with general arguments to the effect that this is not how history happens, because scientific change cannot be conditioned by economic life as the proposal requires. The conduct of this general debate about science and society is obviously relevant to any particular case such as the Darwinian one. But, equally, there are considerations peculiar to this case that need clarifying if that general relevance is not to be misunderstood. One consideration is that we have to be more critical that many writers on Darwinism have been concerning the economic history that may be involved.

I

It is all too easy, for instance, especially in thinking of England, to identify capitalism principally with industry in the sense of manufacturing. However, in looking at Darwin's biogeographical theorising, his Malthusian notions and his selective breeding analogies, it is plain that the relevant economic context is often not the new world of urban factories and mills, with Manchester as its exemplar; rather it is the earlier developments that gave England its new agricultural capitalism and its empire. Malthus after all is principally concerned, as Adam Smith so often was, with food and land, including land as ground won or lost in invasions that lead to struggles between aliens and natives. The animal and plant breeders, after all, were themselves often captains of a new scientific, agricultural industry.

Reflecting on the relevance of the economic history of these developments in agriculture and empire should suggest that Darwin's science drew its cultural resources less from the nineteenth and more from the eighteenth century than is often thought. If so, then the mediation between economic change and scientific theory may be less direct than is sometimes implied. Instead of seeing Darwin as directly reflecting on any innovations in urban and manufacturing life going on up the road from where he sat in his study, so to speak, we should see him drawing on intellectual traditions that had arisen in response to earlier changes in the agricultural and imperial life of his country.

A further way to raise larger questions about Darwin and his science is to ask whether there was, in any reasonably precise sense, a Darwinian revolution in biology, comparable, say, to the Copernican revolution in astronomy, the revolution in the Renaissance whereby the earth ceased to be central and stationary and became a rotating planet in orbit around the sun. The main challenge in maintaining that there was a Darwinian revolution is in showing that there was a shift from one consensus before Darwin to another after him, and that he was the main agent in bringing that shift about. One thing we can say, perhaps, is that in the decades before 1859, many leading biologists took species to be independently initiated in their origins and fixed in their characters, while within a few years of Darwin publishing, very few professionals were defending that position at length as Lyell and Agassiz had, for instance, in their very different ways in the 1850s. What is more, there is plenty of testimony from the time that this change in opinion did occur and that Darwin is responsible.

However, this generalisation hardly shows that there was a Darwinian revolution. On the one hand, there was, before Darwin, a great variety of approaches to questions of form and function, creation and design, progress and development, and life and matter; what is more, Darwin's implicit and explicit teachings on these questions were wholeheartedly embraced by very few biologists even well into this century, let alone in the last. Biologists who have agreed with Darwin that species are not separately originated and fixed perpetuated have often agreed with him on that issue alone, and have disagreed with him on others so fundamental as to make that agreement look almost superficial. To take just one example, there were biologists in the last century who agreed with Darwin that new species do arise from the modification of earlier ones, but who interpreted evolution as a developmental process subject, as individual embryonic developments are, to developmental laws that ensure the progressive realisation of a pre-existing Divine plan for the earth, life and man. In general, then, we have to be very careful not to over-estimate how fully Darwin's proposals were accepted in the last century by professional biologists, let alone people outside that scientific community. Certainly, many considered Darwin's views carefully and some welcomed much of what he said; but only a few, very few, had no serious disagreements with his position as a whole.

6. TWENTIETH-CENTURY EPILOGUE

There is a familiar stereotype as to what happened to evolutionary biology in this century, but it is now discredited by recent historical research. The stereotype says that in 1900 there was agreement that evolution had occurred but no consensus concerning what had caused it, especially, no consensus that natural selection was the cause. However, the stereotype continues, in that year there were rediscovered Mendel's laws of heredity, laws that allowed natural selection to be vindicated against all the objections then current. After two decades or so of misunderstandings, this vindication of Darwin by Mendel began to be appreciated, and a new theory, the new so-called synthesis, or neo-Darwinian synthesis, began to emerge with the integration of Mendel on heredity and Darwin on selection as its principal doctrine. By the 1950s this synthesis had been extended successfully to the reinterpretation of botany, paleontology and so on. Since then the synthesis has been modified but not replaced as a consensus.

There are at least two fatal difficulties with the stereotype. The first is that the characterisation of the state of discussion around 1900 is quite inadequate. The issues dividing the different schools of thought cannot be reduced to the single one of what the causes of evolution was. For a host of prior matters were in dispute too: whether evolution is to be understood as a regular development conforming to laws of development, or whether it is irregular and contingent in its branching and rebranching precisely because it conforms to no such laws; whether evolution is directed mainly by inner forces or, rather, mainly by external conditions and circumstances; whether it is smooth and gradual in its course or jerky and advanced by jumps.[4] These were all obviously complex questions and the bearing of Mendelism on all of them taken together was by no means straightforward.

A second difficulty is that when we see how the new synthesis emerged in the 1930s, it is plain that the integration of the new genetics with the old problems about species and their origins involved far more than demonstrating that Men-

ORIGINS AND SPECIES BEFORE AND AFTER DARWIN

delism can rescue selectionism, by resolving difficulties about the hereditary variation selection needs to accumulate if it is to be effective in causing evolution. A glance at one book can show how much more was involved, and that is the book that did more than any other to convince biologists that new developments in genetics meant that a new synthesis of causal theory in evolution was at hand: Theodosius Dobzhansky's *Genetics and the Origin of Species* (1937). For Dobzhansky's book is a synthesis not merely of Mendel's and Darwin's legacies, but of a Western tradition in theoretical population genetics and a Russian tradition in experimental genetics of populations; of genetics and systematics and of cytology and biogeography. So, even if we confine our attention to how one decisive text came to be written, we can see that the stereotype misleads us thoroughly.[5]

The last two decades have seen various challenges to the new synthesis as it was celebrated, with little dissent, at least in the English-speaking biological world, in 1959, the centennial of the *Origin*. Some writers have even declared that evolutionary biology is now in such a state of uncertainty and disagreement that talk of crisis is appropriate. Others, however, are not at all persuaded that such talk is called for. If there is a crisis, it is not one that is ever likely to be resolved to the satisfaction of all those dissatisfied with the orthodox consensus, for they seem to have little in common beyond that dissatisfaction. Thus some dissenters question the genetical doctrines accepted by the orthodox, but other dissenters do not; again some dissenters want evolutionary and embryological biology unified by a common foundation in developmental laws of form, while others think evolution is even more stochastic or random than the orthodox do.

What has been subject to a more univocal critique is not the new synthesis as such, but a distinctive programme in evolutionary biology, called sociobiology; it is a programme constructed by extending that synthesis to include social behaviour in animals and man so as to subsume topics – such as incest, kinship and altruism – traditionally treated by anthropologists and moral philosophers. Critics, many of them professional biologists on the political left, have found the programme making erroneous and unwholesome assumptions about nature and nurture, competition and adaptation; assumptions that involve, in turn, fallacious presumptions in favour of the *status quo* in class, race and gender politics.

From evangelical Christians, on the other hand – who, in the United States, especially, often tend to right-wing politics – has come a renewed attack on evolution itself, particularly as a dogma imposed by distant, liberal, secular scientists on children whose schools should properly, it is argued, be under local, electoral control in this matter as in any other of moral import.

These various politicisations of evolutionary biology may seem malfunctioning lapses from the proper norms of scientific discourse, but historically considered, as part of the long run of thinking about origins and species, they are far from exceptional; indeed they are what one expects. It is likewise, with the recurrent concern among philosophical analysts of evolutionary biology to compare and contrast the theory of natural selection with one or another branches of physics; at least since Newton and certainly since Darwin, there has been sustained discussion of the precedents set by successes in physics for any efforts to understand the history and diversity of life on earth. That discussion has been continued in this century by a new professional community of philosophers of science, who have taken up the questions raised in asking how far the concepts of evolutionary biology resemble the paradigm examples of scientific concepts taken by philosophers from physics. One issue that is at stake in this discussion is the unity of science itself. If evolutionary biology is really very unlike physics, then one can hardly talk of the structure of scientific theories, or indeed of the nature of science itself, without begging contentious questions.

Many evolutionary biologists may sometimes wish that the ideological critiques and philosophical explications of their theories could be suspended so that they can get on with their work undistracted. But not all feel this way, as is apparent from the way several leading general textbooks on evolution have been written recently.[6] In continuing the tradition of seeing questions about origins and species as involving perennial issues of politics, philosophy and religion, such texts would seem to have history very much on their side.

NOTES

1. See P. R. Sloan, 'Darwin's invertebrate programme', in D. Kohn (ed.), *The Darwinian heritage* (Princeton, 1985) pp. 71–120; M. J. S. Hodge, 'Darwin as a lifelong generation theorist', ibid., pp. 207–44.
2. For differing accounts of how Darwin may have first inclined to transmutation, see F. J. Sulloway, 'Darwin's conversion: the *Beagle* voyage and its aftermath', *Journal of the history of biology*, 15 (1982), 325–96, and M. J. S. Hodge, 'Darwin, species and the theory of natural selection' in S. Atran *et al.*, *Histoire du concept d'éspèce dans les sciences de la vie* (Paris, 1986), pp. 227–52.
3. For a much fuller account, see M. J. S. Hodge and D. Kohn, 'The immediate origins of natural selection', in D. Kohn (ed.), *The Darwinian heritage* (Princeton, 1985).
4. P. J. Bowler, *The eclipse of Darwinism: anti-Darwinian evolution theories in the decades around 1900* (Baltimore, 1983).
5. On the emergence of the new synthesis, see E. Mayr and W. B. Provine (eds.), *The evolutionary synthesis: perspectives on the unification of biology* (Cambridge, Mass., 1980).
6. See, for example, T. Dobzhansky, F. Ayala, G. L. Stebbins and J. W. Valentine, *Evolution* (San Francisco, 1977) and D. J. Futuyma, *Evolutionary biology* (Sunderland, Mass., 1979).

FURTHER READING

P. J. Bowler, *Evolution: the history of an idea* (Berkeley, 1984).
—— *The non-Darwinian evolution: reinterpreting a historical myth* (Baltimore, 1988).
T. F. Glick (ed.), *The comparative reception of darwinism* (Austin, 1974).
J. C. Greene, *The death of Adam. Evolution and its impact on Western thought* (Ames, 1959).
D. L. Hull (ed.), *Darwin and his critics. The reception of Darwin's theory of evolution by the scientific community* (Cambridge, Mass., 1973).

I

ORIGINS AND SPECIES BEFORE AND AFTER DARWIN

D. Kohn (ed.), *The Darwinian heritage* (Princeton, 1985).

E. Mayr, *The growth of biological thought. Diversity, evolution and inheritance* (Cambridge, Mass., 1982).

D. Oldroyd, *Darwinian impacts: an introduction to the Darwinian revolution* (Milton Keynes, 1980).

R. J. Richards, *Darwin and the emergence of evolutionary theories of mind and behaviour* (Chicago, 1987).

M. Ruse, *The Darwinian revolution: science red in tooth and claw* (Chicago, 1979).

E. Sober (ed.), *Conceptual issues in evolutionary theory. An anthology* (Cambridge, Mass., 1984).

R. M. Young, *Darwin's metaphor. Nature's place in Victorian culture* (Cambridge, 1985).

II

Canguilhem
and the history of biology

RÉSUMÉ. — Pour de nombreuses raisons et à juste titre, Georges Canguilhem est considéré comme l'auteur le plus important de notre époque dans le
domaine de l'histoire de la biologie. Son association de l'histoire de la biologie et
de la philosophie fut décisive dans la détermination de son héritage pour notre
discipline. Canguilhem était redevable aussi bien au positivisme historiciste de
Comte qu'à l'idéalisme historiciste de Kant. Sa dette à l'égard de Comte le rend
plus proche de l'empirisme anglais que beaucoup d'historiens des sciences anglo-
phones actuels. La manière dont Canguilhem, comme Kant, se concentre sur les
concepts restreint son programme d'historien. Sa réticence à l'égard de l'analyse
des institutions, des intérêts, des intentions et des influences restreint également
son programme. Si nous souhaitons dépasser les limites attachées à l'héritage his-
toriographique de Canguilhem, nous devons vaincre les obstacles résultant des
diverses divisions du travail universitaire. C'est alors seulement que nous pourrons
disposer d'une historiographie pluraliste et holistique pour la biologie, d'une historiographie qui puisse accorder à l'héritage de Canguilhem sa véritable place.

MOTS-CLÉS. — Historiographie ; concepts ; biologie ; positivisme ; histori-
cisme ; socialisme ; précurseurs ; Comte ; Kant ; Hegel ; intentions ; influences ;
intérêts ; institutions ; holisme ; pluralisme.

SUMMARY. — For many reasons Georges Canguilhem is rightly regarded as
the most prominent writer, in our time, on the history of biology. His association of
biological history with philosophy was decisive in shaping his legacy for our discipline. Canguilhem had debts to Comtean historicist positivism as well as to Kantian
historicist idealism. These Comtean debts make him closer to English empiricism
than many Anglophone historians of science allow themselves to be. Canguilhem's
Kantian concentration on concepts restricts his agenda as a historian. His reluctance
to analyze institutions, interests, intentions and influences also restricts his agenda. If
we are to overcome the limitations in Canguilhem's historiographic legacy, we must
overcome the limitations imposed by various divisions of academic labor. Only then

It is a pleasure to acknowledge my debts to conversations with Jean Gayon, Malcolm
Nicolson, Martin Kusch and Saied Zibakalam.

can we have a pluralistic, holistic historiography for biology that can give a proper place to Canguilhem's legacy.

KEYWORDS. — Historiography ; concepts ; biology ; positivism ; historicism ; socialism ; precursors ; Comte ; Kant ; Hegel ; intentions ; influences ; interests ; institutions ; holism ; pluralism.

I. — A DEBT OF GRATITUDE

Today, any historian of biology and medicine must look back to Georges Canguilhem with a feeling of gratitude. For Canguilhem did more than anyone else in his day to give our subject public, academic prominence, to endow it with status, respectability, serious significance, even with glamour. He was able to do this for at least four reasons. First, he pursued these topics with masterly scholarly ability and with intellectual flair and literary panache. Second, he assimilated the history of the sciences of life to philosophy. This was decisive because philosophy is an endeavour, even outside France, the home of the « philosophes » as the rest of the world calls them, that enjoys a peculiar prestige (along sometimes with ridicule) both within and beyond academic culture. Third, unlike many philosophers writing more recently about science, Canguilhem did not frighten the horses, as we say in English. He did not, that is, threaten the claims of science to be knowledge ; he did not, therefore, subvert the biologists' and doctors' own sources of cognitive self-respect and self-confidence. Fourth, he fathered a succession of disciples who in turn became mentors. So, he founded a school, or more powerfully than that, a tradition, that is now well into its third or fourth generation and that is spreading its influence across national and other boundaries. For spectators looking from a distance, upon the history of the science of life, Canguilhem's protégé Michel Foucault must appear uniquely influential in the dissemination of this legacy. But to practising historians of these sciences it is evident that this disciplinary success is a collective and institutional achievement as well as an individual triumph.

Several characteristics of Canguilhem's historical work would seem to derive fairly directly from his own philosophical education. First, as far as possible he wished to discern not only how science has moved toward its present doctrine, but to see also how it has done so by starting from and then progressing beyond its classical, Greek sources. With the Greeks, philosophy and science are in an alliance, if not forming a single unity. A philosophical historian of science feels at home with the Greeks. Again, Canguilhem chose concepts as his units of analysis and his subjects for narration. Concepts, according to the philosophers' lexicons, are not exactly the same as ideas. But there is a close similarity between the historiography of ideas (originally a German tradition) and Canguilhem's historiography of concepts. Canguilhem's best known writings concern the history of the concept of life, the history of the concepts of the normal and the pathological and the history of the concept of reflex action. Historians of ideas might have taken up these topics as exercises in their historiography without departing very far from Canguilhem's agenda. Where Canguilhem's philosophical ambitions are decisive, surely, is in the choice of concepts. His chooses often, if not always, fundamental concepts, that is concepts that help found a science or can provide foundations for the integration of two sciences. The concepts of life, of the normal and the pathological and of the reflex all concern the foundations of physiology, medicine and psychology as sciences, and as sciences that are related to one another. There resulted from Canguilhem's pursuit of these choices historical scholarship of the highest order. The monograph *La Formation du concept de réflexe aux XVIIe et XVIIIe siècles* (Paris, 1955) contains at its heart a meticulous and sophisticated study of the formation by Thomas Willis (John Locke's English contemporary) of the concept of reflex movement. This study, motivated as it was by Canguilhem's philosophical commitments, stands and will endure as a contribution to the history of biology quite independently of those commitments. Indeed, it presents an exemplary exhibition of the arts of textual explication and contextual exegesis ; while the volume as a whole provides, no less, a model of how to place the physiology and psychology of its two chosen centuries within the *longue durée* from the Greeks to our own time.

This emphasis on the Greek legacy comes naturally to historians of the sixteenth and seventeenth century. But historians of science who think of themselves as historians of biology, and who think of that science as one that only emerged in the nineteenth century, such historians can often feel little obligation to engage Greek authors. That feeling, as Canguilhem would have insisted, is a fallacious one. Elite education continued in that century to integrate ancient with modern thought. Georges Cuvier and Louis Agassiz were not exceptional in keeping faith with Aristotle, and we misunderstand their responses to Lamarck and to Darwin if we forget how such reading habits persisted for centuries after the Renaissance and Reformation. Canguilhem would not forget.

It would be a mistake, therefore, to think that Canguilhem's work as a historian of biology is so closely integrated with his ambitions as a philosopher that historians can only benefit from his histories if they share his philosophical convictions. However, it seems equally true that the responses, made today to his work in the history of biology, will be strongly conditioned by perceptions and evaluations of his philosophy. So, some account of Canguilhem's philosophical alignments and sympathies should be given here before returning to the question of his role as historian.

II. — HISTORY OF SCIENCE AND PHILOSOPHICAL TRADITIONS

It is for philosophical commentators and critics to decide whether Canguilhem has given us a philosophy of science, of reason and of knowledge that is acceptable and persuasive in our time. Is he sufficiently liberated from positivist deference to the authority of science ? Is his historicism about reason compatible with his rationalism about knowledge ? Does his demarcation of science from ideology unhelpfully serve to put the scientific community beyond moral and political criticism ? Such questions, and many others like them, may be of interest to the historian of biology reading Canguilhem at the opening of this century. But they are not perhaps the questions about Canguilhem's philosophy that a historian will naturally take up initially. Rather, it may be more appropriate to begin by placing Canguilhem in relation to familiar trends in the historiography of science, and in relation to the contributions to those trends made by diverse philosophical traditions.

At this point it might seem only proper to acknowledge that in different national academic cultures different trends have been manifested. And, yes, the issue of those national differences must be confronted eventually. However, it may be useful for an English historian, trained partly in the United States, and writing here about a French author to start with a reflection about German (including Austrian) philosophy. The reflection is simple enough, although not perhaps very agreeable to English or French pride : as we all know, all philosophy throughout the West has been predominately German for two centuries now. Indeed most of the differences between French and English philosophy have arisen because those two academic cultures have made different uses of Germanic resources : for instance, the French making, in this century, more use of Nietzsche, Husserl, Marx and Heidegger, the English more use of Frege, Schlick and Wittgenstein. However, generalisations about philosophy do not themselves provide the mappings that we need in understanding the historiography of science. To see why this is so, consider for a moment how it was for a young historian of science around 1964 in any of the English speaking countries. If inclined to philosophical self-consciousness s(he)** was expected to make one decisive commitment. On the one hand were the positivistic views of science and its history. These views were shared by most practising scientists and by most philosophers of science ; and these views led those scientists and those philosophers to write history of science that was naïvely empiricist and uncritically scripted as so many progressive triumphs of the rational over the traditional, the religious, the metaphysical and the superstitious. On the other hand were the views associated with Cassirer, Lovejoy, Koyré and Collingwood, all philosophers with fundamental debts to Kant and Hegel, and all philosophers offering explicit alternatives to empiricism and positivism. The commitment the young historian of science made was, then, to this heritage from Kant and Hegel ; and it was a commitment confirmed by reading in Kuhn or any of the other dissident historicist philosophers of science who offered to liberate historians of science from empiricism and positivism. Indeed this offering was only an exchange. For, at least in Kuhn's case, it was the history of science that was supposed to have liberated philosophy of science from positivism and shown philosophy of science how to draw on the Kantian and Hegelian heritage.

This liberation was itself, however, of short duration. For, within a few years, and certainly by 1970, our maturing Anglo-

(**) S(he) : he or she (N.D.L.R.).

phone historian of science has learned that history of science needed liberating again, this time from philosophy itself. For it needs
to reject Kuhn, Koyré, Kant and the rest in favour of Marx,
Mannheim and a variety of brave new sociologies of knowledge
that brought with them an emphasis on class interests, scientific
institutions and political ideologies, sociologies of knowledge that
were allied with socialist and feminist critiques of the power inherent in all knowledge and so, especially, in scientific knowledge.

So far, obviously, this tale about the historiography of science
has followed standard Anglophone legends. How, then, can this
tale aid us in understanding the role Canguilhem might play in and
beyond the borders of his own country France ?

Consider next a legend about France itself. The legend begins
by insisting, perhaps uncontroversially, that French philosophical
reflection on science never embraced the combination of Mach and
Frege that was consummated by Schlick and others in the *Wiener-
kreis*. Surely, the France of Canguilhem's teacher Bachelard upheld
Kantian if not Hegelian opposition to positivism. But are we wise
to accept this legend in this form ? Are we wise to concentrate
exclusively on Bachelard as the main mentor of Canguilhem, and
to infer that once we have identified Canguilhem as the protégé of
Bachelard, then we have done all the philosophical genealogy that
is required in understanding Canguilhem as a historian ?

As so often we need to recall the *longue durée,* as well as immediate precedents. Canguilhem's work is full of explicit debts to
Comte, the old, historicist positivist, writing in France a century
before the new, logicist positivists in Austria. And who were the
brains behind Comte ? In the eighteenth century, Condorcet and
Condillac were his decisive sources. And who, in the seventeenth
century, inspired them ? Well, Bacon and Locke more than anyone
else. So, perhaps we have a paradox or, at least, an ironic tale to
tell. Historians and philosophers of science working in England are
often told how they must avoid the positivist and empiricist prejudices endemic in their national cultural heritage. And they are told
that the best way to escape these prejudices is to apprentice themselves to teachers on the continent of Europe, teachers such as
Canguilhem, perhaps. However, the advice may not be so sound in
this case. Canguilhem, as descended from Comte, and so more
remotely from Bacon and Locke, may be too English a teacher for
a young English historian of science seeking to transcend his (her)

domestic inheritance. Indeed, having been directed to Cassirer, Koyré, Lovejoy and Collingwood and then to Marx and Mannheim, all in accord with Anglophone routines in the 1960's and 1970's, s(he) has been brought by these German teachings to be less English than any French follower of Comte, such as Canguilhem, is likely to be. Perhaps, indeed it is the French who need liberating from the English Bacon and Locke. And perhaps it is the English historians of science, with their schooling in the German Kantian and Hegelian legacies, who can offer assistance here. Conversely, perhaps Canguilhem is too English for a generation of English historians of science who have been learning from German sources how to not be English. Or, more precisely, learning from Kantian and Hegelian German sources how to repudiate the one Germanic tradition, logical positivism, that drew extensively, albeit remotely, on the heritage from Bacon and Locke.

Maybe, then, English historians have now learned enough from Germany to help liberate not only themselves but their French colleagues from English philosophy. Maybe it is the French who are now more English than the English and so more in need of Germany.

III. — ANALYSIS AND NARRATION IN THE HISTORY OF SCIENCE

On one principal issue, Canguilhem is manifestly indebted to Kant rather than to Comte. His histories are histories of concepts. For Canguilhem is concerned with the history of changing interpretations of phenomena ; with changes in scientific experience, where experience requires the active interpretation of what the senses deliver. Now concepts come in different sizes, as it were. The concepts of time, of space, of matter and of cause are very comprehensive concepts indeed, not restricted to one science rather than another but relevant to many. By contrast, the concept of valency or the concept of the gene are plainly less comprehensive concepts. When Canguilhem studies the concept of the normal, he is contemplating a broad spread of scientific thinking in pathology, physiology, anatomy and so on. A preoccupation with concepts can, therefore, bring with it a broad intellectual view of science.

However, it is surely true that this preoccupation with concepts is associated with certain confinements of his agenda as a historian. For two clusters of enquiries are not strongly represented in Canguilhem's writing. On the one hand, there is little attention to interests and institutions : commerce or imperialism as an interest, for example, or museums and laboratories as institutions. On the other hand, there is little investigation into the influences upon scientists nor of their intentions. Consider, first, this last item. Biographical studies of individual scientists, or indeed of groups or schools of scientists, disclose that their intentions may take a variety of forms. Take two examples. In the work that resulted in his *Principles of geology* (1830-1833) Lyell had a number of intentions, one being to reform the young science of geology. This intention cannot be reduced to one or more intentions about concepts. For the object of his intention was larger and more unitary than any concept analysis would imply. He was aiming to change the whole of a science, and this change was to satisfy a number of very diverse ideals and constraints : epistemological, methodological, ideological, institutional and so on. Any historical analysis of the *Principles of geology* cannot succeed if it restricts itself to an analysis of concepts. Again, recall the moment when Darwin in July 1837 opened his *Notebook* B by writing as a title *Zoonomia,* meaning here not his grandfather's book, but the very search itself for laws of life. Darwin's programme, his goal, his project comprises and integrates several component projects : to revise Lamarck's system as presented and rejected by Lyell ; to introduce a new way of classifying species that incorporates geographical findings, and so allows inferences about the laws of change in species ; a further enquiry into the causes of these changes ; and finally an enquiry into how those causes arise from the most general laws of life. These ambitions, and others formulated in the months ahead, cannot be reduced to an intention to construct one or more concepts. Nor should this surprise us. Scientists are like other people : they have all kinds of goals, ambitions, motivations, even at one time in history. Moreover, when we take history into account it is manifest that there have been radical shifts in the kinds of goals, ambitions and motivations people have had ; even if one considers only those people who have sought to explain and understand animals and plants. A historiography for science that concentrates our attention on concepts cannot do justice to the challenges we face as historians.

Nor, in any case, is the concept a very suitable unit of analysis and narration. For a concept is a product, an achievement not a process or a goal. The concept, then, as a unit of analysis and narration is not adapted to the historian's quest for insight into the processes of enquiry that Lyell or Darwin, or Bernard or Monod, have undertaken, or for insight into the aims that inspired and directed those activities.

But now a larger, philosophical issue confronts us. At least in poststructuralist retrospect, it is tempting to say that Canguilhem did not bring individual, or group, intentions, motives, aims and so on into his history because his thinking belongs to that decentering of the subject, that dissolution of the liberal, humanist conception of the person, that is central to the familiar shift whereby French thinking moved from the age of Sartre in the 1950s to the age of Foucault, Derrida, Barthes, Lacan, and the rest, in the 1970s. To this reflection, a historian could reply in various ways. S(he) could attempt a philosophical refutation of the relevant doctrines of poststructuralism ; with, surely, very little prospect of publicly proclaimed success. Or the historian can choose instead to declare a measure of disciplinary independence ; and to insist that for the purposes of doing history in ways that are not arbitrarily restricted in scope, nor arbitrarily detached from the way most people continue to talk, to write and to understand themselves and each other, for these purposes s(he) will continue undeflected by the philosophical forces of the age. After all does not the very pluralism proclaimed by poststructuralists, when in postmodernist mood, provide the historian with this license to decline to defer to the philosophers.

The power and authority of the historian's response is obviously limited. However, it may serve well when other issues are engaged. Turn next to influences. Canguilhem's histories do not say much about influences. They talk of continuity and discontinuity certainly. But not of the influences, the causes, for these continuities and discontinuities. Again, we have to acknowledge here, also, the transformation of the *Zeitgeist* between the 1950s and 1970s in France. In 1955 no philosopher had yet declared history impossible, history that is with its traditional concern with historical agents and agencies ; and with its traditional concern with the origins and influences of those agents and agencies. By 1975, with some help from Nietzsche's writings, indeed under the influence of Nietzsche, Foucault was proposing that genealogies should replace histories precisely because they do not presume to address those traditional concerns of historians.

After Foucault, therefore, Canguilhem looks, as Sartre does not, a suitable sage for our epoch. However, the historian is once more inclined to remain unpersuaded. S(he) reads, in *Les Mots et les choses* (1966) and in *L'Archéologie du savoir* (1969), Foucault's summary of the traditional historians' tasks that he is not going to pursue, and what does s(he) feel ? S(he) feels, perhaps, that those tasks sound at least as interesting and illuminating as the task Foucault is undertaking. What is more, s(he) may wonder why the choice has to be made between these two kinds of task. Could they not be done together and integrated into some larger venture ? S(he) is familiar with the rhetorical device that Foucault has in play ; for s(he) has often seen authors dramatise the significance of their innovations by declaring what others are doing as traditional, with traditional implying *passé*. Indeed, as the new century begins, s(he) is tempted to adopt a variant of this same rhetorical play. The 1970s have been running now for nearly three decades, s(he) observes. Are the 1970s not themselves a little *passé* ? Is it not time to move on ?

Returning to the business of the historian, one sees that the issues about influences and intentions have an especial pertinence for the historian of science. One challenge for historians of science was rightly emphasised by Canguilhem. Myths and fallacies about *precursors* must be rejected and replaced, he would declare. However, Canguilhem, with his inhibitions about influences and intentions, has surely made it difficult for himself on this matter. For the study of influences and intentions, and especially the study of the influences conditioning intentions, is one of the best antidotes to myths and fallacies about precursors. Obviously, trivially, Lamarck was not influenced by Darwin, nor did he intend to anticipate Darwin. Lamarck had the intentions he had, and to that extent did what he did, because he was the postcursor of Buffon, Newton, Linné and so on, the postcursor of those who influenced him and with whom he was agreeing and disagreeing. A postcursorist historiography of intentions and influences is a good cure for precursorism, because it respects the asymmetries in time and causation. For, to show that Lamarck is a postcursor of Buffon (among others) is not to show that Buffon is the precursor of

Lamarck. Understanding people as postcursors replaces understanding them as precursors. Canguilhem denied himself this subversion of the precursor temptation, because he ignored influences and intentions in order to focus on similarities and differences, focusing therefore on the conceptual dissimilarities between Lamarck and Darwin, in order to show that Lamarck was not a precursor of Darwin. But such a strategy is less than satisfying to a historian of biology. For a historian may want to ask how and why Darwin was deeply influenced by reading about Lamarck, in Lyell's exposition, and yet, also, how and why Darwin's theorising could eventually depart markedly from the structure of Lamarck's system. In answering these questions, judgements about the identities, similarities and differences among concepts may be insufficiently enlightening and discriminating. Continuity and influence do not always coincide. Darwin was deeply influenced by Lyell even while breaking with his views.

IV. — INDIVIDUALS, INSTITUTIONS AND INTERESTS

If we allow interests, economic interests especially, into our history of science, and if we allow institutions and the interests of institutions, then we raise questions about the relations of history of science to philosophy on the one hand and to social theory on the other. To raise these questions is to raise, in turn, questions about the relations between philosophy and social theory. Those relations have obviously changed from one period to another, and from one place to another. In Germany in the 1870s they were very different from what they were in France in the 1970s.

If one likes to study the history of science in a pluralistic, eclectic and holistic manner, then one regrets any disjunction between philosophical and social approaches to the subject. Canguilhem's work is curious in this regard. In his life, a life of political courage and commitment, the role of social ideals is manifest. And yet in his history of biology the political and the economic are kept almost entirely out of the picture. The reasons for this may lie in divisions of labour familiar in French academic life, where philosophical and social studies are often separated. But perhaps other influences are at work as well. Canguilhem, notoriously, was not above dismissing sociological approaches to the history of science as tainted with Marxist associations. The whole topic of Marxism in French philosophical life in the middle decades of the last century is impossibly complex and controversial for quick encapsulation. However, one question can hardly be avoided in any appraisal of Canguilhem's legacy to the history of the life sciences. We can ask why was the entire enterprise of sociological history uniquely associated with Marx ? If a historian does not find the Marxist historiographical tradition congenial, as Canguilhem plainly did not, then why not look to other traditions within sociological history, perhaps those associated, say, with Weber, a social theorist often committed to refuting or at least replacing Marxist legacies rather than perpetuating them ? This question has a special pertinence today. In the shift from the age of Sartre to the age of Foucault, French philosophy appears to have displaced Marx from the centre of the philosophical stage ; for the scientistic treatment of Marx by Althusser can be read as contributing to this tendency, rather than countering it. Canguilhem's resistance to sociological history of science, as too closely associated with unacceptable Marxist history, may seem, then, to make Canguilhem a very appropriate teacher for the current, postMarxist philosophical age. But to accept this verdict is to avoid the larger question about the responsibilities historians of science have to disciplines other than their own. Perhaps, historians of science today should view philosophical culture as impoverished in so far as it has repudiated Marxist or any other social theory, and so perhaps they should conclude that historians of science cannot live by philosophy alone, but must draw on additional academic cultures for historiographical resources and inspiration.

To this reflection, it could well be replied, especially by younger historians of science outside France, that they have long ago turned away from philosophy and toward social theory. However, to describe this tendency in this way may be misleading. If one questions what is involved in making this turn, it can be surprising what one is told, today, in reply. For, one may be told that the emphasis now is on bodies, spaces and practices (or *praxes*). And this does indeed sound very distant from any older, non-Marxist, approach to the history of science such as one finds in Canguilhem. However, further questioning reveals that the current inquiry is into what people were doing and the spaces wherein they were doing it, and in correlating the actions and the spaces. All references to motives and interests, any psychological and economic aspects of the individuals and institutions, are regarded as marginal to the historian's business, perhaps excluded entirely. So, even though we have here a commitment to scientific knowledge as socially constructed, a commitment directly contrary to Canguilhem's teachings, there is a continuity with Canguilhem's disinclination to look to biographical, social and economic analyses as resources in formulating causal and explanatory insights into scientific activity. Anyone who thinks that the new historiography of bodies, spaces and *praxes* is unduly narrow, indeed timid and cautious, in its programme should not, therefore, expect to find a corrective in Canguilhem.

V. — BROAD VISTAS AND INTEGRATIVE PROSPECTS

Duhem famously contrasted the broad and shallow thinking of English physicists in the last century with the narrow and deep thinking of their French contemporaries. National chauvinist that he was, Duhem's contrast was intended to be to the advantage of the French. But if we set aside national chauvinism, surely the ideal of combining breadth and depth has much in its favour. The difficulty in doing this arises, once again, from disciplinary and other divisions and hostilities. A quarter of a century ago, a young historian of biology visiting France was vexed, even intimidated by what he encountered. His reading convinced him that France excelled in grand masters of his subject – Roger, Foucault, Schiller, Canguilhem, Grmek, to name but a few – no other nation presented such an array of authority and accomplishment. And yet he had to learn that there was very little cooperation going on. On the contrary, some ignored others, some disliked others, some spoke very uncharitably of others, and for reasons that were sometimes personal, sometimes ideological, sometimes institutional and so on. In an optimistic mood, he might have concluded that it was easier to learn from the French historians of biology if one was not French. For an outsider could read all of them without taking sides ; whereas a young French student would be expected to assume that valuing and learning from some was inconsistent with appreciating and benefiting from the work of others. In a pessimistic mood he might have concluded that the best scholarship in a field often requires abandoning the ideals of cooperation and coordination. Perhaps French history of biology has moved away from that phase of a quarter of a century ago. Perhaps, it will now be thought that a good way to benefit from the work of Canguilhem is to integrate it, rather than isolate it, from the work of others.

On a larger scale, a foreign historian of biology looked in admiration upon the French historians, especially the *Annales* school, and upon the philosophers' writing about science. But the foreigner observed, with regret, that there seemed no willingness or ability in these two academic communities to learn from one another or to collaborate in contributing to his own field. Again, in an optimistic mood, he might hope today that one way to benefit from the work of Canguilhem would be to attempt to integrate it with the work of Braudel.

All such integrative prospects call for collective as well as individual efforts, and these efforts require us to overcome divisions of academic labour within as well as beyond the history of biology. The chronological divisions, and their professional corollaries, are familiar enough. A specialist on Aristotle's biology is likely to have a philosophical or classical education rather than a historical or scientific training ; whereas a specialist on early molecular biology will have been trained quite differently. No less significant are the divisions among historians of biology studying the very same period. Darwin specialists and Bernard specialists often know little of each others' studies. Indeed, they may feel justified in this isolation on the grounds that Darwin and Bernard themselves had programmes of enquiry and explanation that involved very little engagement with each other's endeavours. One might try to draw up some comprehensive generalisations about the *longue durée* of the sciences of life, generalisations designed to make sense of the gap between Darwin and Bernard. But it is widely agreed that phrases such as « the sciences of life » or « the biological sciences » are as problematic as « biology » itself. One can certainly doubt whether biology exists before the nineteenth century. Indeed one can doubt whether any single, unified science of life has successfully been established even in our own century. In the absence, therefore, of any simple identity for all the enquiries into plants and animals since ancient times, one can only formulate distinctions and groupings to suit a particular historiographical occasion. For the purposes of understanding Canguilhem's work, there may be a case for invoking one rather simple contrast, if only with a view to setting it aside later. The contrast is between two clusters of enquiries. On the one hand are enquiries into the structure and function in health and disease of individual living organisms. Galen and Bernard on the liver or Hippocrates and Sherrington on epilepsy fall into this category. On the other hand are enquiries into kinds of plants and animals, their structural designs and habitual activities. Plato's *Timaeus,* the opening of *Genesis.* Darwin's *Origin* and Monod's *Chance and necessity* (together with Lucretius's *De natura rerum,* Monod's model) fall into this category. Investigations in the first category are, in recent centuries, more closely connected with medicine and microscopes than are those in the second. Those in the second are, in these centuries, more closely connected with museums and voyages. In the middle of the nineteenth century the first category included physiology, the second natural history, at least in some senses of these terms. Now, although he did write some memorable essays on Darwin and on other authors within the second category, the most detailed and influential of Canguilhem's writings deal with those in the first. And when he wrote of Aristotle, whose work falls into both categories, it was Aristotle's contribution to the first rather than the second that was paid most attention.

A comprehensive history of enquiries into plants and animals would obviously seek to do justice to both these categories. This would require, in turn, doing justice to the Hebrew as well as the Hellenic sources of western scientific culture, and to the relations between the legacies from those sources. It would require us to look at the new integrations made by Albert and Thomas, in the thirteenth century, integrations of *Genesis* and of Aristotle on the diversity and designs of plant and animal kinds. In sum, the legacies of both those sources for both categories of enquiry would require an integrated treatment. In this integrative endeavour, certain limitations in Canguilhem's work would have to be circumvented. There is little of the Biblical in Canguilhem and there is little on the parts of biology (to speak anachronistically) that lie furthest from medicine. Nor is that surprising ; with his philosophical and medical commitments it was natural for him to concentrate on the Greeks rather than the Hebrews, and on physiology, anatomy and pathology, rather than palaeontology, biogeography or population genetics.

Here, therefore, as elsewhere, it seems evident that we can make the best use of Canguilhem's legacy to the historical study of the life sciences by trying to clarify, as accurately as we can, what it does and what it does not do. Canguilhem himself had no illusion that he was providing a framework that encompassed the entire history of all of the life sciences throughout all of the centuries. The appropriate way to measure his contribution is not by comparison with some Utopian ideal for the course of future endeavours in this field. Rather, we should ask how his contribution to our discipline took it beyond where it stood before Canguilhem began his long and rewarding career. It is a mark of the value and significance of that career that, in commemorating it, one needs to assemble a team of commentators from a wide array of academic disciplines and cultures ; and that they find themselves taking up an extraordinarily broad range of issues. Historians of biology owe a great debt to the man who has associated their subject with these issues.

BIBLIOGRAPHY

BOWKER G. and LATOUR Bruno
1987   A booming discipline short of discipline : (Social) Studies of science in France, *Social Studies of science,* **17**, 715-748.

CANGUILHEM Georges
1955   *La Formation du concept de réflexe aux XVIIe et XVIIIe siècles* (Paris : Vrin).
1966   *Le Normal et le pathologique* (Paris : PUF) ; this is the expanded edition consulted here of *Essai sur quelques problèmes concernant le normal et le pathologique* (Clermont-Ferrand : Impr. La Montagne, 1943), « Publications de la faculté des lettres de l'université de Strasbourg », fasc. 100.
1968   *Études d'histoire et de philosophie des sciences concernant le vivant et la vie,* 2nd ed. – consulted here – (Paris : Vrin, 1970) ; 1st ed. (Paris, 1968).
1977   *Idéologie et rationalité dans l'histoire des sciences de la vie* (Paris : Vrin, 1977) ; English transl. consulted here : *Ideology and rationality in the history of the life sciences,* transl. by A. Goldhammer (Cambridge : MIT Press, 1988).

1994
  A vital rationalist : Selected writings from Georges Canguilhem, ed. by François Delaporte, transl. by A. Goldhammer with an introd. by Paul Rabinow and a critical bibliography by Camille Limoges (New York : Zone Books).

CARON J. A.
1988
  « Biology » in the life sciences : A historiographical contribution, *History of science,* **26,** 223-268.

GUTTING Gary
1989
  Michael Foucault's Archaeology of scientific reason (Cambridge : Cambridge Univ. Press).

KUSCH Martin
1991
  Foucault's Strata and fields : An investigation into archaeological and genealogical science studies (Dordrecht-Boston-London : Kluwer Academic Publishers).

NICOLSON Malcolm
1991
  The Social and the cognitive : Resources for the sociology of scientific knowledge [a review of the English translation of *Le Normal et le pathologique*], *Studies in the history and philosophy of science,* **22,** 347-369.

ZIBAKALAM S.
1996
  Ideology and rationality in Canguilhem's epistemology, *Physis,* **33,** 266-287.

III

TWO COSMOGONIES
(THEORY OF THE EARTH AND THEORY OF GENERATION)
AND THE UNITY OF BUFFON'S THOUGHT

I
INTERPRETING BUFFON

There are many reasons for 1988 being an appropriate year for a collective examination of Buffon. Two centuries after his death the bicentennial of the ending of
the *Ancien Régime* will soon concern historians around the world. Within the sphere
of Buffon studies, the year of the *Colloque* has had also a special significance for
those of us who first encountered Buffon two decades ago. For it is now a quarter of
a century since the work of Professor Jacques Roger first transformed Buffon
scholarship. From that time on, to study Buffon has been to study with Professor
Roger. Like many of the presentations prepared for the *Colloque*, my own here is
profoundly indebted throughout to the fundamental contributions Professor Roger
has made to the understanding of our subject.[1]

It is a common observation concerning Buffon that his writing and thinking, taken as a whole, is without an obvious, apparent unity of purpose, doctrine or structure. This appearance is undeniable. However, it is no less apparent that there are
unities to be found, unities, indeed, of several kinds. Most manifestly, perhaps, there
are in Buffon persistent problem preoccupations. For there are persistent preoccupations with certain clusters of explanatory challenges. Among these, one finds
very prominently a persistent preoccupation with the cosmogonical problems
(macrocosmogonical and microcosmogonical problems) confronted in the theory of
the earth and the theory of generation. No less conspicuous, there is a persistent
preoccupation with natural philosophy problems concerning matter and force and
our knowledge of matter and force. One way, then, to approach the issue of the
coherence in Buffon's work as a whole, is to ask how these various unities relate to
one another, and to the preoccupations of his contemporaries and of his epoch. A
main aim of this paper is to indicate what may be involved in taking up the
coherence issue through such an approach.

Before proceeding to the pursuit of this aim, however, it will be as well to indicate what possible dangers may lie in our path when we take up any very general
questions concerning the interpretation of Buffon. These dangers often arise because

---

1. See Roger [10], [11] and [12]. My debts are extensive also to the studies of Phillip Sloan [9], [13]
and [14]. John Greene's classic study [6] greatly enhanced understanding of Buffon. For students
confined to English David Goodman [5] has provided a valuable introduction.

we, in the twentieth century, are always liable to perpetuate various judgements made in the nineteenth century concerning the eighteenth century.

Among the most influential of these judgements are those surrounding the question of what it is to think historically. This question has immediate relevance to the coherence issue, because Buffon's thinking attains its greatest measure of unity in his *Époques de la Nature* (1778), and it is there, likewise, that this thinking most obviously has to be designated as historical.

Consider just three judgements made by the nineteenth century about the eighteenth. The first is that Buffon's century continued a seventeenth century inability to write or to theorize, indeed to think, historically. The second is that German authors led the move to historical thinking. The third is that to think historically is to think as an evolutionist, and conversely, that thinking that is not evolutionary is prevented from being so because it is not fully historical.

Now, plainly, the status of these judgements is debatable. Are they definitional stipulations, or empirical generalizations or persuasive redescriptions ? To be critical about these judgements would be to raise such queries. Equally, one could be aided in being critical by asking what ends were served by the making of these judgements in the nineteenth century. Often they seem to be promoting the cause of a new breed of professional academic historians, or the cause of German nationalism or the cause of evolution in biology.

These reflections have a direct bearing on Buffon's historiography. For the matter has been raised as to how far Buffon is an historical thinker concerning species or, indeed, concerning nature. And the assumption has apparently been made that Buffon's thinking cannot be truly historical because it is both deterministic and non-evolutionist. This assumption is, in effect, that one is historical only insofar as one anticipates Lamarck and ultimately Darwin.[2] Again, German sources, particularly Leibniz, as contrasted with English and French sources, have been sought for the historical element in Buffon's conceptualisation of species. Moreover, German authors, such as Kant, have been looked to as the ones most appreciative of the novelty and significance of this conceptualisation. Here the assumption is made that in so far as this eighteenth-century French author, Buffon, succeeds in thinking historically it is because he is mediating between earlier and later participants in a distinctively German tradition.[3]

Now, the view taken in the present paper is that we may need, eventually, to bring such assumptions to the interpretation of Buffon; but that we should see, first, how far we can go, while keeping our critical distance from any nineteenth-century approaches to Buffon and his century. More positively, this paper suggests that there is a coherent position developed in the course of the writing of the *Histoire naturelle*, and one that shows Buffon to be both less haphazard and more willing to learn new things than he has often appeared to be.

II
MICROCOSMOGONIES, MACROCOSMOGONIES, NATURAL PHILOSOPHY AND NATURAL HISTORY

One way to approach one kind of unity in Buffon's work is to start from the

---

2. Professeur Roger seems to come close to such an assumption. See Roger [11] : p. 578.
3. Dr. Sloan seems to come close to such an assumption. See Sloan [13].

common, and correct, observation that the *Époques de la Nature* (1778) was Buffon's unification for much that came before.[4] The unity is such, therefore, that his thinking in earlier decades was unifiable, if not actually unified, in that it could be given unity by the synthesis presented in the *Époques*. Now, one unification that is presented in the *Époques* is the integration of the two cosmogonies. In 1749, the microcosmogony, of the *molécules organiques* and the *moules intérieurs*, is not integrated with the macrocosmogony of the formation of the planets including the earth.[5] In 1778, however, the *molécules*, and hence the *moules*, are depicted as arising within the successive changes on the surface of the cooling earth, and indeed within the successive changes on the surface of any cooling planet.[6]

The centrality of the two cosmogonies for the whole project of the *Histoire naturelle* needs always to be kept in mind. As a genre of writing, natural history was not precisely defined by precedent and convention. However, it would not normally have included either the theory of the earth or the theory of generation. Buffon was, then, making a deliberate and idiosyncratic decision when he opened a work of natural history by writing not only a *Premier Discours* explicitly on natural history, but also treatises devoted to the earth, generation and man.[7]

By engaging the two cosmogonies right from the start, Buffon was refusing to segregate his natural history from his natural philosophy. For his two cosmogonies could only be read at the time as pursuing Cartesian ambitions with Newtonian resources. Most obviously was this true of the theory of the earth. Within medieval Aristotelian natural philosophy there was no theory of the earth; for there was no separating of the earth as a topic from the cosmos as a whole. It was in the new natural philosophy of Descartes that this topic had first emerged as distinct and separable. Since Descartes, the topic had been transformed in the late seventeenth century, first by Burnet and, second, by Whiston.

Burnet had rejected the contrast between an account of how God could have formed the earth, naturally and slowly, in accord with the Cartesian principles of matter in motion, and an account of how He must be believed to have done it, quickly and miraculously, in accord with Scriptural narration. Whiston, declining to reinstate anything like Descartes' contrast, had nevertheless departed from Burnet, both in taking Mosaic chronology literally, and in taking his principles of matter and motion from his mentor Newton rather than from Descartes. Whiston had moreover included in his *New Theory of the Earth* (1696) a commitment to the new *emboîtement*, encasement ('boxes-within-boxes'), theory of generation. Of all the prior theories of the earth, it is Whiston's that receives from Buffon the most extensive portrayal in 1749.[8] It is the one Buffon was most concerned to displace and replace, just as *emboîtement* was the account of generation that he sought to discredit most completely.

In both cases, in doing so, Buffon deployed resources of natural philosophy made available since Descartes and made available, more than anyone else, by Newton. To appreciate the character of this conjunction in Buffon of Cartesian

4. See Roger [10] et [11].

5. Again, see Roger [10] et [11].

6. Professeur Roger [10] explains that 1779 was the year the *Époques* was first published. But I follow here the custom of accepting the date the work itself bore.

7. *Premier Discours* [1749], *in* Buffon [2] : pp. 7-26.

8. *Preuves de la Théorie de la Terre, article II, Du système de M. Whiston* [1749], *in* Buffon [2] : pp. 78-82.

ambitions and Newtonian resources, we have to avoid a misleading stereotype regarding Newtonian science, a stereotype tracing, once again, to nineteenth-century views of the Enlightenment. For the nineteenth century, Newtonian science was principally –it is hardly surprising– what was then still taught : namely, the three laws of motion, the law of gravitation and their use by Laplace and Lagrange in the 1770's, in establishing the stability of the solar system. However, what we now know, thanks to a succession of twentieth-century historians, is that there is vastly more to understanding the Newtonian resources for Buffon's age than understanding the progression from the *Principia* to the *Mécanique Céleste.*[9]

This scholarship allows us to recognize in Buffon a constant commitment to several fundamental theses concerning matter and force that were broadly owing to Newtonian legacies. Three such theses can be recalled briefly here. First is the real existence of forces as powers, dynamical properties of bodies and their particles distinct from geometrical properties; second is the distinction between passive principles of motion such as the inertial force, and active principles such as the forces responsible for gravity, magnetism, electricity and the like; third is the distinction between massive action, action determined by the quantity of matter or mass, and superficial action at the surfaces of bodies.

Now, Buffon was not merely a knowing heir to these theses concerning matter and force in themselves; he was also heir to a further group of theses concerning nature herself that Newton and his followers had grounded in this natural philosophy of matter and force. One such thesis is that nature, as a perpetual worker –a phrase of Newton's often echoed in Buffon– can be understood as constituted by an economy of forces that may or may not be in conservatory balance. Two such balance issues were indeed discussed. First, Newton had taught, if there were only passive principles of motion, then motion would be constantly lost, in particle collisions especially; so the active principles ensure that this loss of motion is compensated by the recruitment of fresh motions. Second, if the attractive force of gravity were the only active force, then nature's work would tend to end in a cold, lifeless, solid clumping of matter with no heat, life or fluidity possible. Conversely, those powers in nature that are expansive and repulsive in action –the powers residing in air and heat– are ultimately responsible for counteracting the gravitational, refrigerational and consolidational stillness and death that would otherwise ensue.

Such conclusions in the understanding of nature as a whole are familiar commonplaces by the end of the 1730's, thanks especially to efforts to integrate Newton's own teachings with Boerhaave's. Buffon's familiarity with such conclusions and his assimilation of them in the years before the *Histoire naturelle* is not in doubt. That he remained committed to the revision, elaboration and vindication of such conclusions through the three decades from 1749 to 1778 is evident when we follow the entire succession of his writings from his translation (1735) of Hales's *Vegetable Staticks* to the article on the forces of nature that appeared in 1788 at the opening of the fifth volume of the *Histoire naturelle des minéraux.*[10]

9. For a recent introduction to the scholarship of Heimann, McGuire, McMullin, Metzger, Schofield and Thackray, see Cantor and Hodge [3].
10. Hanks [7], and Lyon and Sloan [9].

III
THE SUCCESSION OF EVENTS, THE PENETRATION OF FORCES AND THE LIMITATION
OF HUMAN KNOWLEDGE AND ACTION TO THE SURFACES OF THINGS

That such considerations were of constant and decisive significance for Buffon is evident when one sees how they were always related in his thinking to two characteristic concerns, two *idées fixes*, that he returns to again and again throughout his entire career. For these two *idées fixes* are not commonplaces shared by all his contemporaries, or even all followers of Newtonian philosophies of nature. The two lower case *idées* are *succession* and *pénétration*.

For Buffon, succession in facts, *faits* or events, *événements*, is the ground of *certitude* in *physique*, as contrasted with *évidence* in *mathématique*. This succession is therefore the ground for physical science as knowledge of what is physical rather than abstract.

Every reader of Buffon's *Premier Discours* will recall how this contrast is drawn there, and how Buffon uses it to distinguish and distance his natural history from the abstractions of the makers of natural history thesis, most obviously Linnaeus.[11] However, the historiographical interpretation of the succession idea in Buffon is still subject to difficulties. Recently and correctly, Buffon's position on this subject has been associated with similar contrasts then being drawn by several other writers who were also demarcating mathematical and physical science. However, Buffon's succession thesis has sometimes been uniquely associated with Leibniz's insistence that no such abstract entity or absolute time exists, because time is merely the relational order among temporally successive beings.[12] A unique Leibniz association is unlikely to be correct, it seems. For when Buffon is most explicit he does not talk merely of succession, as Leibniz does in talking about time, but of recurrent and uninterrupted succession. Especially does he do so in the *Premier Discours* and, in exactly in the same phrasing, when he makes his famous assertion, in the article on the Ass in 1753, that it is the recurrent and uninterrupted succession of individuals produced by reproduction that constitutes a species as a real, not an abstract, entity.[13] For this reason it is surely promising to relate Buffon on succession to discussions not of time as such, but of the relations constituting causation, including, but not only including, temporal relations.

This interpretation is confirmed by other texts, especially the *Arithmétique Morale* (published in 1777, but written, much of it, it seems, in the 1730's), that indicate that repetition and continuity in successive events were always, for Buffon, the ground not merely of avoiding abstraction in general, but also the ground for demarcating what knowable causes are available for explanatory use in physical science.[14] Most familiarly, it is always Buffon's position that we cannot know primary causes; but we can know very general effects. Such general effects are, then, secondary causes, appropriate to invoke in explanation, as is done in physical but not in mathematical science. The requirement of recurrent and unbroken succession in events is a constraining requirement on physical science as knowledge.

---

11. *Premier Discours* [1749], *in* Buffon [2] : pp. 7-26, and Sloan [14].
12. Lyon and Sloan [9], Sloan [13] and [14].
13. *L'asne* [1753], *in* Buffon [2] : pp. 355-356.
14. *Essai d'arithmétique morale* [1777], *in* Buffon [2] : pp. 456-488, and Lyon and Sloan [9].

III

However, Buffon holds throughout his career, it is a requirement that is successfully met. He is, then, always an optimist about the possibility, indeed the actuality, of a physical science that explains by means of what is not mere abstraction. He is always optimistic in this way, although denying the possibility of knowing the primary causes, the causes of the most general effects. There will be highly significant developments in Buffon's thinking in the 1760's, but one does not have to wait for that time to arrive before Buffon is an optimist about a physical science of natural causes. That optimism is there from the beginning.[15]

There is confirmation of this early and constant optimist interpretation of Buffon in the history of his other *idée fixe : pénétration*. Always, from the 1740's to the 1780's, Buffon insists that man cannot penetrate to the causes of events, if by causation is meant not general effects known by comparing particular effects with one another, but the inner forces and underlying powers that produce those effects. For our knowledge is derived from our senses and our senses only put us in touch, literally, with what any touching touches : that is, surfaces. Now, this commitment to comparison and to the senses as the sources of knowledge is strongly indicative of Buffon's debts, direct and indirect, to Locke; they are debts, moreover, so pervasive and familiar that it is, surely, not at all surprising that Buffon does not feel the need to signal them to his readers by naming Locke as their source. By the time Buffon's generation were writing in the 1740's, it went without saying that such indebtedness to Locke was ubiquitous.

Now, what Buffon does not see as commonplace is his combining of this commitment with the conviction that, thanks to Newton, we do now know that nature works, not superficially in two dimensions, but by means of a force that acts instantaneously in three dimensions. So, Buffon's pessimism about penetration is always combined with and not opposed to his optimism about a physical science that avoids mere abstraction.

In the case of the gravitational force, it is evident, from the 1740's on, why Buffon should see no irresolvable tension between this optimism and this pessimism. Two considerations seem to have been decisive. The first is that he takes Newton to have shown conclusively that the conformity of the planets' motions to Kepler's laws must be due to Newton's inverse square force with its dependency on mass. So, here, one has moved successfully from the knowledge that the senses deliver about the surfaces of the bodies whose motions constitute repeated and continuous successive facts, to a knowledge that these motions are due to a cause that is neither superficial nor successive, but penetrating and instantaneous. The second consideration is that we know from the planets' motions only that this gravitational cause is active but not how its activity operates. However, we can know from other facts, and from the analogy that they provide, that it is a single cause which must conform to the simple law that Newton has specified for it. For Buffon follows others in finding, in the lawful spreading and fading with distance of light issuing from a single source, an analogy for the gravitational action exerted by one body on another. It is for him an analogical resource that allows one to relate what is observably true in two dimensions to what must be true, but not observably so, in three dimensions.[16]

15. Here I am agreeing with Dr. Sloan [14] who was, in turn, disagreeing with Professor Roger [11] and [12].

16. See Buffon's writing on the law of attraction, first published in 1749 and republished in 1774, in Buffon [1], *Supplément...*, T. 1. An English translation is in Lyon and Sloan [9] : pp. 79-85.

When it came to the *moule intérieur*, Buffon did not claim that there was anything equivalent to the proof of gravitational attraction from Kepler's laws or to the analogy with light. There is only the analogy between the *moule* itself and the force of gravity. For this analogy insists that gravity is three dimensional and instantaneous, and therefore beyond the limits of our imagination, since the imagination can only conjoin ideas tracing to the impressions that the senses supply, and the senses receive only impressions of surfaces. So, an inner moulding action is as unimaginable as gravitational attraction is. But, equally, it cannot be physically impossible for that reason alone. It cannot then be objected to as a conjecture or hypothesis about the cause of what the senses acquaint us with, in the recurrent and unbroken succession of events in nutrition and reproduction.[17]

Buffon's reflection in the 1740's on the succession of effects and penetration of causes yielded, then, his most enduring convictions concerning physical science as knowledge, when knowledge was possible, and physical science as analogical conjecture, when it was not. Nor are these convictions ever abandoned. A famous passage in the essay *Des mulets* (1776) urges that our knowledge of nature can be unlimited in the future, provided appropriate observations and experiments are made.[18] But this passage is not concerned with the penetration issue and is not, therefore, retracting the earlier pessimism about penetration. Nor is it retracting, then, the reaffirmation and elaboration of the early teaching on penetration that is given in another famous passage in the article on *Nomenclature des singes* (1766), where Buffon insists that not only is man's knowing limited by the limitation of the senses to the two-dimenstional, his making is limited likewise. For man makes things by acting on material across a surface and along a linear axis, but does not act, as nature does, in three dimensions simultaneously.[19] This preoccupation of Buffon, with the geometric dimensionality of what we know and of what nature does, conditions no less a text of 1783, opening the *Histoire naturelle des minéraux*, the essay *De la figuration des minéraux*; for here he contrasts the growth of crystals, by the laying down of successive layers of material on surfaces, with the simultaneous inner moulding of nutritional material in three dimensions that is distinctive of the growth of living bodies.[20]

IV
PRODUCTION AND REPRODUCTION, INITIATION AND SUCCESSION

It will already be clear that Buffon's convictions concerning succession and penetration are indispensable in understanding his conceptualisation of species. However, in bringing our recognition of these two *idées fixes* to bear on that conceptualisation, we have to distinguish among various questions concerning the production and reproduction of animals and plants.

The *moule* doctrine of 1749 is an account of how parents reproduce, how offspring are produced and so how a species is perpetuated. It is not an account of how

---

17. *Histoire générale des animaux, Chap. II, De la reproduction en général* [1749], *in* Buffon [2] : p. 244.

18. *Des mulets* [1776], *in* Buffon [2] : p. 414.

19. *Nomenclature des singes* [1766], *in* Buffon [2] : pp. 386-393.

20. Buffon [1]. This and other articles in [1] are most easily located by using the invaluable table of contents for the 44 volumes provided in Buffon [2] : pp. 522-524.

a species is produced, is first brought into being. Buffon in 1749 undoubtedly assumes that the earth was once devoid of life; he assumes, then, that every species reproducing today was once produced by some means or other at some time after this planet, the earth, was first formed, even if he, Buffon, was not committing himself in print on how such initiations of species and their reproductive successions were produced. Beyond the rejection of *emboîtement* and what it implies about such original productions, he is silent.

Buffon's successional analysis of the reality and identity of species does then give him a temporal notion of what a species is, in the sense that a species is something constituted by a succession that is temporal. But this notion is not historical, in that there is no implication that the reality and identity of species are owing to the reality and identity of the productive initiations that originated them at particular times and places. It is appropriate, therefore, that Buffon's successional notion of species reality and identity has reminded recent readers both of Aristotle on the *genos*, as a permanently persistent, indeed everlasting, sempiternal kind, and of Hume on the person as constituted by a temporal succession of ideas.[21] In the Buffonian, Aristotelian and Humean cases, time, but not history, is essential to the analysis.

Conversely, therefore, we should not think that once we have found Buffon's succession *idée* fundamental to his early thought about species, then we have found the source of what is historical in his thought, early or late, about species. On the contrary, what we have found, by implication, is that any historical elements in his thinking about species must have other sources.

This conclusion is confirmed when we recognize that bringing species under the quite general doctrine –of succession as the ground of physical knowledge that avoids abstraction– does not in and of itself bring biparental reproduction into the analysis, nor therefore such matters as sterile and fertile crosses. For, after all, uniparental, asexual successions are recurrent and unbroken. Nor does the general doctrine, in itself and as brought to bear on the reproductive successions of species today, raise any issue of descent in past time from common or from distinct stocks, the issue Buffon nevertheless makes decisive to the judgement, in 1753, that the ass and the horse are distinct species. Those issues, like the issues raised by the interaction of two parents in producing fertile offspring, are raised and resolved for Buffon, not by the succession *idée* in and by itself, but by the *moule* doctrine in particular. For, even in 1749, Buffon is assuming that for two putative species to be really distinct is for them to trace to different original *moules*, distinct material structures and powers, that initiated distinct sexual reproductive successions that have been unable to merge into one since.

That the succession *idée*, in and by itself, did not suffice to make Buffon think historically about everything exemplifying that *idée*, is shown also by his macrocosmogonical teaching in 1749. For the hypothesis concerning the formation of the planets –as hot, molten material split off from the sun by a colliding comet– and the theory of the earth are far from integrated with one another in the opening volume of the *Histoire*.[22] Most obviously is this so, because the theory of the earth is principally about the stable, conservative effects worked by the waters of the globe in causing both losses and compensating repairs to the overall sum of rock and land on

21. Compare Sloan [14] and Gayon [4].
22. Roger [10].

the earth's surface. The perpetuation of this stable, conservative action is, therefore, quite independent of any causation conditioned by the manner of this planet's initial production.

Now, it is the repetitive and continuous workings of these aqueous causes of change that Buffon knowingly subsumes within the ideal and the idiom of succession as the ground of any physical science of what is real and not abstract. The formation of the planet and the loss of its early heat are left unintegrated with this exercise in the science of succession, the *Histoire et théorie de la Terre*. Here, then, the history of the earth does not include the macrocosmogonical formation of an orderly solar system and habitable planet or planets from an initial chaos.

Buffon, in 1749, is, then, a long way from the synthesis of the *Époques;* for not only does he not have any integration of the microcosmogony of the *moules* with the macrocosmogony of the formation of the solar system, he also does not have any macrocosmogonical integration of the formation of the earth and the transformation of its surface leading to the present configuration of oceans and continents.

V
GEOGRAPHY, HEAT AND NATURE

It remains, therefore, to inquire into the reasons why Buffon's cosmogonical thinking did not remain unintegrated in these ways. Mere continued commitment to the succession and penetration themes, and mere continued preoccupation with upholding the 1749 theories for generation and the formation of the planets, would not have sufficed, themselves, to shift his thinking to the cosmogonical synthesis of the *Époques*. What helps most in making this shift intelligible to us are, by contrast, three concerns that receive quite new emphases in the 1760s : geography, heat and nature herself. Obviously, none of these topics is, as such, a complete novelty in the 1760s. However, for anyone familiar with Buffon's biography and the sequence of the *Histoire*, there can be no denying that these topics came to prominence as never before in that decade. Geography, the historical geography, that is, of the animals of the old and new worlds, comes into its own not only in the three essays of 1761, the *Animaux de l'ancien continent*, the *Animaux du nouveau monde*, and the *Animaux communs aux deux continens*, but also in a text that is the direct successor to those, the *De la dégénération des animaux* of 1766, a text that includes, in turn, a celebrated phrase providing a precedent for the title of the *Époques*.[23] As for heat, it was in 1766-67 that Buffon followed his readings in the writings of Dortous de Mairan with experiments on heat designed to elucidate the thermal history of our globe.[24] By this time, the famous essays on nature –*De la nature. Première vue* (1764) and *De la nature, Seconde vue* (1765)– had already gone beyond any earlier claims for nature in extolling the principal powers of nature, gravity and heat, as able to produce all bodily phenomena in the living world no less than in the world of brute matter.[25]

Returning a moment to Buffon's geography, the first point to make is a quite general historiographical one. For the chronicler of Western thought, geography has, one suspects, an image that is in every sense rather mundane. One does not, there-

---

23. Roger [10] : p. XXVI.
24. See Roger [10] : p. XXVII.
25. The two *«Vues»* are in Buffon [2] : pp. 31-44.

fore, expect such a chronicler to be nearly as excited by any growth of geographical consciousness as by the rise of historical consciousness. However, this indifference turns out to be misleading. For, as the very case of Buffon shows, in certain ways the rise of historical consciousness –if such an impressionistic *cliché* be allowable here– was decisively fostered by geographical concerns, since the explanatory strategies exercised in the face of geographical matters were often historical ones. It is, then, no accident at all that Buffon's historical writing about species comes principally in writing about the historical geography of the Old and the New Continents.

Thus his central concern was to explain present spatial differentiation; here he seeks to decide how much of this differentiation is due to prior spatial separation, and how much to prior spatial dispersion. Some of the differences in the faunas of the two continents, he decides, are due to the spatial separation of the places of origin of species native to the New World, as distinct from those native to the Old World. Some, however, are due to disparities in the migrations of Northern species in the New and Old Worlds; there being much less migration in the New because of the mountains of Central America.

In itself the deployment of such a dual explanatory conjecture may seem neither novel nor consequential. But Buffon did not confine himself to this conjecture alone. He eventually argued, in the *Époques*, that the South American native species are distinct from and inferior to their African counterparts because, thanks to these mountains, South America was not supplied with organic molecules by migration of life from the North, as Africa was.

No matter how one chooses criteria for what is to count as historical thinking about the natural world, what Buffon is doing here will surely have to count as genuinely historical, and to that extent will have to count as a distinctive element in the general rise of historical consciousness. Moreover, there is a significant lesson to be learned from contemplating the wider significance of this distinctive element. If historical thinking is sometimes developed in making sense of geography, and if geography is often motivated by the colonial and trading interests of all the European nations since the Renaissance, then we do not always need to find the sources of historical thinking about man and nature in such locations as a distinctive German philosophy of man and nature.

Now, Buffon's historical geography for the South American species was not initially integrated with his Newtonian philosophy of nature as a perpetual gravitational and thermal worker, nor with his succession and penetration convictions. However, such an integration is made in the *Époques*, and it is an integration that depends, more than anything else, on the new narrative and chronology for the loss of the earth's original heat, and the consequent degeneration of terrestrial nature as a lawful system of unfading gravity and fading heat. For, with this additional explanatory resource, Buffon now, in the *Époques*, ascribes the difference and the deficiency of the South American fauna to the younger, cooler age of the original formation of that land; for from that recency follows an inevitable feebleness of nature there, entailing a corresponding feebleness in the living matter produced by nature in that land.[26].

Buffon's most general convictions, concerning the constitution of nature as a perpetual worker, did not have to be rethought in making the move to this integra-

26. *Des Époques de la Nature, Cinquième époque* [1778], *in* Buffon [2] : pp. 174-175.

tion of geography and thermal history. The two *«Vues»* should be read, on the interpretation offered here, as indicating a growth in confidence in a Newtonian nature's powers, a growth on Buffon's part, and on the part of those others in his generation who were impressed with such contemporary developments as the greater place now accorded to electrical, magnetic and thermal activity in the overall theory of attractive and repulsive forces. For someone of Buffon's deistic, unchristian, Enlightenment, *philosophe* outlook it was to be expected that such developments should raise the prospect of reaffirming the possibility of a complete cosmogonical scheme, one not only freed from Moses, providence and miracles, but also deliberately harking back to that ultimate precedent in ateleological worldmaking and lifemaking : Lucretius's *De Rerum Natura*. To Buffon, then, it would seem that such a prospect arises not so much from a newly enhanced confidence in man's ability to achieve scientific knowledge of the natural order, but rather from a newly enhanced confidence in nature's powers to produce originally, no less than to perpetuate subsequently, the order already discerned in the solar system, in the earth and in its animal and plant species.

That the power of nature is a decisive consideration in this move to the *Époques* is shown, especially, by the insistence in this treatise on one corollary of nature's thermal decline : namely, that many productions which are today not possible were earlier within the power of causes tracing ultimately to ordinary matter and ubiquitous forces acting with greater intensity in more favourable circumstances in olden times. Most particularly is this consideration evident when one compares the *Époques* with the emphatic affirmation of spontaneous generation in the additional article on generation composed for the fourth supplementary volume of 1777.[27] These two texts taken together confirm that Buffon intended to be read as implying that when the largest quadruped and mollusc species now found as fossils were first produced, all such species could be originated, quite naturally, by spontaneous generations such as are now beyond nature's powers on earth. So, here, the question of a knowable natural order is the question of whether such ancient productions have arisen within an order that is knowable today. It is Buffon's attitude to this question that reveals so vividly what has been constant, and what not, in his thinking over the three decades since composing the opening volumes of the *Histoire*. For he is now prepared to include such productions within the scope of a physical science grounded in the recurrent and unbroken succession of observable events.

VI
SCIENCE, NATURE, COSMOGONY AND COSMOGRAPHY

He can include such productions in such a science because the coolings of spheres –whether of small ones in his own hand, where the primacy of touch is respected, or a large one in the heavens, reachable only by analogy– all exhibit a repeatable and continuous succession of temperatures; even if, in any one case, the succession can not be repeated in a particular sphere, say the earth, but is to be repeated in another. For this regular cooling succession entails a regular succession of organisation, of life and so of species, provided one makes, as Buffon is prepared

---

27. Buffon [1], *Supplément...*, T. IV.

to, one further assumption, and that is an assumption of thermal organisational determinism that he announces quite explicitly. The assumption, as already formulated in 1775, is that the same temperature produces everywhere the same living beings; and this Buffon evidently takes to hold true of all the planets, not merely the earth alone.[28]

This assumption came easily enough to Buffon in the 1770's. In his historical geography, he had already supposed that the New and Old worlds would have had exactly similar native organisms, where their climates and so on were exactly similar, had it not been for the differential access of migrations from the North. A main rationale for the migrational conjectures was to explain what was otherwise anomalous to a thermal determinist. Moreover, that degrees of heat should be the ultimate influences, in determining the organisation produced at different times and places, was a natural corollary of the role ascribed, in the two *Vues*, to heat as a cause of the properties of living organisms.

Buffon's thermal determinism is, therefore, a central element in a cosmogonical synthesis that it is easy for us to misunderstand. As Professor Roger has brought out so valuably, it is tempting but incorrect for us to work with a simple dichotomy.[29] On the one hand, there would be the Aristotelian type of science where species, as forms, constitute the very order of nature. Nature is constituted by everlasting, specific natures. Here nature has no history because species, as natures, have none, and because the natural order is constituted by the orderliness of natures. On the other hand would be the Darwinian type of science where species are contingent products of the historical course of nature, with each species owing its singular origin and unique properties to the consequences of particular local circumstances.

Buffon shows us that all of Western thought cannot be mapped on to such a dichotomous scheme. Species are not, for Buffon, as intrinsic to the ultimate orderliness of nature as the gravitational force and law are.[30] And yet the thermal determinism makes any specific organisational structure an invariable consequence of what is a repeatable stage in nature's inevitable decline on one planet after another. Are we, then, to conclude that Buffon's Newtonian philosophy of nature as a perpetual worker, acting through persistent gravity and evanescent heat, has somehow kept him from being less than fully historical in his thinking about species in relation to the orderliness of nature ? Surely we are not required to draw this conclusion. Where geographical explanatory challenges and their historical resolution were integrated with the Newtonian account of the dynamic economy of nature, that account did not prevent the possibility, or indeed the actuality, of historical thinking. It was only where the geographical challenges had no equivalent –in the consideration of interplanetary comparisons– that Buffon's science of successional changes ceased to sustain a historical understanding of species origins.

Our conclusion must be, then, that it would be unhistorical for us, as historians of Buffon's epoch, to presume that in all cases Newtonian science is inevitably and intrinsically opposed to a historical understanding of nature. More positively, we may perhaps agree that, for any individual such as Buffon, we cannot decide in advance what will prove to be indispensable to our understanding of the relationship between macrocosmogonical and microcosmogonical questions. For Buffon, we

28. Roger [11] : p. 580.
29. Roger [11] : pp. 580-582.
30. Roger [11] : pp. 580.

find that, in addition to the philosophy of nature, there are two indispensable keys –geography and heat– that form a peculiar conjunction that is decisive for very specific reasons. The character and role of this conjunction in Buffon we can only recognize if we do not restrict ourselves to the usual historiographical abstractions : the conceptualisation of time, the philosophy of nature, the theory of knowledge and so on. To be sure, those abstractions are requisite, but equally they are not sufficient in the present case. Study of the present case must be allowed to show us what else is requisite, because peculiar to this case.

An enquiry into the unity of Buffon's cosmogonical thinking has, therefore, to be a study also of its singularity. For the conclusions must confirm what we already know : namely, that Buffon as unified macrocosmogonical and microsmogonical theorist had no followers. There were no *Buffoniens*. This outcome for our enquiry is, then, confirmed in turn by a glance at what happened in the next generation. None of the new breed of professional naturalists and natural philosophers –nor any amateur either– was prepared to emulate the full range of Buffon's neo-Lucretian cosmogoniacal-theoretic ambitions. Only one person, Buffon's personal *protégé*, Lamarck, worked out his own position by addressing the full scope of Buffon's. And Lamarck was sufficiently in disagreement with Buffon that we cannot count even him as a cosmogonical-theoretic successor to Buffon.

For Lamarck's most general claim about nature and life, as made quite explicit after 1800, has to be read as a response to Buffon; but it is by no means a mere re-vision or development of Buffon's own final teaching.[31] For Lamarck's most general claim is that, yes, nature –which is to say the collaboration of attractive and re-pulsive forces of gravity and heat– has indeed produced all bodies, living and lifeless, organised and otherwise, the animals and plants no less than the minerals. However, among living bodies nature can only make the simplest directly, in direct spontaneous generations from lifeless matter; the more complex have therefore been produced indirectly through the gradual complexification of these over vast eons of time. Where Buffon has nature making organic molecules from brute matter and then these molecules suddenly making animals and plants, simple and complex, Lamarck has the simplest organisms formed as intermediaries and the rest formed gradually from them. Buffon, in Lamarck's view, although correctly Newtonian about the nature of nature's agencies, is incorrect about their powers and their history. The earth's heat is constantly revived by the circulation of fire to and from an unfading sun; so nature now can do no more, and no less, than at any time in our planet's apparently unlimited past. Always obliged to begin the production of living bodies with the simplest, she has always to produce the higher forms of life by modifying the simplest in a progressive, successive production of the classes forming the two or more series of animal and plant organisation. This successive production of the several class series involves a successive production of species that are unlimitedly mutable.

Here, significantly, the historical geography of our planet and its irreversible cooling are both missing. So, paradoxically, perhaps, the two elements that did most to make Buffon's thinking about nature and life count as historical are lost on the way to Lamarck. To that extent, Lamarck may be a less historical thinker about nature and life than was Buffon. Whether or not this heretical historiographical suggestion is ultimately sustainable, we can at least agree, surely, that an inquiry

31. For this interpretation of Lamarck and his relation to Buffon, see Hodge [8].

into the unity and singularity of Buffon's thinking may help to free us from the temptation of perpetuating many things the nineteenth century had to say about Buffon's century.

BIBLIOGRAPHY *

(1)†   BUFFON (G. L. Leclerc de), *Histoire naturelle, générale et particulière*, Paris, Imprimerie Royale, puis Plassan, 1749-1804, 44 vol.

(2)†   BUFFON (G. L. Leclerc de), *Œuvres philosophiques*, J. Piveteau ed., Paris, Presses Universitaires de France, 1954.

(3)   CANTOR (G.N.) and HODGE (M.J.S.) eds., *Conceptions of Ether. Studies in the History of Ether Theories. 1740-1900*, Cambridge, Cambridge University Press, 1981.

(4)   GAYON (J.), «L'individualité de l'espèce : une thèse tranformiste ?», this volume.

(5)   GOODMAN (D.), *Buffon's Natural History*, Milton Keynes, The Open University Press, 1959.

(6)   GREENE (J.C.), *The Death of Adam. Evolution and its Impact on Western Thought*, Ames, Iowa State University Press, 1959.

(7)   HANKS (L.), *Buffon avant l'Histoire naturelle*, Paris, Presses Universitaires de France, 1966.

(8)   HODGE (M.J.S.), «Lamarck's science of living bodies», *British Journal for the History of Science*, 5 (1971) : pp. 323-352.

(9)   LYON (J.) and SLOAN (P.R.), *From Natural History to the History of Nature : Readings from Buffon and his Critics*, Notre Dame, Indiana, University of Notre Dame Press, 1981.

(10)   ROGER (J.), *Buffon : Les Époques de la Nature. Édition critique. In :Mémoires du Muséum National d'Histoire naturelle, Série C, Sciences de la Terre*, T. X (1962). Reprinted in Paris, Éditions du Muséum, 1988.

(11)   ROGER (J.), *Les Sciences de la vie dans la pensée française du XVIIIᵉ siècle*, seconde édition, Armand Colin, Paris, 1971.

(12)   ROGER (J.), «Buffon» in *Dictionary of Scientific Biography*, C.C. Gillispie ed., New York, Scribner, 1970-80, 16 vols.

(13)   SLOAN (P.R.), «Buffon, German biology and the historical interpretation of biological species», *British Journal for the History of Science*, 12 (1979) : pp. 109-153.

(14)   SLOAN (P.R.), «From logical universals to historical individuals : Buffon's idea of biological species», *in Histoire du concept d'Espèce dans les Sciences de la vie*, Colloque international (J.L. Fischer and J. Roger eds.), Paris, éditions de la Fondation Singer-Polignac, 1986 : pp. 101-140.

Note added in proof : the Buffon scholarship of the late Professor Roger is accessible now also in his biographical study : *Buffon. Un philosophe au Jardin du Roi*, Paris, Fayard, 1989.

* Sources imprimées et études. Les sources sont distinguées par le signe †.

IV

LAMARCK'S SCIENCE OF LIVING BODIES

I. THE DIRECT AND INDIRECT PRODUCTION OF LIVING BODIES

(i) *The Interpretation of Lamarck*

As a historical figure, Lamarck proves a rather difficult subject. His writings give us few explicit leads to his intellectual debts; nor do they present his theories as the outcome of any sustained course of observations or experimental research; and, what is equally frustrating, it is hard to see how his personal development as a scientific theorist was affected by the dramatic political and social upheavals of the period, in which he took an active and lively interest.[1] And so, with his importance for later writers much clearer than his relationship to those of his own and earlier ages, historians have repeatedly interpreted his works as prophetic of doctrines developed more fully by subsequent generations. No less surprisingly, this facile tactic has provoked a reaction; we have been offered Lamarck as a Stoic, a romantic, harking back to Heraclitus.[2]

---

[1] For Lamarck's life, works and a bibliography of his writings, the standard source remains M. Landrieu, *Lamarck: Le Fondateur du Transformisme* (Paris, 1909). The only truly important and extensive study since is in the three authoritative volumes, contributed to the series, *Études d'Histoire des Sciences Naturelles*, by H. Daudin: *De Linné à Jussieu: Méthodes de la Classification et Idée de Série en Botanique et en Zoologie (1740-90)* (Paris, 1927) and *Cuvier et Lamarck: Les Classes Zoologique et l'Idée de Série Animale (1790-1830)* (2 vols. Paris, 1926). My own discussion of Lamarck has benefited from comments made on earlier versions of this paper by Dr. E. Mayr, Dr. T. Levere, Miss M. P. Winsor and Dr. J. Greene. The paper is based on a section of a dissertation, accepted by Harvard University in April 1970: *Origins and Species: A Study of the Historical Sources of Darwinism and of the Contexts of Some Other Accounts of Organic Diversity from Plato and Aristotle on*. There are two recent discussions of the manuscript writings of Lamarck now in the library of the Muséum d'Histoire Naturelle, Paris, but there is as yet no evidence that this material throws any new light on the sources and development of Lamarck's speculations. See the English article: I. M. Polyakov, "The Lamarck Manuscripts", *Russian Biological Review*, xlviii (1959), 289-296, and W. Coleman, "Lamarck Manuscripts", *The Mendel News Letter: Archival Resources for the History of Genetics and Allied Sciences (Issued by the Library of the American Philosophical Society)*, Number 4 (Nov. 1969), 3-4. For a fuller description of manuscripts, see M. Vachon, "A propos de manuscrits de Lamarck conservés a la Bibliothèque Centrale du Muséum National d'Histoire Naturelle de Paris (Note préliminaire)", *Bull. Mus. Nat. d'Hist. Nat.*, xxxix (1967), 1023-1027, where future publication of manuscripts was promised and M. Vachon, G. Rousseau and Y. Laissus, "Liste complète des manuscrits conservés à la Bibliothèque Centrale du Muséum Nationale d'Histoire Naturelle de Paris", *Bull. Mus. Nat. d'Hist. Nat.*, xl (1968), 1093-1102. Three items, all from very late in Lamarck's career and containing no sure clues to his earlier intellectual development, have been published in Russian translation in: I. M. Polyakov and N. I. Nuzhdina (eds.), *J-B. Lamark: Izbrannye Proizvedeniia v Dvukh* (2 vols. Moscow, 1955-59), ii, 573-673. I am grateful to my assistant Mr. M. Siekierski for examining these translated items.

[2] For the romantic Lamarck, see C. C. Gillispie, "The Formation of Lamarck's Evolutionary Theory", *Archives Internationales d'Histoire des Sciences*, ix (1956), 323-338, "Lamarck and Darwin in the History of Science", in B. Glass, O. Temkin, W. L. Straus (eds.), *Forerunners of Darwin: 1745-1859* (Baltimore, 1959), 265-291, and most recently, Chapter vii of Gillispie's book, *The Edge of Objectivity: An Essay in the History of Scientific Ideas* (Princeton, New Jersey, 1960). Greene has rightly criticized Gillispie's interpretation of Lamarck as plainly at odds with the texts and seriously misleading. See J. C. Greene, *The Death of Adam: Evolution and Its Impact on Western Thought* (New York, 1961), 350. Greene's own discussion of Lamarck deals only with his later writings and stays close to those texts. But, after some sound general remarks on Lamarck's affinities with Buffon, Greene slips into the erroneous orthodoxy: "Thus the mutability of species,

As history, neither of these interpretations can satisfy, of course. It is influences which make sense of history, and, to put it mildly (without forgetting the important use made by the French Enlightenment of ancient authors) it is hardly likely that Lamarck was extensively influenced, even indirectly, by Heraclitus, and—popular "historiography of ideas" notwithstanding—impossible that he was indebted to the as yet unborn Darwin.

There is never—to recite another truism—a historiographical substitute for two tasks: analysing an author's arguments strictly in his own terms, and relating those arguments, as far as possible, to the problems dominating the pertinent theoretical discussions of the day. My purpose here is to pursue the first as a prolegomenon to any future efforts at the second of these tasks. My strategy will be to start by exploring in considerable detail the main claims of Lamarck's "biologie". Then, in the second part of the paper, I want to exhibit the influence of the early Lamarck on the late Lamarck. Finally, my general conclusion will be that Lamarck's "biologie" was a natural philosophy project thoroughly characteristic of the late Enlightenment. For, like his mentor Buffon and others, he was in his theoretical writings on animals and plants primarily concerned with the venerable problem of the *generation* or *production* of these organized bodies; like Buffon, he sought especially for an alternative to the *emboîtement* theory as an answer to this problem, and, like Buffon, he deliberately proposed an alternative that was, in a definite sense, mechanistic and, in an equally definite sense, Newtonian. Certain views, now current, as to Lamarck's relationship to Buffon and to Newton are, therefore, erroneous.

To begin to appreciate the importance of these points, some deliberate forgetfulness is needed. Forget the words "evolution" and "transmutation"; Lamarck uses neither. Forget questions about the "origin of species"; the phrase is nowhere in Lamarck. Forget common ancestors; Lamarck's is not a theory of common ancestry. And, finally, put out of mind all thoughts of a geological succession and geographical distribution of species. All such topics must be kept out of the discussion until we have Lamarck's explanatory arguments before us and can see just how various kinds of facts bear on his assumptions and inferences.

Once we clear our heads of distracting anachronistic notions, Lamarck himself proves fairly helpful as a guide to his motives and intentions. But he does not remove all our perplexities. In particular, we would like to know more about how he came to propose, in 1800, what

---

a conception toward which Buffon had slowly groped his way, became Lamarck's starting point, the postulate which determined his approach to natural history." *Ibid.*, 162. Cf. a similar discussion of Buffon and Lamarck, in J. S. Wilkie, "Buffon, Lamarck and Darwin: The Originality of Darwin's Theory of Evolution", in P. R. Bell (ed.), *Darwin's Biological Work: Some Aspects Reconsidered* (Cambridge, 1959), 262-343.

were for him entirely new theories—the ones he is famous for—concerning the natural production of higher animals and plants from more simple ones. But even here we can reconstruct the main features of his thinking. For we have two strong lines of evidence, one fairly direct, the other less so. First, Lamarck tells us in two places why he gave up his former opinions and how he decided on his later ones. Second, his earlier writings make it plain what his old views were on a wide range of subjects, so that we can easily see what convictions he retained as well as which were abandoned. Significantly, his chemical, physiological and taxonomic teachings survived largely unchanged as part of his new biology. As long as we stick strictly to Lamarck's terms, we can, then, establish most of the context in which his mature account of organic diversity arose and why its arguments took the form they did.

Lamarck always stressed the connections among his conclusions on different subjects. Roughly speaking, his scientific thinking was concentrated in what he considered four intimately related fields of inquiry. These were (a) comparative and taxonomic botany and zoology, (b) the physics, chemistry, physiology and psychology of living and non-living bodies, (c) meteorology, and (d) geology or the theory of the earth. To understand his "biologie" we have therefore to appreciate what his various conclusions in each of these fields contributed to it. But we must first identify the central arguments which made up the core of this new "biologie".

(ii) *The central biological claim*

Lamarck defines his "biologie" as "conclusions concerning the nature, the faculties, the developments and the origin of living bodies".[3] These conclusions were a truly novel departure for Lamarck. For what we may call his central biological claim was a dramatic and complete reversal of his previous stand on the origin of living bodies.

For more than twenty years, Lamarck taught that non-living, inorganic, mineral, compound bodies cannot be produced by nature, for they can only be produced by living bodies whose existence likewise "in no way depends on nature", since "all that is meant by 'nature' cannot

---

[3] The title of a manuscript outline for a book Lamarck never published reads: "Biologie, ou Considérations sur la nature, les facultés, les développemens et l'origine des Corps vivans." See P. P. Grassé, " 'La Biologie' Texte inédite de Lamarck", *Revue Scientifique*, v (1944), 267-276. Apparently ignorant of its publication by Grassé, M. Klein also wrote of this MS.: "Sur l'origine du vocable 'biologie' ", *Archives d'Anatomie, d'Histologie et d'Embryologie*, xxxvii (1954), 105-114. Grassé, Klein and Polyakov (in his English article) all discuss the dating of this sketch but agree only that it was written after 1802 and before 1815. It seems clear, however, that Klein is right and that the sketch dates from 1802 or 1803, and is the "Équisse d'une *philosophie zoologique*", also referred to by Lamarck as "ma Biologie", which he told readers of his 1802 lecture they would find printed at the end of that lecture but which was not of course there. See "Discours d'Ouverture de l'An XI", reprinted by A. Giard, in "J. B. Lamarck, Discours d'Ouverture des Cours de Zoologie, Donnés dans le Muséum d'Histoire Naturelle (An VII, An X, An XI et 1806)", *Bulletin Scientifique de la France et de la Belgique*, xl (1906), 539. All subsequent references to these four lectures are to this handy edition of them.

give life".[4] Starting in 1800 he urges that, on the contrary, organic and inorganic bodies are all "productions of nature". Living bodies produce mineral bodies, as he always said. But nature can, Lamarck is now convinced, produce all living bodies, some "directly", the others "indirectly".[5]

The claim central to his new teaching is then this: in spontaneous generation, nature produces, directly, a living being, an organized body; the distinctive essence of living bodies is their possession of contained fluid parts moving in a containing cellular matrix; only the simplest living bodies can be produced directly, but given the motions of the vital fluids and the influence of changing circumstances, these simplest of organized bodies can give rise, in sufficient time, to all the more complex animals and plants, which have thus been indirectly produced by nature; these more organized bodies, and so the different species among them, have therefore been produced of necessity successively, starting with the simplest, and hence are not all of them "as old as nature herself".[6]

(iii) *Lamarck's route to his new biology*

Lamarck's *Philosophie Zoologique* of 1809 opens with an account—invariably, but for no good reason, ignored by historians—of how he arrived at the new doctrines of his nineteenth-century writings.[7] Faced with establishing general zoological principles for use in his teaching, he studied the diversity of animal organization "in each family, each order and above all in each class". He not only compared the various "faculties" due to the different degree of organization in "each race", but investigated, too, "the most general phenomena" of animals. Thus was he led to "difficult zoological questions" and to a first crucial step. Seeing

---

[4] *Recherches sur les Causes des Principaux Faits Physiques* (2 vols., Paris, 1794), ii, 213. These *Recherches* were written in 1776; see the "Avertissement" prefacing the work.

[5] The significance of the phrase, "production de la nature" will become evident shortly. Although Lamarck occasionally uses it of animals and plants in his writings of the 1780s and 1790s, from 1800 on its frequency rises dramatically when in his lecture of that year he announces his new views. See "Discours d'Ouverture, Prononcé le 21 Floréal An 8", *loc. cit.*, 443-595. The 1800 lecture was first published by Lamarck as an introduction to his *Système des Animaux sans Vertèbres* (Paris, 1801), 1-48. Lamarck complained that although many adopt it to refer to animals and plants, writers attach "no positive idea" to the general term "productions of nature". *Philosophie Zoologique* (2 vols., Paris, 1830), i, 90. There was only ever one edition of the *Philosophie*; this so-called "nouvelle edition" is only a re-issue of the sheets of the 1809 printing, as was pointed out by S. Smith, "The Origin of 'The Origin'", *The Advancement of Science*, lxiv (March 1960), 393-403.

[6] Another favourite phrase. See especially *Philosophie Zoologique*, i, 66.

[7] Sceptics will remember that scientists' later recollections of the steps leading up to their major insights have often proved inaccurate when checked against the letters and notebooks written at the time. But, by itself, that fact is no reason for discounting any autobiographical sources in a case like Lamarck's where there is as yet no such check available. Sceptics may also suggest that Lamarck only cast his book's preface in autobiographical form for rhetorical effect. But there is no shred of confirming evidence for this suggestion, and Lamarck's well-known honesty to weigh against it. Furthermore, even granted the force of the sceptics two warnings, the historian would still be mistaken who did not pay close attention to what Lamarck thought it proper to give as a public reconstruction of the steps which had led him to his new views.

a "singular degradation" in the complexity of animals, in passing "from the most to the least perfect of them", was he "not forced to think that nature had produced the different living bodies successively", beginning with the simplest?[8]

Next came an equally important step and one surely taken, like the first, as a direct if delayed result of Lamarck's new appointment, in 1793, to the study of the National Museum's invertebrate animals. Successive production seemed most certain, he tells us, when he discovered that the simplest animals have no peculiar organs or faculties, only those common to all living beings; some infusoria, for example, like higher animals, can move, but they lack any special locomotory organs. It was these conclusions, Lamarck tells us, which led him to "examine what life really consists of and what were the conditions" for its "production and maintenance in a body". Lamarck, that is to say, sought the conditions for the production—in either spontaneous or sexual generation—of a viable organic body. As the simplest organization offered just those conditions, and nothing else to distract him, "it held the solution to this problem". So far, so good: clearly, what he "next needed to know" was "how this simplest of all organizations, as a result of some changes, had been able to give rise to others less simple", indeed, to the entire animal scale.

Here, then, is Lamarck's own analysis of his two problems. His mature views are usually presented as springing from a shift in his interpretation of organic species, and have even been taken as little more than an easy extension to all species of his long-standing conviction of the mutability of mineral species. But "espèce" and its synonyms are quite absent from Lamarck's delineation of his two key problems. Obviously a change of opinion about living species was involved. In due course, we will examine Lamarck's own exegesis of this change. Meanwhile, however, we must recognize that Lamarck was not merely able to but chose to define the questions his new theories were to answer without mentioning species at all, and without drawing on any facts outside comparative anatomy and physiology.

(iv) *The two causes of diverse organization*

In two principles Lamarck "saw the solution" to his second problem. First, continued use by successive individuals of an organ leads to its development, while disuse has the contrary effect. Hence, "with a change of circumstances forcing the individuals to change their habits", organs

---

[8] These and the quotations in the next dozen paragraphs are all from the "Avertissement" prefacing the *Philosophie*. The translation of Hugh Elliot—always free and often anachronistic—contains a misleading 'howler' on its first page where Lamarck's "en procédant du plus simple vers le plus composé" is rendered "from the simplest worm upwards". See J. B. Lamarck, *Zoological Philosophy*, trans. H. Elliot (London, 1914), 1-2. Belief in inherited effects of habits had, of course, been prevalent since antiquity. See C. Zirkle, "The Early History of the Ideas of the Inheritance of Acquired Characters and of Pangenesis", *Transactions of the American Philosophical Society*, xxxv (1946), 91-151.

would develop or decline according to the changes in their use. Second, "reflecting on the power of the motion of the fluids in the very supple parts containing them", Lamarck was quickly convinced that, when accelerated, "the fluids modify the cellular tissue, open up passages, form various canals there and finally create different organs depending on the condition of the organization". Certainly, then, "the *motion of fluids* inside animals, and *the influence of new circumstances* on animals as they spread throughout habitable areas, were the two general causes which brought the different animals to the present state".

Everywhere in Lamarck's works we meet with motion and with fluids; there are vital fluids pushing open passages in plant and animal tissue, ocean waters scouring out the seabed, while the moon draws not only our planet's seas but also its atmosphere back and forth in tireless tides which, Lamarck contends, bring us some of the changes in our weather. But Lamarck always distinguished two kinds of motions: imparted, "communicated" motions like those of colliding billiard balls, and "excited" motions like those of muscles which are stimulated, not pushed, into action.

The preface to the *Philosophie* continues by describing how his earlier awareness of this distinction enabled him to see how the two causes of animal and plant organization combine in producing their effects. Turning from the simplest creatures to those which have "feeling [sentiment]", Lamarck sought the physical causes of this "faculty". Being already convinced that feeling could not be "the property" of "any matter" but must be due to "the workings of a system of organs", he looked for "le mécanisme organique" that could produce it. He decided that a fairly complex nervous system is a condition for any "feeling arousable by physical and mental needs [besoins physiques et moraux]", which can thus be a "source" and "means" of excited motions. But plants and the simplest animals, having no nervous system, lack this source of excited motions. How, then, do they stay alive? Lamarck asked himself. Both, he supposed, owe their vital motions solely to their penetration and animation by "subtle constantly moving fluids in the environment". Here was a "flood of light", Lamarck recalls. For "by joining this with the other two principles"—concerning fluid motions and changed circumstances—he "could seize the thread" connecting "the numerous causes of the phenomena presented by animal organization in its developments and diversity".[9]

It was subtle fluids, then, which allowed Lamarck finally to fit all this together. Plants get their motions always from the outside. All

---

[9] Cf. Lamarck's earlier enthusiastic comments concerning these subtle fluids: "to renounce research which can lead to knowledge of these subtle matters is to refuse to seize the only thread which nature offers to lead us to the truth". *Recherches sur l'Organisation des Corps Vivans* (Paris [1802]), 159-160.

irregularities in the successively produced scale of plants are therefore due to the uneven distribution of active subtle fluids around the world, especially the fluid of heat, caloric, which is the crucial feature obviously distinguishing various climates. It is the same with the simplest animals. But with the nervous system came a new source of irregularities. For fluid motions have the power not only to effect a general gain in the complexity of a growing individual: if given an extra acceleration in any portion of the cellular matrix, they produce extra complexity there. And so, when animals with a nervous system respond to changed circumstances by adopting new habits involving increased stimulation and use of any organ, that organ must gain in development as long as succeeding generations continue those habits, for all gains in organization, general or special, are conserved in reproduction by the transmission of the necessary motions from parents to offspring.

(v) *Organic diversity and the mutability of species*

When his central biological claim was completed, Lamarck had a theory in which for two reasons the mutability of species in changing conditions occupied a logically quite secondary place in the explanation of organic diversity. For, in the first place, the graduated scale of organization is not a serial ranking of species; on the contrary, only the largest divisions, the classes, are held to constitute such a series. Thus, in a lecture of 1806, Lamarck insisted that, thanks to "progress in comparative anatomy", we now know "the principal systems of organization" among animals, and by ranking these classes according to complexity we can draw up "an order which is in no way arbitrary and which we can take as the very order of nature". And in these conclusions, Lamarck finds nothing less than "very adequate means" for uncovering "the greatest secret of nature, the *origin of all natural bodies*".[10] Not just living bodies, of course; the italics are Lamarck's.

Secondly, there would never have been this enormous diversity exhibited in the series of classes were it not for the power of vital motions to produce the vast gains of organization involved in the indirect production of mammals from spontaneously generated infusorians. It is essential to recognize this obvious feature of Lamarck's position. For his claim implies not only that species are mutable, it implies also that a mere ability of species to be changed by altered circumstances could not, by itself, begin to explain the great range of diversity present in the whole animal and plant scales. That is why Lamarck developed his explanation for the ability of changed habits to alter animal organs as strictly a special case of the general power of fluid motions to produce local as well as general increases in the individual animal's organization.

[10] "Discours d'Ouverture de 1806", 558.

330

This analysis of the place of species' mutability in Lamarck's theorizing does not, of course, settle just why Lamarck first decided in favour of this mutability. And, fortunately, we do not have to guess on this matter; Lamarck himself has told us why he abandoned the constancy of species and so how he came to believe that any account of organic diversity must admit their mutability in altered conditions.

(vi) *The endless sources of changed circumstances*

Here geology and meteorology proved decisive, but once again in ways that only remind us more forcibly than ever how totally Lamarck's theorizing was conditioned by eighteenth- rather than nineteenth-century issues.

In a famous passage, he tells us he gave up the common error of thinking species constant when he realized that even though the conditions in which any species lives are constant for long periods of time, relative to our own observations, in fact everything on the globe's surface is slowly but steadily changing. Elevated places are always being worn away; "the shores of streams, of rivers, even of seas are insensibly displaced and so likewise are climates", he says, referring us to his *Hydrogéologie* for proof of "these great truths".[11] Sure enough, the parallel passages in the *Hydrogéologie* of 1802, and in a paper of three years later, introduce us to a fully developed theory of perpetually shifting climates, in which Lamarck's old interest in the weather is coupled with his more recent concern with invertebrate fossils and the general theory of the earth.[12]

We must notice right away what questions Lamarck's geology does and, even more importantly, does not try to answer. The theory is not concerned to discuss how the earth may have first got into its present shape or how it first came to be a fit abode for animals and plants as different as polyps, ferns and mammals. Lamarck's silence on these traditional topics for theorists of the earth was quite likely due to reflection on Lagrange and Laplace's recent demonstration of the stability of the solar system.[13] Anyway, Lamarck's geological arguments all take as given an oblately spheroid earth, with land and sea providing a complete range of suitable environments for all types of living things.

His geology is confined still further; it is a hydrogeology dealing principally, therefore, with the effects of water on the earth's surface.

---

[11] *Recherches* (1802), 141-142. This passage is among the pages reprinted by Giard, *loc. cit.*, 517-521.

[12] *Hydrogéologie* (Paris, 1802), 52 ff. The original pagination of the book is given in J. B. Lamarck, *Hydrogeology*, trans. Albert V. Carozzi (Urbana, Illinois, 1964). The most systematic presentation of his geological and climatic theories was given by Lamarck in: "Considérations sur quelques faits applicables à la théorie du globe, observés par M. Peron dans son voyage aux terres australes, et sur quelques questions géologiques qui naissent de la connaissance de ces faits", *Annales du Muséum National d'Histoire Naturelle*, vi (1805), 26-52.

[13] For this conclusion of the two astronomers, see R. Grant, *History of Physical Astronomy* (London, 1852), 52-56.

One effect is that the climate at any spot on the globe is slowly changing. For the oceans are constantly moving westward as eastern shores are worn away and western ones built up. Although the gravitational action of the sun and moon keeps the seas oscillating sufficiently to scour out the ocean beds which would otherwise silt up, the westward motion of the oceans takes millions of centuries to make a full revolution; nevertheless, it has probably completed several already. The movement is not precisely symmetrical, being deflected by irregular land masses. Hence there are changes in the distribution of elevated and depressed areas on either side of the equator, causing shifts in the earth's centre of gravity and so in the axis of its rotation. These extremely slow displacements of the earth's polar points are, in turn, "responsible for simultaneous displacement of the *equatorial bulges* and of the climates".[14]

(vii) *The three conclusions from fossils*

Fossils come into Lamarck's arguments in three ways, none of which marks any departure from what earlier writers, particularly Buffon, had been saying. First, he points to marine fossil deposits now far from the sea and at high elevations as evidence for the shifting of the ocean basins and the elevation of former ocean beds. Second, he notes that these fossils are often remains of animals or plants which live in a climate quite different from that now prevailing where they are found. They are evidence, then, for the displacement of climatic zones. Third, Lamarck compares fossilized remains with living animals and discerns in some cases a lack of "perfect resemblance" between them. Buffon and Cuvier, among others, had, of course, concluded that these fossils were relics of extinct, "lost" species.[15]

Tacitly referring to Cuvier, Lamarck's memoir of 1801, "Sur les Fossiles", discusses two possible inferences from this not uncommon discrepancy between the fossil and the living.[16] On the interpretation he rejects, the dissimilarity is evidence for widespread catastrophes which have annihilated certain species. Lamarck implicitly accepts that there are but two alternatives: species must change or go extinct. But, denying extinction, he concludes that these fossils are simply the remains of animals whose descendants have been forced by changing circumstances to undergo gradual modification in a long succession of generations. The comparison

---

[14] See the entry, "Points polaires", in Lamarck's analytical index to the *Hydrogéologie*. Cf. his paper of 1805, cited in note 12. Werner's attempt to find the origin of mineral strata in a primeval ocean was, of course, a quite different project, based on assumptions about mineral compounds that Lamarck's chemical theories ruled out altogether.

[15] *Hydrogéologie* and 1805 paper, *passim*. For a comprehensive survey of opinions on "lost species" from Ray to Cuvier, see J. Greene, *Death of Adam*, 96-133. Lamarck's place in speculations on the age of the habitable earth is discussed by F. C. Haber in his study "Fossils and the Idea of a Process of Time in Natural History", in *Forerunners of Darwin*, 222-262, and in *The Age of the World: Moses to Darwin* (Baltimore, 1962).

[16] Lamarck read a memoir, "Sur les fossiles", to the Institut National in 1799. It was presumably this memoir which he published, in 1801, as an appendix to the *Système des Animaux sans Vertèbres*, 403-411. See Landrieu, *Lamarck*, 186.

of fossils and living animals provides, therefore, just one more piece of evidence for the power of conditions to modify species in long stretches of time.

This is an important point, one to be grasped firmly. Lamarck's theorizing originated before, and never took into account or even mentioned the later conclusion from the identification of peculiar sets of fossils confined to particular strata: that there has been a succession of faunas and floras in which many extinct species have been replaced by many new ones. The successive production of species which appears in Lamarck's biology is never presented as an inference from stratigraphical palaeontology; it is deduced from the inability of nature to produce directly any but the simplest living bodies and the thesis that all living bodies, simple and complex, are truly "productions of nature". Nor does Lamarck ever argue that a proper appreciation of the earth's age must lead to a correct view of animal and plant production. Quite the converse: the *Hydrogéologie* predicts that when people recognize how long it takes to make complex organisms from simple ones, then old errors about the earth's youth will be rejected. We cannot be sure, but it looks as though Lamarck's own intellectual progress may be reflected here. Rather than having searched originally for a theory of organic production to complement newly enlarged views of terrestrial chronology, his enquiry had perhaps proceeded in precisely the reverse direction.

(viii) *The unimportance of geography*

Just as palaeontological data came into Lamarck's theorizing in a distinctive and characteristically eighteenth-century way, so do the facts of zoological and botanical geography. In his one brief reference to the geographical studies of Buffon and Lacépède, Lamarck cites the differences in animal organization in different climates merely as additional evidence for the susceptibility of organization to external influences, heat in particular.[17]

The striking irrelevance of further geographical questions to Lamarck's arguments is not surprising when we remember once again the questions his general biological claim was constructed to answer. Lamarck never even discusses whether a species can originate more than once, nor does he once take up the question, central to the geographical theorizing of the period, whether each species originates at just one spot or can on the contrary start up at several different places independently.

The reason for Lamarck's disinterest in these matters is fundamental. His theories were not attempts to understand why species appeared when and where they did. In elaborating his theory of animal and plant production, he is only concerned to argue that whenever and wherever lions and

[17] "Discours" of 1802, 515-516.

oak trees have turned up, they have owed their organization to the ability of nature to produce directly the simplest creatures and to the power of vital fluids to produce, given eons of time, any higher degrees of complexity. Certainly Lamarck's claims imply that there were at one time only simple animals on earth and as yet no lions or tigers. But he has no reason to discuss that moment in the earth's past, much less to seek traces of it in the rocks. All such questions were so far beside his main biological points that he gave them no attention at all.

Thus in a passage stressing the difference between "indirect" and "direct" production and the great time needed for the former, Lamarck says that nature cannot, of course, make a lion, an eagle, an oak or butterfly directly. They are of necessity indirect productions. Were these "species" annihilated in an area by some misfortune, it would, Lamarck observes, be a very long time before nature could replace them.[18] The implication is clearly that the very same species would eventually appear. But it is characteristic of Lamarck's position that his discussion ends there, without any explicit comment on the specific identity of the old with the new lions and butterflies; it is for his purposes a purely peripheral issue.

(ix) *Species among living bodies*

It is routine in discussing Lamarck's interpretation of species to go, as Charles Lyell did, straight to the chapter, "On species among living bodies", in the *Philosophie Zoologique*. But that book was, as Lamarck says, just a new edition of the 1802 *Recherches*.[19] And the structure of the *Recherches* cannot be overlooked.

There the general claim for the two causes—fluid motions and changing circumstances—is set out in summary. The conclusions are then supported in two ways. The first part of the book establishes a scale in the dozen or so main types of organization: mammal, fish, insect, and so on. The second part moves from sexual reproduction to spontaneous generation and the effects of life in a body. It is only in an appendix that Lamarck deals with "species among living bodies" and gives us the account of how he came to think that just as minerals, such as chalk or

---

[18] *Philosophie Zoologique*, i, 368

[19] *Philosophie Zoologique*, i, 14. Lyell's long exposition of Lamarck's views is in his *Principles of Geology* (5th edn., 4 vols., London, 1837), ii, 360-421. Lyell's interpretation of Lamarck is mistaken in three fundamental respects. He takes Lamarck to be proposing primarily a theory of the origin of species, he presents it as a theory of common ancestry and as an attempt to explain the fossil record. This allowed Lyell to separate the question of common ancestry from Lamarck's discussions of spontaneous generation and the alleged invariable progress from more simple organisms to those above them in the scale of perfection. This radical misrepresentation of Lamarck's position (which was followed by Whewell too) was of the greatest importance for the development of Darwin's and Wallace's thinking on the origin of species. Lyell's version of Lamarck influenced them as Lamarck's own writings could never have done. I have discussed this point at length in the dissertation cited in the first footnote. For Whewell on Lamarck, see his *History of the Inductive Sciences* (3rd edn., with additions. 2 vols., New York, 1890), ii, 561 ff. Whewell's section on Lamarck first appeared in the original edition of 1837 and was familiar to both Darwin and Wallace.

granite, can slowly change in certain conditions into other species by loss or gain of the elements, air, earth, fire or water, so organic species can in long periods become modified by external influences.

The same title, "species among living bodies", is used for Lamarck's introductory lecture of 1803, where he gives the topic a fuller treatment than elsewhere; the chapter in the *Philosophie* is just a rewriting of this 1803 lecture without its important opening remarks and with an additional note, "On species said to be lost", which simply argues that species' powers of preservation ensure that extinction does not occur, except perhaps as a result of human activity.

The 1803 lecture brings out, even more clearly than the *Philosophie* chapter, how derivative Lamarck's interpretation of species was from his general thesis about organized bodies. He begins by setting out that thesis in three "*considérations*", and then asks, "What is the species among living bodies?" He admits immediately that individuals always spring from ones like them. Offspring resemble parents, certainly. But naturalists go beyond this observation to assert that this likeness is constantly preserved and that the characters and so species are thus strictly unchanging and "as old as nature herself".

Against this conclusion, Lamarck makes several moves which come to this. In changed circumstances, a succession of individuals can in time take on new characters sufficiently different to mark them off as a new species. Even, therefore, if distinguishable but similar stocks were to be present at any time, the distinction of character would soon be destroyed by successive generations approximating in character closer and closer to one another. In the best-known genera we have, according to Lamarck, excellent evidence for this constant elimination of slight distinctions. For only in lesser-known genera can easily discernible groups be made out. Clearly, we have yet to find the intermediate individuals. To support this claim, Lamarck cites cases where species, introduced by gardeners, farmers and others, to new conditions, have taken on characters differing from their former characters as much as the characters of two recognized species differ from one another. And so ends the preliminary argument for the power of conditions to produce, not only the minor differences recognized among varieties but also those large differences used to mark species off from one another.

Lamarck now plays his ace. To see whether the traditional idea of constant species has any real justification, we will return, he says, to the "*considérations*" already established. These conclusions, he says, show six propositions to be true. The first is that all living bodies are truly products of nature. The next four complete the general claim for vital motions. Only in the final proposition are species even mentioned. Living bodies being subject to changes in organization, "species among them have been insensibly and successively formed, have only a relative constancy and cannot be as old as nature". Needless to say, fossils are not mentioned.

Nor is there here, any more than elsewhere in Lamarck's writings, the slightest hint of any argument for common ancestry. At no point does he extrapolate from the widely accepted common ancestry of conspecific *varieties* to a common ancestry for congeneric and confamilial *species*. Lamarck's theory neither arose as such an extrapolatory proposal nor, once that theory was fully worked out, was there any place in it for this extrapolation. His new view of species did not even require him to establish the common descent of all the members of a species from a single original stock. What his new view could use was any evidence of the power of changing conditions to alter slowly the characters of successive generations of individuals. But this supposition, as we have just seen, could be supported quite independently of any consideration of the Linnean thesis of an original Adam and Eve in each animal and plant species. Lamarck never discussed that thesis; he did not need to.

II. GOD, NATURE, MATTER AND LIFE

(i) *Earlier chemistry and later biology*

So far we have been discussing the account of organic diversity set forth in Lamarck's new "biologie", particularly the ways in which different kinds of facts came into his leading arguments. The overwhelming impression that Lamarck's theories sprang from the preoccupations of eighteenth- rather than nineteenth-century natural history and natural philosophy is amply confirmed when we turn to his earliest writings and find there the presuppositions which, more than any others, determined the final form of his biology.

Most importantly, Lamarck never changed his mind on what was involved in finding the physical origin or establishing the natural production of something. A footnote neatly summarized the burden of several longer passages written in 1776. "To explain physically the origin and mechanism of the universe", we would need to know three things which we do not know: first, "the productive cause of matter with all its qualities and powers", second, the cause of living beings, since matter cannot produce them, and third, the cause of "the activity distributed throughout the universe".[20] Taking as given, then, matter, the general motion in the world and the distinctive motions which are the peculiar and essential property of living bodies, Lamarck set out to explain everything else. His procedure shaped all his subsequent theorizing.

Lamarck was especially concerned to combat the thesis (he could have found it in Diderot, among others) that motion or activity is innate, natural or proper to any matter. Accordingly, he insisted on explaining chemical changes by tracing the motions not to the *nature* but to the *state* of the matters involved. Activity is not natural to the matter of fire, for

[20] *Recherches* (1794), ii, 26-27.

example, but the friction between two bodies can put any available fire into an active state, thereby producing a local source of activity. Lamarck's main and repeated objection to his contemporaries' chemical theories is thus not merely that they abandon the traditional four elements, earth, air, fire and water, but that they credit their proposed elements with affinities which are supposed to function as "active and special forces". These theories therefore postulate a source of activity which effects chemical combination but survives the combining undiminished. Now, Lamarck admits diverse affinities among his elements, but holds them to be purely passive "fitnesses [convenances]", determining the readiness but not supplying the activity with which any combination proceeds. However, while no matter in the natural state can modify itself, once brought far from this state by a suitable modifying cause, any matter, according to Lamarck, tends of its own power or faculty to return to its natural state. And, when combined in compounds, the elements are in highly modified states which only the special activity in living beings can bring about. All mineral compounds have hence been formed either "immediately", in and by animals and plants, or indirectly, in the subsequent decomposition of those minerals directly produced. Further, while all compounds are composed, some are more composed than others. The more "perfect", the more composed the compound, the greater "intimacy of union" and restraint among its elements, and so the stronger its tendency to that destructive decomposition which leads to the formation of "less perfect compounds" below it in the mineral scale.

Such are the foundations for that curious chemistry to which Lamarck was steadfastly loyal from his first to his last writings. The arguments central to his later biology were, of course, to match exactly these mineralogical maxims. In living bodies, Lamarck came to assert, a strictly contrary tendency toward higher degrees of organization follows just as necessarily from *their* essential properties: contained fluids moving in a cellular matrix; the more perfect the organization, the more active the motion, and so the stronger that tendency whose effects are completely conserved in reproduction.[21]

(ii) *The effects of life in a body*

Nor did Lamarck's analysis of what life is ever change. The first

---

[21] See Lamarck's own summary of his chemistry, in *Recherches* (1794), ii, 290-400. On affinities, see *Recherches* (1794), ii 387; *Memoires* 7-20, and *Philosophie Zoologique*, ii, 95. Lamarck talks of "tissu cellulaire" as do several of his contemporaries, but the phrase indicates nothing approaching the cell theory of the eighteen-thirties. See A. Hughes, *A History of Cytology* (London and New York, 1959), 33-34; and T. S. Hall, *Ideas of Life and Matter, Studies in the History of General Physiology*, 600 B.C.-1900 A.D. (2 vols., Chicago and London, 1969), ii, 139. Hall gives a useful discussion of Lamarck's physiological views. Lamarck's distinction between contained and containing parts goes back to antiquity; see William Harvey, *The Anatomical Lectures*, ed. and trans. by G. Whitteridge (Edinburgh and London, 1964), 8. See further, J. Schiller, "Physiologie et classification dans l'oeuvre de Lamarck", *Histoire et Biologie*, ii (1969), 35-57.

Recherches (1794) declared that "the essence of an organic being is constituted by the existence in it of a vital principle or motion which gives it the faculty to develop, to grow and finally to reproduce its like by means of appropriate organs".[22] As an explanatory manifesto, this may remind us of Aristotle. It should, for there is a logic shared by all comparative physiologies which proceed from the definition of the common essence to the elucidation of the many different faculties. But we are in the eighteenth century now; spirits and souls have been replaced by subtle fluids, which are, according to Lamarck, but the various active states of the ordinary element fire. As Lamarck was quick to observe, there were striking Newtonian precedents for such proposals. And, if he seems to recall Heraclitus, too, that is only because the distinction between matter and motion, stuff and its movements, also has a permanent logic. Immediate impressions can yield no precise historical clues here, then. We must descend gradually from the general features to the specific items of Lamarck's early chemical and physiological theories, if these are to cast light on his later account of plant and animal origins.

For let us watch Lamarck in explanatory action. Consider the phenomena of growth, the assimilation of nutrients, and death. The first *Recherches* (1794) argue that, in any alteration of a compound body, the first "principles" to escape are largely water and air—the most elastic, active and least fixed of the elements. Since earthy principles disengage with greater reluctance, assimilation constantly accumulates these so that growth is limited and death inevitable as the vital motions are gradually slowed and the organism turns to stone. The motions in more perfect organisms are more lively; their elements are consequently combined under a tighter restraint, producing a stronger tendency to disengage, so that these organisms need more food. For Lamarck, the phenomena of living beings are solved, then, when he has passed from identifying the essence of these bodies to explain a continuous scale in the intensity of its physical and chemical effects.[23]

(iii) *The production and reproduction of living bodies*

Lamarck's problems are always posed or declared insoluble in terms of matter and motion. The early *Recherches* (1794) tell us repeatedly that no physical origin, no natural cause, which is to say no source in any matter "in any circumstances", can be found for the peculiar motion present in living things. It is a motion "perpetuated by reproduction" and so is not transmitted by "any impact proportional to the masses involved". Furthermore, since this motion requires only organization for its maintenance and not any particular organ, living beings can differ

---

[22] *Recherches* (1794), ii, 188.
[23] *Recherches* (1794), ii, 398-399.

"in the number and perfection of the organs and so in the faculties peculiar to each species".[24]

From the impossibility of finding the physical origin of such motion, Lamarck derives the impossibility of finding a physical cause for the first individual of any species. Each such individual, as the first member of a distinct species, would be endowed with a specific version of the general vital motion which it could pass on in the generation of similar descendants. He is, then, convinced that "all which is meant by the word nature cannot give life—namely, that all the faculties of matter, added to all possible circumstances, and even to the activity spread throughout the universe cannot produce a being endowed with organic movement, capable of reproducing its like, and subject to death". For, he continues, "all the existing individuals of this nature come from similar individuals, which, all together, constitute the entire species" and "it is as impossible for man to know the physical cause of the first individual of each species as to determine physically the cause of the existence of matter or of the whole universe". The passage is well-known because Lamarck ends it with a denial that species can be changed by altered conditions.[25] But although Lamarck has given us this clear statement of this, the position he will have abandoned by 1800, the historical significance of his switch has been consistently missed.

The key to understanding his switch is to recognize that it was a shift from one side to another in a venerable eighteenth-century debate which, indeed, went back to the days of Descartes and Boyle. The question was, in the phrasing of the day, Are the laws of nature able to *produce* an animal or plant? Can a living, organized body be formed by mere matter in lawful motion? The terms of the question were in part mediaeval and in part an innovation of a much later period. Scholastic theologians had distinguished between God's unique power to *create* matter *ex nihilo* and his creatures' power to *generate* or *produce* other creatures from pre-existent material. The word "production" like "eduction" was thus originally a piece of technical terminology used in the Christian exegesis of Aristotelian accounts of causation. But the general questions as to what universal laws of motion there are and what matter in lawful motion can or cannot

---

[24] *Ibid.*, 185-215.

[25] *Ibid.*, 214. Lamarck was, as taxonomists say, a "lumper" rather than a "splitter". Thus when he says, here: "If there exist any varieties produced by the action of circumstances, these varieties do not change the nature of the species; but doubtless one often errs in indicating as a species what is only a variety; and I perceive that this error may be of consequence in reasoning upon this subject", he means that, in alleged cases of changes in specific characters, altered conditions are really only causing new varieties. For the several earlier statements of Lamarck, from 1778 to 1793, asserting the fixity of species, see Landrieu, *op. cit.* (1), 290. Note also that in 1794 Lamarck rejected a reported "transformation, du *lichen pyxidatus* en *lichen digitalus*"; *Voyages du Professeur Pallas dans Plusieurs Provinces de l'Empire de Russie et dans l'Asie Septentrionale, traduits de l'Allemand par C. Gaultier de la Peyronnerie. Nouvelle édition, revue et enrichie de notes per Lamarck, Langlès et Billecoq* (8 vols., Paris, 1794), ii, 459.

produce, date, of course, from the establishment of the new "mechanical philosophy" of the seventeenth century.

The pre-existence of germs doctrine, the box-within-a-box theory, emerged late in that century as a direct response to these new issues.[26] George Cheyne, for example, explicitly denied, in adopting that theory, "that a Plant or an Animal can be produced by Mechanism, i.e. Nature, or the Laws of Motion", and sided therefore with those "modern philosophers", who, as Issac Watts wrote in 1733, supposing "that the Formation of Plants and animals is beyond the Reach and Power of the Laws of Nature", have therefore conceived "that the Creator himself in the first individuals of every kind included all the Future Plants and Animals", and that "the daily Productions of Nature are nothing else but the unveiling of these little Plants and Animals in continual Successions".[27]

Lamarck's awareness of his place in the history of this debate is evident throughout his writings. It was heightened by his knowing, as he could not have failed to know, that Georges Cuvier was a firm although somewhat reticent supporter of the *emboîtement* doctrine.[28] Once one grasps that Lamarck's new "biologie" was, above all else, an attempt to replace that doctrine, then many things in it make sense which otherwise do not.

In particular Lamarck's general relationship to Buffon becomes plain. The usual view of this relationship is, of course, that Lamarck must have been most influenced by what he must have interpreted as a half-hearted attempt by Buffon to base a new zoology on a denial that species were as rigidly immutable as most naturalists held. Encouraged by this Lamarck, the legend has it, proceeded to bring to fruition what his master had so presciently sown. But this legend is simply an error followed by an unwarranted guess. For not only is its view as to what was central to Buffon's zoology quite mistaken, there is no evidence that Lamarck shared that mistaken view. We can see this by considering briefly what was central to Buffon's general treatment of organic beings: his inveterate opposition and deliberate alternative to the theory of *emboîtement*.

[26] For a definitive history of these and related issues, see the masterly work by J. Roger, *Les Sciences de la Vie dans la Pensée Française du XVIIIe Siècle: La Génération des Animaux de Descartes à l'Encyclopédie* (Paris, 1963). Although this magnificent study stops short of Lamarck it provides the indispensable background for any analysis of his writings. My discussion of Buffon and hence of Lamarck is greatly indebted to Roger's insights and scholarship. For a full commentary on the term *productio* and its cognates in modern languages, see H. B. Adelmann, *Marcello Malpighi and the Evolution of Embryology* (5 vols., Ithaca, New York, 1966), ii, 934-935.

[27] George Cheyne, *Philosophical Principles of Religion: Natural and Revealed* (4th edn., corrected. London, 1734), 131; Isaac Watts. *Philosophical Essays on Various Subjects* (6th edn. corrected. London, 1763), 209-210. The first edition appeared in 1733 and gives an excellent summary of the various positions then current, regarding the production of animals and plants.

[28] For Lamarck's explicit rejections of pre-existent germs, see his *Mémoires de Physique et d'Histoire Naturelle* (Paris, 1797), 272, and *Système Analytique des Connaissances Positives de L'Homme* (Paris, 1820), 123. For Cuvier's cautious endorsement of pre-existing germs, see the passages quoted in W. Coleman, *Georges Cuvier, Zoologist: A Study in the History of Evolution Theory* (Cambridge, Massachusetts, 1964), 163-164.

340

More than anyone, Buffon made familiar the term "reproduction", which Lamarck adopted in turn. Buffon insisted that parents do truly reproduce themselves and do not simply nourish and release a preformed germ with whose organization, formation and production they have had nothing to do. But even more importantly, Buffon's celebrated chapters on animal generation in the second volume of the *Histoire Naturelle* had argued that the new Newtonian understanding of matter and its forces allowed a new definition of the sense in which "mechanical principles" *could* account for the production of organized bodies. Descartes' exclusion of all but mechanical principles was fine as a programme, said Buffon. But his list of mechanical principles was too short. Newton has not only added to the list, but given us a better way of arriving at it. Descartes derived his list of mechanical principles from an *a priori* analysis of the essences of matter, of motion and of God's perfection. Newton tells us to infer our mechanical principles *a posteriori*. In addition to extension and inertia, matter must be credited with solidity and gravity. Gravity acts not according to the surfaces of the bodies, as does bumping, but, Buffon stresses, according to their masses. The active force of gravity must, then, penetrate all the matter of the attracting bodies. May we not think further that there are other active penetrating forces possessed by matter? We may, and confronted with animals and plants, we should. For their component particles are clearly acting on one another in ways, and so, we may conclude, with forces, analogous to gravitational attraction. To explain how these particles and forces act in reproduction and spontaneous generation is therefore to trace these phenomena to what Buffon insists can fairly be called "mechanical principles".[29]

Lamarck's new biology embodied exactly the same general explanatory ideals as those inspiring Buffon's opposition to the theory of pre-existent germs. But he did not accept Buffon's alternative to that theory, and for good reason; his chemical and physiological principles not only ruled out Buffon's proposal, they clearly suggested another solution to the problem.

(iv) *Sexual generation and direct and indirect production*

Lamarck deals with reproduction at length in the *Mémoires* of 1797 and again in the *Recherches* of 1802. As Cuvier recognized in his notorious *éloge* of Lamarck, the extensive continuity between the two books is the proper introduction to the leading theses of the new "biologie", for they all appear to date from the few years separating these works.

Buffon's alternative to pre-existent germs was "organic molecules", possessing active, penetrating and organizing forces. An animal or plant is an assemblage of these "living molecules" which are dispersed but not

---

[29] *Histoire Naturelle, Générale et Particulière* (44 vols., Paris, 1749-1804), ii, 18-53.

destroyed when the organism dies. Free organic molecules arise also from the action of heat on certain inanimate materials. In spontaneous generation, free molecules gather themselves together to form a living body which matures by assimilating further organic molecules in its food. In sexual reproduction, the new individual is produced by molecules provided from the parents' bodies in their respective seminal contributions.

Lamarck explicitly rejects Buffon's "organic molecules". He says he cannot see how they could be assimilated in nutrition without blending into one another; and, being composite bodies, they would tend to destruction once outside a living organism. Apparently still opposed, in 1797, to those like Buffon who had accepted spontaneous generation, Lamarck says that it is with good reason that "everything having life is said to come from an egg" (although his position at this time may just possibly have been that of Harvey, who reconciled this dictum with spontaneous generation by holding that there, as in sexual generation, the new individual always starts as an egg). In any case, we are told by Lamarck only that an egg arises in a female parent by growth, just like any other organ. It is then furnished with assimilated and prepared matter by the male's seminal liquor in which is a subtle vapour, an *aura vitalis*. This active vapour or fluid provides the stimulus which excites the orderly, organic motions constituting the beginning life of the new individual.[30]

The discussion is elaborated in the *Recherches* of 1802. The unfertilized egg is unreceptive to vital motions. No amount of incubation will animate it. The *aura vitalis* prepares the egg for animation by penetrating it and reducing its chaotic matter to order. The vital motions succeed immediately in viviparous animals, and after incubation in oviparous animals. In spontaneous generation, nature imitates these processes, achieving the production of a living body without the aid of organization or vital motions already existing. Heat or caloric, which is just the element fire in an active state, is everywhere and is analogous to the *aura vitalis*, perhaps identical with it. Heat, this "material soul of living bodies" as the *Philosophie* dubs it, can prepare a "gelatinous body" and then excite in it the appropriate motions of its fluids.[31]

The rest of the argument is now familiar to us. Arguing from the power of vital motions to increase the amount and the specialization of organization in a growing *individual*, and the ability of reproduction to conserve every increment of organization, Lamarck concludes that an infusorian can give rise after enough generations and sufficient ages of time to organisms of all the classes constituting the natural order of the animal and plant kingdom.

---

[30] *Mémoires*, 272-274. As additional evidence that, in 1797, Lamarck still has not reached the central convictions of his later biology, see his rejection, once again, of the possibility that mineral compounds can be produced "by nature" and his reassertion of the "contrary" thesis that they are all produced by "living beings"; *Mémoires*, 319-320.

[31] *Recherches* (1802), 92 ff. Cf. *Philosophie Zoologique*, ii, 82.

(v) *The natural order*

In natural history Lamarck had two mentors, who bequeathed very different, even conflicting, legacies to the next generation. Linnaeus insisted on the paramount value of classification. Each species should be assigned a place in a classificatory system drawn up by introducing a sequence of divisions and sub-divisions into the animal and plant kingdoms. Buffon, especially in his earlier writings, had been scornful of the "*nomenclateurs*", mainly on the ground that they arbitrarily introduced divisions into what is a completely continuous series of animals.[32]

Lamarck, by following both the Swede and his own countryman, early took up a novel position on these matters. He gave a full statement of this position in the first, 1778, edition of his *Flore Françoise* and never departed from it, except of course in giving up the conviction that species were marked off by real and permanent differences. He always agreed with Buffon that all divisions above the species are artificial and with Linnaeus that groupings of species can be made natural.[33]

Start with any species. There will be, according to Lamarck, others which can be grouped with it quite naturally because they resemble it in numerous ways. Between such species, there are, in Lamarck's words, "*rapports naturels*". Start with another species and, the same being true, one can assemble another group joined by natural affinities. But the demarcation between neighbouring groups can only be drawn arbitrarily. There will always be borderline cases, species which have affinities with those in more than one group. Gaps between groups exist only because intermediate species have still to be discovered.

Following these assumptions, Lamarck separated the task of providing a set of artificial divisions and sub-divisions enabling a student to identify the plants in his area, from the task of establishing the "natural order" among species. Linnaeus had distinguished between an artificial system of classification and a natural one, but Lamarck does not present his "natural order" as a classification at all. It is not a classification, he observes, only an arrangement of continuous groups. Thus, if genera are being arranged according to their affinities, the arrangement recognizes intermediate species linking any two genera.

The essential step in arriving at an arrangement which displays the

[32] For a more extensive account of Lamarck's taxonomic writings and their relation to those of other naturalists, see Daudin's exemplary studies, cited in the first note to this paper.

[33] See the "Discours Préliminaire" which appeared unchanged in all three editions (1778, 1795, 1805 of Lamarck's *Flore Françoise*. As cited here, the "Discours" is in *Flore Françoise* (2nd edn., 3 vols., Paris, 1795), i, i-cxix. See also, for Lamarck's early statements of his taxonomic views, various general articles—especially, "Analyse", "Botanique", "Caractères", "Classes", "Espèce", "Familles", "Genres"—published in the first three volumes (1783, 1786 and 1789 respectively) of the *Encyclopédie Méthodique: Botanique. Dictionnaire de Botanique* (13 vols. Paris, 1783-1817); the preface and introduction to his *Tableau Encyclopédique et Méthodique Des Trois Règnes De La Nature: Botanique. Illustration des Genres* (2 vols., Paris, 1791-93); and the many articles (listed in Landrieu's bibliography) contributed by Lamarck to the *Journal d'Histoire Naturelle* (2 vols., Paris, 1791-92).

natural order is, Lamarck always maintained, deciding on two indispensable landmarks. Between the simplest and the most complex plant
there is, he held, the smallest of all degrees of affinity. With these landmarks identified, the job of arranging the rest can proceed. The identification is simple enough in principle. If you know what a plant is, then,
Lamarck assumes—and here of course, he follows a long tradition in the
interpretation of the *scala naturae*—you know what makes one plant more
perfect than another. He has difficulty choosing a candidate, but is
satisfied his scale should be headed by "the most alive, the best organized,
in a word, the most perfect plant".[34] And so, in 1778 Lamarck offers his
readers a natural arrangement of plant genera in which each genus
(except the two at either end of the series), has just two natural neighbours,
one above and one below it in the scale of perfection which links the most
with the least alive and organized.

By 1785 he was arguing for parallels between the plant and the
animal series. He had already abandoned genera as the groups which
form a continuous animal series and had settled on six classes—quadrupeds,
birds, amphibians (including reptiles), fishes, insects and worms—as the
constituents of his linear scale.[35] As the years went by, and particularly
after his zoological work at the National Museum began, this list of six
classes increased and was eventually more than doubled. But in spite of
Cuvier's objections, Lamarck continued to insist on a linear series in the
scale of classes. Among smaller groupings—orders, families, genera and
species—there were certainly affinities which could only be represented
by branched and reticulate arrangements, but not among the classes.
On that continuous, linear series of classes the new "biologie" was, of
course, founded.

(vi) *The very order of nature and the order of production*

With his new theory constructed, Lamarck eventually found a
further distinction of value. At first he simply identified the two linear
series of classes as the actual orders followed in producing the higher
from the lowest animals and plants. Later he came to distinguish these
linear series of organization from the order of production.

The grounds for this distinction were straightforward. First, among
invertebrates there seemed to be two quite separate series of classes.
Second, the terrestrial mammals clearly had affinities with the amphibian
mammals, which in turn had affinities, Lamarck thought, with reptiles,
as do birds. Rather than land-mammals coming from birds, then, it

---

34 *Flore Françoise* (1795), i, xciii.
35 "Mémoire sur les classes les plus convenables à établir parmi des végétaux, et sur l'analogie
de leur nombre avec celles determinées dans le règne animal, ayant égard de part et d'autre
à la perfection graduée des organes", *Mémoires de l'Académie des Sciences*, 1785, 437-453. This
memoir was largely reprinted as the article "Classes" (1786) cited in note 33.

344

made more sense to suppose that the chelonian reptiles gave rise to the birds, while saurians led to amphibian mammals, and these to three "branches" of higher mammals—the cetaceans, unguiculates and ungulates.[36]

Lamarck's use of the word "branche", and his famous branched diagram illustrating these speculations in a hastily added appendix to the *Philosophie*, have invariably led commentators quite mistakenly to take this scheme as a genealogical reconstruction of the common ancestries joining the divergent limbs of a tree of descent. A look, however, at Lamarck's final and much more systematic and explicit treatment of the scheme, given six years later, in a *"Supplément"* to the first volume of the *Histoire Naturelle*, shows at once how entirely remote were his presuppositions from those of any theory of common ancestry.[37]

The ranking of animal classes gives us a linear arrangement which, says Lamarck, not only facilitates our zoological studies, it presents "the very order of nature, that is to say, that order which nature would have carried out if accidental causes had not modified her operations". This arrangement is the "simple series". The "order of production", however, differs from this simple linear series in two ways, both due to accidental causes. First just as there is a plant and an animal scale whose simplest spontaneously generated members differ as a result of accidental differences in chemical composition, so there are two animal scales whose lowest extremities owe their differences to accidents of chemistry. Second, there is branching in the order of production, due to the operation of an "accidental and consequently variable cause".

This language is highly significant. Branching, according to Lamarck, is not due to the *essential* properties common to all living bodies. It is due to things which are in the every-day, and in the school logician's sense, *accidental*, things which happen to be true of some animals and plants. Because of their exposure to uneven and changing circumstances and their responses to such changes, the organization of animals and plants has not always made steady gains in complexity.

The result is that two classes which are adjacent in the simple animal series, drawn up by comparison with human structure, may have no members which resemble each other very closely. For example, the crustaceans are rightly placed just below the annelids in the simple series, but there is no "truly nuanced transition" between them. The difficulty is removed, Lamarck contends, by supposing that the annelids came, not from the crustaceans immediately below them, but that they "came originally" from intestinal worms, and that another branch from the same source led successively to the insects and crustaceans. Thus, although close enough in their *degree* of organization, annelids and crustaceans are too

36 *Philosophie Zoologique*, ii, 451-466.
37 *Histoire Naturelle des Animaux sans Vertèbres* (7 vols., Paris, 1815-22), i, 451-462.

unlike in *structure* for a transition to be presumed. But lateral branching, due to the modifying accidental cause, allows the existence of a simple scale in the organization of the classes to be reconciled with a lack of close resemblance or transition between any two classes. And that is the end of Lamarck's discussion. We note again that fossils are not mentioned. Nor is there any reason why they should be. All of them fell clearly into one of Lamarck's dozen or so "masses" or classes. They simply do not raise any additional questions.

The contrast with a theory of common descent is now obvious. Such a theory ascribes the common characters shared by all the mammals to their descent from a common ancestral species possessing those characters. For Lamarck, on the contrary, all the mammals are alike, not because they have a common mammal ancestor, but because they have all resulted from the same complexification of some earlier reptiles. What is more, of course, Lamarck's account also supposes a constant elevation of reptiles into mammals, and so, while the first mammals ever to appear and the ones now living both came from reptiles, they did not come from the same stock of reptilian individuals.

(vii) *Lamarckian and Buffonian science*

We can now see that these features of Lamarck's theory are not, to use his own phrasing, the results of mere accidental causes. They arise directly from the assumptions, explicit and implicit, which define his whole explanatory project. At its most general, his biology sought to square the essential properties of all matter with the distinctive and essential properties common to all living bodies, and then to square the latter with the special properties peculiar to different animals and plants. The powers present in activated subtle fluids turn the inanimate into the simplest of the living, and the powers of life in these, in turn, provide the origin of all the diverse organization, and hence faculties, of higher organisms.

The great explanatory achievement made possible by this synthesis was, Lamarck early claimed, the demonstration that gravitational attraction and heat were adequate to account for all the properties of all bodies. A fundamental misconception, shared by Cuvier and Bichat, can thus be exposed. These two, Lamarck reports, have supposed that a living body possesses peculiar forces by which it resists the destructive effects of the interplay of the elements outside it. Death, according to their view, comes only when this battle is lost. But, Lamarck counters, there are no special active forces peculiar to chemical elements. The organism has no external destructive powers to contend with and no peculiar vital powers with which to postpone death. Death is an inevitable consequence of life; in assimilating food, the organism accumulates a higher and higher proportion of earth and so petrifies itself. There are no special forces responsible for any properties of living bodies, irritability, sensation and thought not excepted. Irritability is due to a tension or orgasm in the parts of the cellular tissue, a tension produced by the attraction of gravity balanced by the repulsion of heat. In touching an irritable body, which is to say an animal, we cause some caloric to escape, thereby allowing a local gravitational contraction of the tissue.[38]

The explanation is wholly typical of Lamarck. It is required by a striking and familiar fact of experience, the peculiar irritability of even the lowest animals. But no experimentation is referred to. The phenomenon is accounted for by showing its consistency with the first principles of chemistry and physiology. No verification or testing of the proposal is recorded or even recommended. That is not Lamarck's way. Even in his epistemological apologetics, where he displays his new biology as being all that sound reasoning from "positive" facts should be, there is no suggestion that, in the face of rival explanations for the same phenomena, one should deliberately seek those further facts fatal to one conjecture but not to another.[39]

But it would be a mistake to let Lamarck's distaste for discussing rival theories mislead us; we have here no grounds for concluding that his principal explanatory goals required him first to dissociate himself decisively from all the presuppositions acceptable to the previous generation. For, on the contrary, the central claim of Lamarck's biology was in a real sense just a revision of the main proposal of Buffon's natural science.

As his voluminous work proceeded, Buffon had championed, ever more explicitly, gravitational attraction and heat as fully adequate to produce all the phenomena presented by the living no less than the lifeless. As in cosmogonical astronomy, so in chemistry, Newtonian physics suffices. There is no need, Buffon told chemists, to seek special laws of chemical affinity. For with the inverse square law, there will be differences in gravitational attraction due to a diversity of figures in their respective constituent particles which can cause all the various affinities elements or compounds have for one another.[40] And so, with "living, organic molecules" arising simply from the action of heat on certain kinds of inanimate matter, heat aiding gravity is responsible for all the additional active qualities or forces peculiar to these special molecules; thus, seeing that in a hot enough earlier phase of any planet's cooling history new species of complex animals and plants can originate as readily as the simplest organisms do today, in spontaneous generations involving only

[38] *Recherches* (1802), 71-83.
[39] These epistemological apologetics get their fullest statement in the *Histoire Naturelle*, i, 334-341, and in the *Système Analytique*. See also the manuscript sketch, "Aperçu Analytique des Connaissance Humaines", published, with an English translation, in W. M. Wheeler and T. Barbour, *The Lamarck Manuscripts at Harvard* (Cambridge, Mass., 1933), 75-83 and 175-183.
[40] See Buffon's "De la Nature. Seconde Vue", of 1765, in the *Histoire Naturelle, Générale et Particulière* (44 vols., Paris, 1749-1804), xiii, i-xx. For an account of Buffon's place in the history of Newtonian approaches to chemical theory, see A. Thackray, *Atoms and Powers: An Essay on Newtonian Matter-Theory and the Development of Chemistry* (Cambridge, Mass., 1970), 155-160.

the normal interaction of these molecules, then no part or faculty of any creature is due, ultimately, to any cause but these two ubiquitous natural agents.

In constructing his new "biologie" Lamarck's response to this Buffonian thesis was, in effect, to say, "Yes, nature does produce all organization and does it by heat and gravity. But the intermediary is not any organic molecules able to assemble themselves into a mammoth or an oak no less easily than into an intestinal worm. The intermediary is the organization of the simplest animals and plants. Only these can be produced by heat and gravity. The rest must, therefore, be generated successively, over enormous stretches of time, from these simplest ones. And so species are not 'as ancient as nature herself' as you, Buffon, say in your celebrated 'Second View of Nature'. Nor, therefore—to take the later suggestion of your 'Epochs of Nature'—has there been that succession of ever less complex species which you deduced there, by joining your theory of heat-caused 'organic molecules' with your estimates of the dwindling degrees of temperature supposedly accompanying the aging of the earth. For, quite the reverse, the correct view of how heat causes organization entails that the more complex must have come long after the more simple organisms, whatever temperature changes, if any, the earth may have undergone."

(viii) *Lamarck in a sense a mechanist and in a sense a Newtonian*

But did Lamarck share Buffon's commitment to "mechanical principles"? Was Lamarck, too, deliberately working to bring organic diversity within the ambit of Newtonian science?

On one recent view, of course, there are easy answers to these questions. For according to this proposal, Lamarck was a tender-minded romantic, who, along with Goethe, Schelling and their like, recoiled from the heartless march of modern science. Spurning especially Newton and all he stood for, these men constructed a quaint, comforting, organismic, even anthropocentric, cosmology as an intellectual wrapping against the chilling emotional prospect and encroaching objectivity of the Galilean-Newtonian tradition.

This account of Lamarck cannot, however, survive any confrontation with his writings. His chemistry was archaic certainly; it was also reactionary, although we must respect chronology here; the new Lavoisierian chemistry Lamarck attacked so relentlessly in the 1790's was hardly a comprehensive and sytematic doctrine twenty years before, when Lamarck first took up his position in this field. Nor was Lamarck trying to discredit any attempt to do for chemistry what Newton had done for physics, whatever that might involve. The short "Discours préliminaire" of the first *Recherches* (1794) says merely that the physicists and chemists still differ over several points, and "the immortal Newton himself, who knew and fixed the certain laws of attraction, was not everywhere equally fortunate in explaining many facts he sought causes for". When Lamarck mentions the heroes of the scientific movement again, it is, first, to praise Descartes for introducing "freedom of judgment" into science; and, second, to disclaim any intention of explaining "the formation of the universe", by proceeding directly to first causes and universal laws. Views so sublime can only occupy "those rare and superior geniuses, who, like the Leibnitz's, the Descartes', the Newtons", rising, as it were "above human faculties", seem made to uncover natural secrets that would otherwise remain forever unopened.

Nor does Lamarck anywhere object that such moderns have wrongly subverted man's traditional position at nature's centre. His final view of our species was, of course, that we are strictly nature's productions. We are indeed, the *Hydrogéologie* declares, her most impressive product to date. Accordingly, Lamarck recommends that, as the most complex, human structure should be taken as the standard of perfection when we draw up a natural arrangement of animals by their affinities. But the quite distinct plant scale should be arranged according to its own independent criteria. Further, when using these two separate natural orders to chart the course of natural production, we will find even animal lines, like that now headed by the birds, which point no longer at man.

There is, then, no *a priori* anthropocentrism presupposed here. It is, for Lamarck, a purely contingent matter of fact that the species doing zoology also has the most complicated organization. He would explain the fact causally; as with all species, our faculties arise from purely bodily structures; that is how nature produces reason and irritability alike.

Thus Lamarck seeks nothing resembling the anthropocentrism found in German romantics like Schelling and Oken. For, in his philosophy of nature, there is no place for their transcendental deductions from epistemological and ontological principles, showing, as these idealists put it, that man is nature grown up and that the human mind is nature become conscious of herself.[41] Lamarck himself made this clear in several passages, especially the section of the *Histoire Naturelle* entitled "On Nature or the power in some way mechanical, which has given rise to the animals and necessarily made them what they are". Here he argues that his new "biologie" has the great merit of allowing him a consistent mechanistic and deistic position.

Phenomena such as life, feeling and thought are neither the properties of any matter nor are they things, we are told repeatedly. They are the effects of an order of things present in a body; they are the effects of

---

[41] For these features of German romantic idealism, see A. G. F. Gode-von Aesch, *Natural Science in German Romanticism* (New York, 1941), *passim*.

"mechanical causes, regulated by laws". Life, feeling and thought are therefore "physical, mechanical and organic phenomena".[42] The three adjectives are here strictly complementary. "Organique" confirms "mécanique" in Lamarck because it means arising from organization rather than from spirits or animation—a fact which should make us very wary of accepting from historians any impressionistic distinction between organic and mechanical "world-views" at this period.

The distinction which Lamarck himself regards as crucial is something quite different. It is present in his earliest writings and gets explicit treatment in the later works. The *universe*, he says, is all matters which are quite inert, powerless and motionless; *nature* comprises not physical things but the order of things; space, time, motion and laws of motion. Nature is thus an active power created by God which goes to work on matter (also created by God) to produce all bodies and their properties, necessarily. Like nature, life is a power but is itself a production of nature, not a direct creation of God.[43]

Lamarck accordingly castigates those of his contemporaries who have revived the *anima mundi* of the Greeks. Nature, he says, includes no such universal soul power working toward one end. Lamarck is resolutely necessitarian. There is no intention in nature or in life, only in God. What appear to indicate intention in natural powers are in truth only the necessary consequences of the laws of motion laid down by God.[44] God is the prime mover. We reach him in tracing back activity, which is to say, motion, to its source. But this "hierarchy of powers" need not be fully delineated. The motion of the earth round the sun keeps plenty of terrestrial caloric agitated and active. There is no need for the chemist or biologist to look any further for his sources, his principles of motion. God is responsible for the original properties of matter but the properties of bodies, living and lifeless, are due to the subsequent motions of matter.

All those motions and so, therefore, the properties of all bodies are, Lamarck concludes, due to "two opposed forces"—the universal gravitational attraction which is "regular" in its action and the irregularly acting repulsive power present in the ubiquitous subtle fluids.[45] This was of course a conclusion common among natural philosophers upholding, as they saw it, the true Newtonian tradition. But labels are in themselves very limited sources of historical insight. Many and various are the eighteenth-century intellectual traditions claiming to trace their ancestry to Newton; we need always, therefore, to specify carefully the taxonomic

[42] *Histoire Naturelle*, i, 304-341; *ibid.*, 12-13.
[43] *Ibid.*, i, 307-322. Cf. *Recherches* (1794), i, 19-27 and parallel passages in the *Recherches* (1802), *Philosophie Zoologique* and *Système Analytique*.
[44] *Histoire Naturelle*, i, 323-334.
[45] *Ibid.*, 44-45, and 168-169.

grounds for calling any particular project, such as Lamarck's "biologie", Newtonian.[46]

The first need is, obviously, to recognize that Newton and his followers had no monopoly on "mechanical" ideals of explanation. This was, of course, acknowledged by Buffon and other Newtonians at the time. And we do best here to follow his example, making our first distinction that between the Newtonian and the Cartesian traditions. Again, as Buffon suggested, there are two quite different kinds of *differentiae* pertinent here. One can contrast the broadly "rationalistic" way, in which Descartes claimed to establish the essential properties and universal rules of matter in motion, with the more "empiricist" stance taken by Newton in defence of his alternative theses. Or one can turn from contrasts in style, procedure and philosophical apologetics to consider the physics, the forces and the laws, proposed by the two schools' founders.

Now, when we apply the first consideration to the case of Lamarck's "biologie", he appears rather more Cartesian than Newtonian. Not only do his explanations for particular organic phenomena usually start with an appeal to what constitutes the essence of this type of body, but it seems implied, at least, that any sound science will make use of distinctions, such as that between the universe and nature, and assumptions, such as that of the natural inactivity of matter, for which sufficient evidence can be found and should be sought, not in experience, but in the clarity and distinctness of the conceptions themselves.

However, on applying the second of our two considerations, we can only conclude that Lamarck's "biologie" was in obvious respects a Newtonian rather than Cartesian project. For Lamarck does not try, as Descartes did, to exhibit all corporeal phenomena as the outcome of matter bumping into matter in accord with universal laws of inertia and impact. He accepts the Newtonian reformation of physics as having shown that forces and laws unheard of in the old Cartesian philosophy are everywhere in nature; and that these universal forces and laws of attraction and repulsion promise to explain all we can trace to their effects. Indeed, in this regard, Lamarck might have said: "there are no Cartesians left; we are all Newtonians now."

But now we need a further distinction. For one can find by Lamarck's generation a clear alternative to his insistence that all qualities can be traced to these two forces. Rather than assume that heat, for instance, is

[46] The literature on Newton's influence grows apace. In what follows I have drawn especially from R. E. Schofield, *Mechanism and Materialism: British Natural Philosophy in An Age of Reason* (Princeton, New Jersey, 1970) and conversations with Dr. Schofield and Dr. J. L. Heilbron. An anonymous reviewer of Schofield's book recently provided an excellent orientation in this difficult area. See "Cartesians and Newtonians", *Times Literary Supplement*, No. 3, 576 (11 Sept. 1970), 1000. For an incisive survey of different Newtonian physiological theories in the eighteenth century, see T. S. Hall, "On Biological Analogs of Newtonian Paradigms", *Philosophy of Science*, xxxv, No. 1 (March 1968), 6-27, and the relevant sections of Hall's valuable book cited in note 21.

ultimately the outcome of attractive and repulsive action between ordinary particles, many people took it to arise from the presence of a special kind of matter, the matter of heat.

Lamarck's strange account of heat was a deviant version of the older tradition. Only when activated by a prior motion does the element fire yield heat and this sensible heat appears not as a new material substance but merely as the repulsive action exerted by fire in its activated subtle fluid condition. Like Buffon, Lamarck always takes heat as gravity's less regular partner in natural productions.

As a Newtonian, Lamarck was emulating not the geometrician of forces following out the Euclidean format of the *Principia Mathematica*, but the natural philosopher who found in a subtle ether the active principles whereby God recruits and sustains the variety of motion in his material creation.[47] It was, we saw, as a source of such activity that Lamarck's new "biologie" first appealed to the ubiquitous subtle fluids exciting the motions in plants and the simplest animals. An exact reconstruction of Lamarck's route to this insight is obviously impossible, but of the importance of these Newtonian precedents there can be no doubt. In a rarely-read memoir, significantly of 1799, "On the Matter of Sound", Lamarck repeatedly quotes the illustrious Newton on the ether and identifies with it the active "ethereal fire" so central to the eccentric chemistry and physiology already expounded in his *Mémoires* and elsewhere.[48]

As in the subsequent expositions of his new "biologie", then, so, perhaps, in its first origins, Lamarck may well have been seeking deliberately to exclude all but mechanical causes; but like Buffon, his view on what could fairly be counted a mechanical cause departed widely from Descartes' own proposal. Here Lamarck was to follow Buffon in claiming Newton's work as a precedent for extending the understanding of mechanical principles to include any active as well as passive principles of regularly observed motions. Newton, we can be fairly sure, would have objected; but then Newton would have complained about a great deal that was said and done in his name, especially by Frenchmen. And, however much Lamarck deviated from Newton on other points, there remains the instructive respect in which he, like Cavendish and other contemporary natural philosophers, followed Newton very strictly. The exhibition of a wide range of phenomena as the result of attractive and repelling forces was, Newton had declared, in the final pages of the *Opticks*, the proper explanatory programme for progress in the natural

---

[47] For Newton's changing views, published and unpublished, on active principles, see J. E. McGuire, "Force, Active Principles, and Newton's Invisible Realm", *Ambix*, xv (1968), 154-208.

[48] "Mémoire sur la Matière du Son", *Journal de Physique, de Chimie et d'Histoire Naturelle*, xlix (1799), 397-412.

philosophy of living no less than lifeless bodies.[49] Newton had, of course, seen this programme as marking a radical break with the refuted fables of Cartesian physics. If Buffon and Lamarck saw it as a way to vindicate, in a revised form, their countryman's original faith in mechanical principles, that surely makes them none the less Newtonian.

[49] For loyalty to attractive and repulsive forces as defining one kind of Newtonian natural philosophy in the late eighteenth century, see R. McCormmach, "Henry Cavendish: A Study of Rational Empiricism in Eighteenth-Century Natural Philosophy", *Isis*, lx (1969), 293-306. In introducing his "Biologie", Lamarck offered an explicit analogy between the great synthesis Newton's law of gravitation made possible in astronomy, and the long search for a single law behind all organic diversity and developments which had guided his own researches in physics and natural history. See "La Biologie", 270 (f.n. 3).

V

Lamarck's Great Change of Mind[1]

1. A change and its interpretation

Jacques Roger's magnificent studies of enlightenment science have set a shining example for younger historians writing on many and varied subjects. In two ways, especially, he has inspired those of us who have been writing about Jean-Baptiste Lamarck (1744–1829). First, Lamarckian scholarship has now taken on the international character exemplified by Jacque Roger's own career. Second, as with Jacque Roger's masterly monographs on Buffon, Lamarck is now increasingly understood in the context of his own time, rather than in relation to developments that would only come later in the nineteenth century.[2]

It is with these two precedents in mind that I turn here to consider the question of Lamarck's great change of mind, the striking shift in his thought about the living world whereby he came to hold those views familiar in his books from 1802 on, most familiarly the *Recherches sur l'organisation des corps vivants* of 1802 and its larger elaboration, the *Philosophie Zoologique* of 1809.

Let me next, therefore, simply epitomise this change of mind, using only terms that Lamarck himself used. There will be, then, no dispute about the content of this change of mind, nor about the biographical judgement that it took place, whether gradually or suddenly, sometime between 1794 – the date of Lamarck's book *Recherches sur les causes des principaux faits physiques* – and 1802. I will accordingly call the Lamarck of 1794 Lamarck I and the Lamarck of 1802 Lamarck II.

Lamarck I held that compound, inorganic bodies are not produced by nature. For the four elements, earth, air, fire and water, have no tendency of themselves to combine into compound, mineral bodies. All such bodies are

---

[1] First published as, 'Lamarck: un grand changement de cadre conceptual', in Claude Blanckaert, Jean-Louis Fischer and Roselyne Rey (eds.) *Nature, Histoire, Société: Essais en homage à Jacques Roger*, Paris: Klincksieck, 1995, pp. 227–239.

[2] Roger, Jacques. *Les sciences de la vie dans la pensée française du XVIIIe siècle*, Paris: Armand Colin, 2nd edn., 1971; *Buffon: Les Époques de la nature, édition critique*, Paris: Édition du Muséum, 1988; *Buffon, un philosophe au Jardin du Roi*, Paris: Fayard, 1989.

produced rather by the peculiar activity, the peculiar vital motions, present in organic, living bodies. All the less complex ones are therefore produced indirectly in the degradation of these most complex ones. This degradation entails, then, a successive production of less and less complex mineral kinds or sorts, a degradation following from the very essence of mineral composition – the constrained union of those combined elements that cause the degradation by exerting their tendency to revert to the natural uncombined state; for, the elements being unequal in their tendency to free themselves, there ensue changes in their proportions.

Now, Lamarck I believes that nature cannot produce organic bodies either. Nature can never produce such a body, which is to say that in no circumstances can the four elements act on one another so as to produce an organised living body. These bodies perpetuate themselves by reproduction in fixed species. Each species embodies peculiar vital motions. Nature, which is to say the four elements and their distinctive powers, cannot initiate such motion in any matter. Nor can nature alter the pattern of such motions; so there is no spontaneous generation possible, nor any change of one species into another.

Switch next to Lamarck II. He, Lamarck II, continues to hold all that Lamarck I taught concerning mineral compounds. But Lamarck II, contra Lamarck I, holds that all living bodies are nature's products. Here too, however, he makes a distinction between direct and indirect production. Nature, which is to say the four elements in suitable circumstances, can produce the simplest plants and animals directly, in direct or spontaneous generations. But that is all that she can produce directly. It is possible for complex organic bodies to be nature's products because they can be made indirectly by the complexification, over myriad generations and eons of times, of the simplest. Here, too, indirect production results from a tendency arising from the essence of the bodies concerned. The essence of living bodies includes possession of contained fluids in solid containing parts. From the movement of the fluids follows the complexification involved in indirect production. Just as degradation follows from the essence of mineral bodies to give a successive production of lower and lower mineral kinds; so in living bodies there is a successive production of more and more complex organised bodies thanks to the essence distinctive of living, organised bodies.

Manifestly, there are many questions that we might pose about this shift in Lamarck's thinking. For instance, we could try to give it an accurate chronology; we could seek for it an initiating stimulus; we could identify what resources – people, readings, experiences – Lamarck drew on in making this transition; we could establish how the transition was conditioned by the institutional and ideological life Lamarck and others were then living; and we could analyse its ultimate, broader significance, whether for natural philosophy, religion, politics or whatever. Naturally, in taking up these questions we would have to avoid citing things in answer to one question that were more appropriate to another question; but, equally, we would hope to integrate our answers to various questions, so as to achieve a coherent history of the entire matter.

Obviously, the present paper can make but a slight, partial contribution to the clarification of what such a history should include. In any case, a great deal has already been achieved, as can be discerned most easily by consulting two recent authoritative monographic studies, those by Richard Burkhardt and by Pietro Corsi, studies I am much indebted to in coming to my own conclusions.[3]

The principal proposal to be made here is that what is needed most, at this time, is to return to the intellectual content of Lamarck's great change of mind, in order, above all, to reconstruct it in a less anachronistic way than is still currently the common practice. For, when this avoidance of anachronism is consistently pursued, we can then consider new chronological and contextual possibilities that allow, in turn, for new insights concerning those broader, more ultimate questions.

To see, in advance, what historiographical revision is called for, consider next, therefore, the place of one issue – the issue of the mutability of living species – in Lamarck's great change of mind. Those writing most recently on the origins of Lamarck II's views, have not only continued to make this issue a central issue, but they have done so for a very telling reason. Lamarck, they have said, moved to a belief in mutable species because he wanted to uphold certain further conclusions concerning species, and he could not see how to preserve those conclusions in any other way. More precisely, it is said that Lamarck embraced mutable species when he became convinced that without species mutability he would have to admit species extinction. Likewise, in an analogous fashion, it has been said that Lamarck came to accept the spontaneous generation of simple organisms because he became convinced that otherwise he would have to admit the extinction of such species. Accordingly, through its bearing on a species mutability and on spontaneous generation, the issue of the

---

³ Burkhardt, R. *The Spirit of System: Lamarck and Evolutionary Biology*, Cambridge, MA: Harvard University Press, 1977; Corsi, P. *The Age of Lamarck: Evolutionary theories in France, 1790–1830*, Berkeley, Los Angeles and London: University of California Press, 1988. I am much indebted to discussions with these two authors. Corsi's book supplies very complete references to the literature on Lamarck. There is now a revised edition: *Lamarck: Genèse et enjeux du transformisme, 1770–1830*, Paris: CNRS Éditions, 2001. A website on Lamarck, http://www.lamarck.cnrs.fr/, supervised by Pietro Corsi, has become indispensable.

extinction of species – whether of complex or of simple organisms – has been made central to the shift from the Lamarck of 1794 to the Lamarck of 1802.[4]

Now, lest there be any misunderstanding, let it be said right away that there is no doubt at all that Lamarck II did indeed appeal to the extinction issue as an explicit reason for accepting mutable species, and he did indeed appear to see that same issue as a reason for accepting spontaneous generation. Why, then, should a biographer of Lamarck not conclude that this extinction issue was the decisive issue of bringing Lamarck to his new views, seeing that it bore so evidently on his new belief in species mutability and spontaneous generation? The answer is – as followers of the extinction issue historiography have emphasized themselves – that there is manifestly far more to the story of the great change of mind. However, the trouble is that the other issues involved bring in species mutability in a way that can be difficult to recognise and appreciate fully. It can be difficult because in these cases Lamarck is not moving to mutable species because he wants to achieve or preserve some further conclusions about species. For, to consider one line of thought only here, recall what he does with the series of animal organisation. The series is one of classes or large families. And, familiarly enough, Lamarck II interprets it as an order of production. Now, this new interpretation of the series requires species to be mutable, in that indefinite change must be possible in successive generations of animals, in a race, to use Lamarck's own term. Here, therefore, species mutability is embraced in order to uphold a conclusion that is not about species at all, but about a series of complexity that is not a series of species but of classes or large families.[5]

But why make so much fuss about such an analytic element in the biographical challenge? It is necessary to do so because a very consequential and general historiographical principle is at stake. If any author is approached, as Lamarck has continued to be, as a contributor to 'evolutionary biology' or to 'transformism', or indeed any such enterprise delineated in any such way, then a certain biographical quest inevitably takes the centre of the stage. That quest is the quest for some convictions about species that could have provided the original and principal rationale for that author first embracing species mutability. However, as the very case of Lamarck shows, this whole approach may lead us seriously astray in our attempts to understand the most fundamental shift in our author's thinking. For, to stay for the moment with the issue of the animal series

---

[4]   Burkhardt, R., op. cit. supra n. 2; Corsi, P., op. cit. supra n. 2.

[5]   For an interpretation, similar to my own, of the series and of the primacy of the thesis of graduated production, see Schiller, J. *La notion d'organisation dans l'histoire de la biologie*, Paris: Maloine, 1978; *Physiology and classification. Historical Relations*, Paris: Maloine, 1980.

as an order of production, it may lead us to make uncritical assumptions as to what were the primary and what were the secondary issues involved in that shift; so, we fail to make central to our biographical quest questions about when and why Lamarck first came to interpret that series as an order of production. Once again, then, despite all our resolutions to avoid anachronism, we would indeed have allowed an 'evolutionism' or 'transformism' historiography to mislead us into approaching a pre-Darwinian author as just that, a pre-Darwinian author, rather than a post-Buffonian or post-Cartesian one, as Jacques Roger tried to teach us to do.

What comes next is, therefore, an attempt to show that we could have an alternative approach to Lamarck's great change of mind, an approach that does not make the presuppositions made by any 'evolutionism' or 'transformism' historiography. The advantages in avoiding inappropriate anachronisms are, as always, that we can recover and understand more of an author's own rationale for doing what he did. Only a historian committed to some self-defeating form of post-structuralism can consistently insist that such advantages are illusory.

2. The primacy of progressive production for the new views

The alternative approach taken here starts from the relationship between two elements in Lamarck's new views: the two elements being the series of classes, on the one hand, and the effects of life in any individual organized body on the other. Notice, right away, therefore, that we are concentrating our attention on the relationship, as Lamarck sees it, between the very grand, general, comparative order that delineates the organisational series and the miniscule motions constituting the life present in the particular matter comprised in any organised body, however small it may be. That we should concentrate our attention in this way is surely appropriate. The writings of Lamarck, both before and after his change of mind, are obviously distinguished by just such a pre-occupation – to relate any larger order to the most minute activities.

Indeed, Lamarck himself, makes just such a pre-occupation primary when, in the very opening of the *Philosophie Zoologique*, he appears to be remembering and recalling the considerations that had first led him to the views taught in that book. I say that this is what he appears to do in this text, because there has been a reluctance by Lamarck scholars to read it as a genuine attempt at an accurate, retrospective narrative; for, it has seemed too much like an idealised rationalisation rather than a reliable, autobiographical history. However, since, as everyone would evidently agree, there is no way either to confirm or discredit it as a trustworthy story, one surely does well not to dismiss it altogether in any efforts we may make to avoid anachronism in our own approach to Lamarck's change of mind. So, we should at least allow that text to suggest to us what biographical possibilities we should be investigating.

One such possibility is that Lamarck did indeed make two decisive steps away from his 1794 views in roughly the same way that that text implies. The first step would have consisted in coming to appreciate the striking degradation found in the organisational series of animals as one passes from the most to the least perfect, and concluding that nature had in fact produced living bodies successively starting with the simplest. A second step would have consisted in asking how life can be produced in a body, and looking to the simplest organisations for an answer on the ground that these present the minimal conditions for the existence of life.[6]

Now, the first step, especially, as here represented, following Lamarck's 'Avertissement', may seem quite implausible, biographically, insofar as it implies that natural, progressive production was originally inferred merely from reflection on the serial ordering of animal organisation. Surely, one is inclined to object, Lamarck I had long recognised such an animal series without seeing in it any grounds for inferring progressive production. When understood in abstraction from any further chronological and contextual considerations, this objection has a manifest force; but once we bring in those further considerations, then this force is greatly lessened, as we can see by recalling briefly Lamarck's life in the years immediately following his appointment, in 1793, to his *Muséum* professorship.

Before turning to those years, however, there is a general challenge to be faced regarding the biography of Lamarck during the whole period after he publishes the last Lamarck I text, the 1794 *Recherches*, and before he writes his first Lamarck II text, the 'Discours d'overture' of 1800, first printed in the *Système des Animaux sans Vertèbres* of 1801. We have various texts from these years, some published, such as the *Mémoires de Physique et d'Histoire Naturelle* of 1797, and some in manuscript and only printed recently, such as a 'Discours préliminaire' read on the 3rd of May, 1799.[7] Now, all who have studied these

---

[6]   Lamarck, J.B. *Philosophie Zoologique*, Paris: Dentu, 1809, vol. I, pp. I–III. Note that in Lamarck's mineralogy there is a decisive precedent for distinguishing direct from indirect productions in a graduated series of productions. The mineralogy of Lamarck I did not need completion by the later biology of Lamarck II. His conception of species among living beings was different from his conception of different sorts of minerals; on this point I agree with Burkhardt, op. cit. *supra* n. 2. However, the explanations of Lamarck I for the production of the mineral series were a decisive influence on Lamark II's account of the graduated production of animal and plant classes. Once again, to appreciate this influence, one must avoid taking questions about species as the principal questions.

[7]   Vachon, M., Rousseau, G., and Laissus, Y. *Inédits de Lamarck*, Paris: Masson, 1972, pp. 153–164.

texts would agree that they do not contain either an explicit and unambiguous reaffirmation of Lamarck I's denial that nature can produce living bodies, nor, conversely, any such affirmation of the Lamarck II insistence that living bodies are, all of them, nature's products. We face, then, a dilemma. We can either conclude that the Lamarck of these years, 1795–1799, must have moved a long way from the position of Lamarck I, on the ground that he is no longer prepared to make those old denials; or we can conclude that he must still be a long way from the Lamarck II position because he is, as yet, not enunciating any of the theses that will be expounded after 1800. My own view is that, when we consider the public character of all these 1795–1799 texts, we should put the emphasis more on the first conclusion. After all, it is quite consistent with all the evidence to suppose that Lamarck had reached many of his Lamarck II views soon after publishing his 1794 *Recherches* – perhaps, then, by 1796 when he was presumably writing the *Mémoires* – but refrained from going public with them. What is more, on this supposition, we have an explanation for what is otherwise very hard to explain; namely, why is there no clear affirmation, after 1794, of the old Lamarck I denials that living bodies are nature's productions.

Let us turn next, then, to contemplate the possibility that Lamarck, in 1795–1796, took that first step of embracing a natural, progressive production of living bodies, principally as an inference from the degradation of organisation from the top to the bottom of the animal series. A comment to be made right away is that to embrace that conclusion entailed, obviously, a commitment to mutable species. Less obviously, perhaps, it may or may not have entailed a presumption of spontaneous generation, in some form or other; for Lamarck could have embraced progressive production of the more complex, starting with the simplest, while leaving open the question of just how the simplest were themselves produced.

Now it is well known that, from the middle of the 1780s on, Lamarck had insisted on the existence among animals of the series of classes arranged in order of degrees of organisational perfection. Moreover, these degrees are degrees of inner organisational complexity. For classes among animals, unlike plants, are distinguished by internal organisational differences, rather than by differences in external organs. What new, general conclusions, then, has Lamarck reached by 1795–1796, about the scale of animal classes, that he has not accepted since the previous decade, new conclusions that might have prompted, in turn, a new belief in progressive production? Not surprisingly, the new conclusions most likely to have done this for him are conclusions deriving from his new studies of the invertebrate classes. If we look at the *Mémoires* of 1797, we find that one conclusion, in particular, is now in place that was not available to him in the 1780s. This is that the animal series goes

V

all the way down to a group of animals called Monades. These belong in the group of microscopical Polyps that forms part of the Polyp class, the ninth and lowest class in the table given in the *Mémoires*. Moreover, Lamarck is already convinced of the extreme simplicity of their organisation. They have, notably, no specialised reproductive organs, not even gemmae, but are fissiparous.[8]

Generally, then, the serial arrangement in the *Mémoires* table does display just those features that Lamarck, after 1800, will always emphasize in moving from an order of complexity to a progressive production; namely, a loss of special internal organs as one goes down the series, terminating in a complete absence of special internal organs in the simplest animals of all. But, surely, that finding alone, in and of itself, could not have sufficed to move Lamarck to abandon all his Lamarck I principles and embrace progressive production? No, to suppose that it did is quite implausible. There is, however, no need for us to make such an implausible supposition. For, we know that Lamarck was reaching other, new and relevant conclusions in these 1795–1796 years.

One such conclusion concerns what goes on during formation of internal organs within the individual animal body. In the *Philosophie Zoologique*, Lamarck expounds his thesis that cellular tissue is the matrix in which all the organs of living bodies have been successively formed, and that the movement of fluids in this tissue is the means nature employs in gradually creating and developing these organs out of this same tissue. And he recollects, in a footnote, that since the year 1796 he has expounded these principles in the first lessons of his course.[9] Now, these principles, as stated in the *Philosophie*, and so presumably in those lessons, concern the formation of organs within an individual animal body as it grows and matures following conception. Nevertheless, it is surely reasonable to conjecture that Lamarck – not in public, but privately – may have been applying these principles to the organisational series in 1796; especially when we see that he explicitly notes, in the *Mémoires* table of animal classes, that whereas the Radiarians have radiating vessels outside their intestinal canal, those Polyps with an intestinal canal have outside it only a cellular or tubular system.

An obvious objection will occur at this moment, however. For Lamarck to have shifted to progressive production around 1796 would have entailed his moving to a new understanding of life as within, not beyond, nature's powers; and where, it may be objected, is the evidence that this naturalism is there so early? Again, the objection is a fair one, but not decisive. It is well known

---

[8]   Lamarck, J.B. *Mémoires de Physique et d'Histoire naturelle*, Paris: published by the author, 1797, p. 314.

[9]   Lamarck, op. cit. *supra* n. 5, t. 2, pp. 46–47.

that the entire discussion of the definition of life in the *Mémoires* is strikingly and explicitly more naturalistic than the equivalent discussion in the *Recherches* of 1794.[10] The most general inhibitions about finding an explanation for life within nature have, then, broken down during the years since the *Muséum* appointment. And that, we might add, is far from surprising. A new professor in a scientific institution that owes its new form to the Revolution might be expected to move towards more rather than less naturalistic interpretations of the material he is responsible for in his research and teaching duties. But are there not equally well-known passages in the *Mémoires* themselves that deny such particular naturalistic proposals as spontaneous generation and mutable species? Yes, there are indeed passages that have been read that way (even by the present author some years ago); but a close examination of their precise content and context would show that those readings are far from manifestly correct.

Three clusters of passages are of immediate concern. First, Lamarck twice says that, in reproduction, an individual alone or collaborating with another, produces a new individual of its species.[11] But this commonplace, about what happens next in the very short run from one generation to the next, could well have been seen by Lamarck as consistent with a gradual mutability of species as they change in the long run as required by progressive production. Second, Lamarck writes that it is said 'with good reason [avec raison]' that everything with life comes from an egg. But notice that he does not say that this is truly or correctly said, only that people have had good grounds for asserting it.[12] There is, then, here no deliberate, overt exclusion of the possibility of spontaneous generation. What is more, in 1802, when Lamarck does give an account of spontaneous generation, he honours an extended version of the egg principle, in that the new living body is supposed to arise from a gelatinous mass that is like an egg in being capable of vivification following a preparation for life that is analogous to fertilisation. Third, in at least one place Lamarck writes, in the *Mémoires*, that mineral compounds cannot be produced by nature but are, on the contrary, produced by living bodies. This contrast looks, then, like an old, Lamarck I, contrast; but twice before, in the *Mémoires*, Lamarck has been

---

[10]   Burkhardt, R., op. cit. *supra* n. 2; Roger, J. 'Chimie et biologie: des molecules organiques de Buffon à la physico-chimie de Lamarck', *History and Philosophy of the Life Sciences*, I, 1979, pp. 41–64. For the Newtonian natural philosophy invoked by Lamarck's naturalism see: Hodge, M. 'Lamarck's science of living bodies', *British Journal for the History of Science*, V, 1971, pp. 323–352 and Conry, Y. 'Une lecture newtonienne de Lamarck est-elle possible?' in *Lamarck et son temps, Lamarck et notre temps*, Paris: Vrin, 1981, pp. 35–57.

[11]   Lamarck, J.B., op. cit. *supra* n. 7, pp. 250 and 270.

[12]   *Ibid.*, p. 272.

more precise and specified that it is nature in abstraction from the peculiar activity in living bodies that cannot produce mineral compounds[13]. These more careful statements, far from opposing nature and vital activity, seem rather to include that activity within nature, and so to mark a distance from the views of Lamarck I.

Similarly, if one reads carefully in the manuscript writings from the late 1790's, those passages that have sometimes seemed to imply a rejection of spontaneous generation will be seen not to address that issue directly, much less to resolve it decisively against spontaneous generation. Thus, in the 'Discours préliminaire' of the 3rd of May, 1799, Lamarck urges that the 'animated molecules' populating the waters have reproductive organs just as higher animals do. But, as the context confirms, he is not here talking about the very simplest 'monades' of all – those he might have called atoms – but of any very small organisms, including those that are more complex than such 'monades'. When he talks, therefore, of these organisms having a rank in nature as little equivocal as have higher animals, he is referring not to the possibility or otherwise of an equivocal generation for them, but to the way their reproductive and other organs give them a place in the organisational ranking as definite as any higher animals have. For he has just warned that our imagination fails us in comprehending all the nuanced degradations of organisation that confront us when we consider lowly as well as higher animals.[14]

There is, then, no textual ground for resisting the biographical conjecture proposed here: that one of the first major steps Lamarck took away from his old, Lamarck I views was to interpret the animal series as an order of progressive production, and that he may have taken this step as early as 1795–1796.

Now, once we take seriously this biographical, narrative, conjecture regarding Lamarck's great change of mind, then we get many further historiographical dividends. First, we can modify and then welcome the proposals made as to the part played by Lamarck's new reflections, apparently in 1799, on the extinction of species. For let us distinguish, first, progressive mutability as the mutability required by the primary thesis of progressive production and, second, circumstantial mutability as the mutability in changing circumstances that Lamarck saw as required by his extinction reflections. This circumstantial mutability thesis could well have been joined with the progressive mutability thesis at this later stage, in 1799, as a secondary thesis. That it was so is certainly

[13] Ibid., pp. 240, 241 and 320. For a different interpretation of these three paragraphs, see Burkhardt, R. 'The Zoological Philosophy of J.B. Lamarck' in Lamarck, J.B. *Zoological Philosophy*, Chicago and London: University of Chicago Press, 1984, pp. xv–xxxix.

[14] Vachon, M. et al., op. cit. *supra* n. 6, p. 154. For a different interpretation of this text, see Corsi, P., op. cit. *supra* n. 2, p. 67.

consonant with the way Lamarck II relates progression, on the one hand, with adaptive diversification through circumstantial mutability on the other.[15]

Obviously, the details of an intellectual biography for Lamarck during the 1795–1799 period will always remain elusive because of the paucity of relevant documentary evidence. However, the present paper will have succeeded if it has done two things: first, to show that, with Lamarck, as with Buffon and others of their period, we have always to avoid that begging of questions that inevitably accompanies any 'evolution' or 'transformism' historiography; second, that there are good reasons for taking the question of progressive production as primary in Lamarck's great change of mind; and that when we do that, we are led to narrative conjectures that allow us to pay appropriate attention, not only to what Lamarck was moving toward, but also to what it was he was moving from in that remarkable shift in his thinking between 1794 and 1802.

---

[15] R. Burkhardt recognises explicitly that focussing on the problem of species can distract from what is most fundamental in Lamarck's thought in 'Lamarck and Species' in: Atran, S. et al. *Histoire du concept d'espèce dans les Sciences de la vie*, Paris: Fondation Singer-Polignac, 1985, pp. 161–181. See also Hodge, M. 'Species in Lamarck' in Schiller, Joseph (ed.) *Colloque International Lamarck*, Paris: Blanchard, 1971, pp. 31–46; Corsi, P., op. cit. *supra* n. 2, p. 67 cites the opinion of G. Barsanti on the change of Lamarck's conceptions in Barsanti, G. *Dalla storia naturale alla storia della natura: saggio su Lamarck*, Milan: Feltrinelli, 1979, a book that I have not had the opportunity to consult.

The History of the Earth, Life, and Man: Whewell and Palaetiological Science

1. Introduction: The Older and the Newer Historiographies

ANY part of Whewell brings us to confront the whole. But Whewell as a whole is a daunting prospect. He was a very big man was Whewell, perhaps a great one; and they do not make them like that any more. Only by grafting Cassirer, the historical philosopher, on to Conant, the scientific educator, could we have in our day a comparable combination of intellectual stride and institutional clout, of polymathic savant and academic supremo. But no such reflections can help us as historians. Great man theories of history, let alone big man theories, are in no better repute than myths of giants who walked the earth, or at least the quadrangles, in distant days.

What can help us is the Whig tradition in the writing of history, and the reactions that tradition has provoked in recent decades; especially we can be helped if we can learn to be as critical of the fashionable, newer reactions as we are of the unfashionable, older tradition itself.[1]

Whig history has taken many forms, obviously; but most of them require the science of the early nineteenth century to be inspired by the same spirit of rational progress that inspired the partisans of reform in religion, education, and government. Whewell has, then, always seemed easy to place within such a historiography. Even the dutiful volumes of Todhunter and Douglas make implicit concessions to the Whig tradition in the muted, apologetic hagiography they offer for their Tory, Anglican, Idealist hero.[2] Early triumphs are seen to give way to the failings of a later loser: most conspicuously a loser to Mill and to Darwin, and even to young men bent on transforming Trinity College itself. Meanwhile, more willing guardians of the Whig heritage, from Huxley on,

---

My understanding of Whewell and his views on palaetiological science has benefited greatly from suggestions made by Geoffrey Cantor, John Christie, Menachem Fisch, Larry Laudan, Rachel Laudan, Michael Ruse, and Simon Schaffer.

[1] A good introduction to the Whig tradition is provided by Chris Wilde in the entry on 'Whig history' in Bynum (1981).

[2] Todhunter (1876); Stair-Douglas (1881).

M.J.S. Hodge, 'The history of the earth, life, and man: Whewell and palaetiological science, *A Composite Portrait*, eds. M. Fisch and S. Schaffer (Oxford: Clarendon Press, 1991), pp. 255–288. By permission of Oxford University Press.

have found in Whewell a natural foil for the vaunted Victorian vindications of reason and reform in empiricism, evolutionism, and liberalism.[3]

It is here that we can so easily go wrong. A strong sense of the indefensible preconceptions in all Whig views of Whewell can make it hard for us to see the limitations in those interpretations of the last three decades, such as Cannon's, that have been adopted in deliberate opposition to the historiographies of Huxley and the rest.[4] Anything, we may presume, anything so resolutely opposed to something so thoroughly unacceptable must itself need no re-appraisal. However, the fallacy in this presumption, once the presumption itself is caricatured quite tactlessly, is obviously resistible. There should, then, be agreement that three decades of anti-Whig historiographies for Whewell ought now to feel too long for uncritical comfort.

One reason for taking a wary stance, at this moment, is that the invaluable achievements of the last three decades may otherwise become indelible idols of a nascent profession. Historical, philosophical, and sociological studies of science have exhibited many of the customary concomitants of professionaliza-tion only since the 1950s. As a corollary of this timing, any viewpoint dating from this period is associated with the new, enhanced status of these disciplines; while, conversely, any expression of dissatisfaction with such a viewpoint may be taken, once again quite fallaciously, to be an invitation to abandon what has accompanied professionalization, and to return to the unsubtle errors once perpetrated by naïve amateurs.

Such fallacies are especially pertinent to any analysis of Whewell's writings on the history of the earth, life, and man. Because of their relevance to an understanding of what some have called the Darwinian revolution, these writings have been given close attention since 1959, the Darwin centennial year. On this ground, all of us who write today are massively indebted to Cannon and his ally, Hooykaas, and to those others who have learned from them.[5] What we have to learn now is to combine that indebtedness with an appreciation of what they cannot do for us, especially in understanding Whewell in relation to Darwin and, earlier, in relation to Darwin's mentor Lyell.

Cannon was not only repudiating the Whig tradition in general, but also, specifically, the views of Lyell and Huxley about the history of geology and biology. Unsurprisingly, Cannon had no trouble in refuting Lyell's legends about how geology had developed, nor in discrediting Huxley's contrasts between Lyell and Whewell. However, Cannon went on to propose that it was 'catastrophists' such as Whewell, and not the 'uniformitarians' led by Lyell, who were being reasonable and empirical rather than dogmatic and arbitrary about the course and causes of change at the earth's surface. What is more, Cannon argued that the orthodox Anglican, natural theology of a catastrophist such as Whewell made him more a 'precursor' of Darwin than any early

---

[3] See, especially, Huxley (1887).
[4] W. Cannon (1960); id. (1960a); id. (1961a); id. (1978).   [5] Hooykaas (1959).

'evolutionist' such as Lamarck. Faced with this historiographical alternative to Huxley, we have to conclude, then, that Cannon did not move completely away from distinguishing good guys, who were undogmatic and empirical and were precursors of Darwin, from bad guys who were not. Hats have been switched around, but the old questions, as to who should wear the black and who the white, are still being asked and decided on the old criteria.[6]

Now, Cannon was perfectly aware that these dialectical switches were often yielding not a radical Tory history, much less a Socialist one, but only mirrored inversions of the old Whig epics. For a principal aim of Cannon in overturning longstanding wisdom about the Whewells and the Lyells of the 1830s was to prepare us for the irony in the tragedy, as Cannon saw it, of what happened in 1859. On Cannon's account, Darwin was eventually to confront a venerable consensus about the coherence of science, morals, and religion, as upheld by the likes of Whewell; and Darwin, according to Cannon, was to confront this consensus with its very own universe, but now completed, in accord with its very own constitutive principles, by natural selection; it was an addition, however, that made this universe abhorrent to its original possessors because manifestly fatal to the coherence cherished in their consensus.[7]

There is, then, a further reason for being wary of this comprehensive Cannonical thesis as to the inception and reception of Darwinian science. For it continues the old Huxleyan habit of concentrating on those issues in the 1830s that are supposed to teach us most about the debates of the 1860s. Regardless, therefore, of whether we think Cannon right—about the relationship between Darwin's *Origin of Species* and Whewell's Bridgewater Treatise on *Astronomy and General Physics, Considered with Reference to Natural Theology* (1833) a quarter of a century before—we should still pause, before following an approach to the 1830s that is so often in the service of a thesis about what was to happen a generation later.

There are equally compelling reasons, although of a very different kind, for being cautious concerning any interpretations of Whewell deriving from his recent rehabilitation by philosophers of science. For this rehabilitation has often been encouraged by developments in philosophy of science itself, developments that can easily prompt inappropriate assumptions about the character of Whewell's work.

According to a familiar stereotypical Anglophonic narrative, three decades ago the philosophy of science was dominated by a combined commitment to logic and to experience marked by the label logical empiricism. Within this commitment there was, it seems, no place for history and much more sympathy for Hume than for Kant. Not surprisingly, therefore, we find Whewell benefiting, in attention and evaluation, from the widespread disenchantment, dating from the 1960s, with logical empiricist views. It is a disenchantment often associated, as for instance in Butts, with Kantian sympathies; or, as in

---

[6] W. Cannon (1961a).   [7] S. Cannon (1978).

Curtis more recently, with a neo-Popperian concern for the rationality of progress as exhibited in the historical development of science.[8]

The trouble is that such philosophical rehabilitations of Whewell may bring inappropriately problematic contrasts—history versus formal logic, Kantian rationalism versus Humean empiricism—to the characterization of his work. These, and other contrasts, as currently influential, can divert us from what was concerning Whewell as he took up the issues his philosophy of science was designed to resolve. For the project consummated in the two volumes on the *Philosophy of the Inductive Sciences, Founded upon their History* (1840) was not originally aimed at deploying Kant and history to oppose any Humean legacy, nor was it aimed at dislodging the formal logic of the day from the analysis of science.

Quite generally, then, as we in the 1980s go back a century and a half to the 1830s, we should beware of perpetuating uncritically too much from several rewarding legacies from the sixties—the 1860s, that is, and the 1960s.

2. The French Revolution,
the Industrial Revolution and a Cambridge Nationalist

One way to integrate our reading of Whewell with an understanding of his epoch is to start our thinking about the man with our knowledge of his strikingly successful career, and to start our thinking about his epoch by recalling what two revolutions, the French and the Industrial, entailed for England as a state and a nation. Such an integration has to be instructive because Whewell's career was very much an individual success conditioned and facilitated by national institutions and cultural ideals that were distinctive of England (where England excludes Scotland). For recall Whewell at his career apogee in 1841. With the five volumes of the *magnum opus* (the *History* and the *Philosophy*) now published, he marries, and is elected Master of Trinity College, a Crown and therefore a national appointment and second to none in its power and prestige in the English academic realm. It is the year he is President of the British Association for the Advancement of Science, thus becoming ringmaster of the annual state circus of science assembled that year at Plymouth. Consider, too, Whewell in the following year, writing a most revealing letter to a friend, a letter that moves, quite naturally, from opening remarks about examinations, to talk of God, truth, and the Church, the Nation and its Constitution. These six, including that is the examinations, although not necessarily in that order, were for Whewell nothing less than what he believed in as objects of belief essential to civilization in its highest form. Nor does he leave us in any doubt as to the leading connections among them. Most familiarly, he was one of those defending the distinctive traditions of English

---

[8] Butts (1968); Curtis (1986); id. (1987).

university education as essential to those established relationships between the Church and the Nation that were enshrined in the Constitution, and thereby made essential to the continued securing of liberty under King or Queen and Parliament.[9]

Now it is easy to see that such commitments served, for Whewell and many another like him, as alternatives to the unreformed Church and revolutionary politics of France, and so served to contain the challenge, the threat from that quarter. What is perhaps less easy to see is that they served no less to contain challenges and threats arising from the industrialization of England. Here it helps to view Whewell as the hyperassimilated emigré within his own country. For, in the course of his progressive career successes, he became assimilated entirely to the older national culture of metropolitan England, as entrenched in London, Oxford, and Cambridge, a culture that was—thanks to the burgeoning strengths of its agricultural, financial, and imperial capitalism—still able to marginalize the provincial alternatives associated with the new manufacturing interests represented most vividly by the exemplar of Manchester's transformation. Whewell, in rising from scholarship winner to grandee don, had indeed circumvented Manchester and all it stood for, in his passage from a Lancaster carpenter's household to the Master's lodge. It tells us much about England then—and before and since—that success, in becoming the accomplished academic and ecclesiastical *arriviste* that Whewell was, required such circumvention as a condition for hyperassimilation.

That nationalism, in the sense here indicated, should be a decisive element in the ideology of anyone with Whewell's career trajectory allows us, therefore, to make sense of two recurrent sources of difficulty for a savant in his position: science itself and Scotland. The difficulty with science was that, abstractly considered, it was professedly free from nationality. The difficulty could be resolved however by respecting that ideal while at the same time insisting that different nations could make distinctive contributions to science, so that, for instance, there could be a case made for a particular foreign, not to say alien, import such as French physical astronomy, if it could be represented as fulfilling earlier Newtonian hopes or improving existing tripos practices.

The difficulty posed by Scotland could not be so resolved, obviously. Scotland as viewed from Oxford and Cambridge was habitually disrespectful of the distinctively English commitment to the National Church, the National Constitution and the contribution made to the maintenance of both by a system of university teaching that made classics and mathematics primary, and had the clergy not in special seminaries but as the most prominent presence in a scheme of higher education equally adapted to the landed gentry. It is no mere happenstance at all that Whewell was as disconcerted by his friend the smooth deist Lyell praising American universities in comparison with Whewell's own,

---

[9] The letter, to J. G. Marshall, is printed in Stair-Douglas (1881), 279–83. Morrell and Thackray (1981), *passim*, is excellent on Whewell's career.

as by his enemy, the abrasive and pious Brewster.[10] Both were Scots knowingly continuing decades of Edinburgh chastisement of Oxford and Cambridge. To reduce such conflicts to personality clashes or regional rivalries is to miss how they involved an issue made unavoidable, for anyone with Whewell's outlook, by the very existence of Scotland within Britain: the issue of reconciling the ultimate sources of authority for the British people with the presence of another national tradition within the nation's bounds.

A main suggestion here will be, therefore, that the character of Whewell's encounter with geology is often distinctive for the same reasons that his entire career was distinctive: namely that the goals he was pursuing as an individual were largely set for him by a context of nationalistic notions as to what counted as success in advancing civilization and so, too, science.

3. Locations for Geology

To understand the various locations geology had for Whewell, including the location, within his career, of his own encounter with this science, one has to recognize the broad scope and deep significance the science held for him. One might think his views on the origin and duration of the earth, including its stocking with plants and animals, man among them, would be spread over several sciences. However, for Whewell, all such matters are directly connected with geology if not subsumed within it. For these are all matters of historical causation, of causes, that is, acting occasionally and successively in past times. All fall, then, within the palaetiological sciences, the sciences of causes active in the past. Now, although several other sciences—glossology, for example, or philology as others were already calling it—count as palaetiological, it was geology that was Whewell's original exemplar of the genre, and the only one that he treats of at any length. Its structure and development are, moreover, exemplary in that the structure and development of any other palaetiological science are understood by comparison and contrast with geology. So geology, broadly construed as Whewell construed it—so as to comprehend climate changes, biogeography, and the origin of species, including man, no less than earthquakes and mountain formation—is what has to be considered.

Whewell's life and his writings both exhibit sufficient restless movement that anything within them has a location *en route*, a location where his various trajectories take him through whatever the subject is and on to something else beyond. There is the movement apparent from any survey of his writings over three decades from the 1830s: a movement from mathematics, pure and mixed, to morality, theoretical and practical. Geology here takes its place in the 1830s, the decade dominated, from his review of Herschel's *Preliminary*

---

[10] K. Lyell (1881), especially ii. 127. On Scottish critiques of Oxford, see Rupke (1983).

Discourse (1830) to the *Philosophy* in 1840, by writing on induction and the inductive sciences. In this context, the principal challenge was to understand geology as an inductive science, and to adjudicate between the rival accounts given, by Lyell and Herschel on the one hand and by Sedgwick on the other, of what it is for geology to have a secure position among the inductive sciences. There is another movement, in the trajectory that is Whewell's career climb from undergraduate to Vice Chancellor, via a tutorship in Mathematics and Professorship in Mineralogy (1828). Within this movement, there is a progress from one election to another; the first is the election, in 1827, to Fellowship in the Geological Society of London, an election appropriate for one who was already a dedicated, publishing mineralogist; the second is the election to President in 1837, an election appropriate not because of any pretension to original geological research, but because Whewell had come to be seen by many Fellows as an excellent alternative to less progressive elder statesmen who might otherwise gain the office.

Engaging geology in the course of these several trajectories, Whewell delineates various placings for the science. In the *History* and the *Philosophy*, both adopting the same ordering, geology finds its place at the very end of a movement that begins with mathematics and then proceeds through the inductive sciences, much in accord with the distinctions embodied in the BAAS sections. Geology provides, then, the transition to topics that lie beyond the 'physical sciences', topics such as classical studies, civil history, the First Cause as an object of theological science, and 'tradition', including the Scripture narrative. Although this transitional siting, for palaetiological science in general and for geology in particular, may seem natural, there was no inevitability about it. As Whewell remarks in the *Philosophy*, at the end of the previous section on Biology, the palaetiological sciences were not typical when considered as instances of the scheme of ideas, the scheme that the *History* and the *Philosophy* are designed to elaborate. For these sciences do not have their own Fundamental Idea or Ideas, only a conception—historical cause—a version of the Idea of Cause. When he has done with biology, he has then done with the circle of sciences as founded on distinctive Ideas, as he admits. What he does not admit is that geology, if not other palaetiological sciences, might well have been treated much earlier on under the heading of causal science. But such a decision would have forfeited the positioning of geology as a conduit to the sciences, the moral sciences, beyond the physical.[11]

As so positioned, geology is delimited by Whewell, in accord with the position and boundaries of its assigned object: a post-nebular and pre-human positioning between, that is, the incipient earth speculated upon in nebular hypotheses and that same earth inhabited finally by man, and so coming within the history of God's providence as revealed in the special dispensation for our species.

---

[11] Whewell (1847). See i. 635–6.

262

These positionings are, therefore, Whewell's own work. They are not given by unavoidable prior intellectual or institutional precedents. There was no consensus concerning geology as a site on the map of knowledge or as a division in the classification of the sciences. Nor was there any customary curricular niche for the science. In Whewell's life at Cambridge geology was familiar, even prominent, from early on, following the appointment in 1818 of his Trinity friend, Sedgwick, to the Woodwardian chair. But its prominence was due to the conjunction of Sedgwick's energy and reputation and the current topicality of the science, not to the statutory claims it could make on undergraduates' time and effort. For curricular purposes it was no closer to being joined with the mathematics, classics, and divinity required for a degree than was botany, say, or mineralogy. Until late in Whewell's life, geology at Cambridge was largely living in the limbo of professorial lectures, outside the rigmarole of examinations and set texts overseen by the Schools and Senate and serviced by the colleges and tutors, private and public. Where Cambridge is reflected in Whewell's positionings of geology is not through any curricular role, but in the prospect of advancing physical geology by bringing mixed mathematics to it, and in the promise that geology, in so far as it studies purposes, will lead to Final Causes, and in so far as it studies origins, will lead to the First Cause.

Within geology itself, when exhibited as an inductive science in the *History* and the *Philosophy*, the main movement is one typical of all induction: from facts to laws without causes, and so onward and upward to laws of causes. For, within this inductive ascent, Whewell distinguishes the phenomenal from the aetiological members in any palaetiological science. Here the exemplar of Newtonian astronomy is both emulated and transcended. It is emulated in that there are, Whewell insists, laws and causes of geological dynamics corresponding to the equivalent branch of the science of mechanics. It is departed from however, in that geology is not a science of mechanical causes, but rather of historical causation. It is not a science of forces that are permanent causes of motion, such as gravitational attraction is, causing motion whenever not opposed by other forces; but of the actions of causes that have worked their effects in a temporal succession, as the causes have, for instance, that have raised mountain ranges of different ages at various times.[12]

Such comparisons and contrasts among distinct sciences are decisive for the higher order inductive science of science—a moral science of physical science—that is pursued throughout the *History* and *Philosophy*. Only phenomenal geology has been completed in passing through an inductive epoch to its sequel. The perfecting of geology as a science that will make it yield true knowledge, aetiological no less than phenomenal, has hardly begun; requiring as it does much more inductive progress than has so far been made in geological dynamics and in the application of that dynamics in arriving at the true theory

---

[12] Whewell (1837a). See especially iii. 481 ff. Also id. (1847), i. 637–8.

of the earth. The marks of truth as manifested in a consilience of inductions and exhibited in inductive tables are not demonstrable here; not because geology comprises counterinstances to such higher order conclusions from the history of other sciences, but because the necessary inductive progress has not yet been made.

This historical and philosophical judgement was reached by Whewell at a particular moment in the history of geology and in his own life, the very early 1830s, and he never felt a need to retract or even revise it. Whewell's treatment of geology is not, then, like his treatment of most sciences; for it is not an exercise in generalizing from the history of attained success; nor, conversely, of applying to the particular case of geology the general doctrines constituting a philosophy of science designed to fit all inductive sciences as such. His philosophy of geology is not, therefore, a straightforward outcome of bringing together a particular science and a general philosophy of science. It was, rather, a philosophical response to geology conditioned by contingencies of timing at its inception, around 1830, before it was further constrained by the structure of exposition given it in 1837 and 1840.

4. Philosophical and Theological Outcomes

Whewell's arguments in his treatment of palaetiological science in general, and geology in particular, are sufficiently complex and convoluted that we do best to invert the order of his own exposition, and to begin with his conclusions. There is a clear advantage in beginning with his conclusions, because we can then avoid one drawback to most of the accounts historians have given of Whewell as a writer on geology. For most historians, understandably enough, have gone to Whewell to see where he stands on the contested issues within the geology of the time, most obviously the debate that Whewell himself taught the world to call the uniformitarian–catastrophist debate. But to read Whewell mainly for instruction on such matter is often to miss the significance for Whewell of the higher lessons of philosophical principle that he is explicitly and insistently concerned to draw from his examination of palaetiological science, lessons that were peculiarly of Whewell's own formulating and that did not become elements in any public discussion.

The first thing to notice about these lessons is that they are nearly all negative. Most evidently is this true of his two leading maxims: the first, 'That no palaetiological science has been able to arrive at a beginning which is homogeneous with the known course of events', which is to say, '*No Natural Origin discoverable*'; and the second—explicitly a corollary of the first—that '*Science tells us nothing concerning Creation*'; for since 'science can teach us nothing positive respecting the beginning of things, she can neither contradict nor confirm what is taught by Scripture on that subject; and thus, as it is unworthy timidity in the love of Scripture to fear contradiction, so it is ungrounded presumption to look for confirmation, in such cases'.[13]

That Whewell is travelling a *via negativa* throughout his examination of palaetiological science is evident to anyone who reads to the end of his presentation. What is less evident is that his negational ambitions are conditioning all of his prior discussion of the development and nature of this genre of science. For, on the face of it, Whewell seems often to be writing in a very hospitable, welcoming, and so positive way. All sorts of proposals are introduced respectfully, even sympathetically, including contested current speculations such as terrestrial cooling from decline of primitive central heat, nebular hypotheses, the transmutation of species, and so on. What is more, the uniformitarian and the catastrophist positions are likewise both given extended, respectful, and sympathetic hearings.

However, if one asks what, for Whewell, do all such scrutinies of more or less general proposals and positions yield, as overall conclusions, the answers are consistently negative. The extent of this negativity in the outcome is easy to miss, because Whewell often seems to be taking sides as he discusses rival views, at least in so far as he indicates that, in his judgement, one rather than another has the better arguments going for it. But this apparent decisiveness over particular matters should not mislead us. For if we ask, quite simply, what does Whewell show himself to believe, to accept, about the earth's past and the changes in life upon it, we notice at once that is very hard to discern the answer to this seemingly straightforward exegetical question.

One reason for this difficulty is that Whewell is adamant that there is as yet no true physical theory of the earth, no true theory, that is, of the course and causes of the changes undergone by the earth and its inhabitants. There is much secure knowledge of geological phenomena, to be sure, and some acceptable, if not certain, knowledge of geological dynamics. But when it comes to the deployment of that dynamics in constructing a true physical theory, there is no candidate worthy of discussion. There are the 'fanciful' theories of Burnet and others to recall, and there are the 'premature' theories of Werner and Hutton also; but there is no successor to these that is neither fanciful nor premature but true. Indeed, a main thesis of Whewell's is that there cannot be, because there is no resolution available, as yet, of the issues dividing the uniformitarian from the catastrophist doctrines, and those doctrines consist principally of incompatible views as to how one should move from geological dynamics to physical theory.[14]

Whewell's attitude here is telling in its implications for our whole interpretation of his place in the history of geological science. For it would be easy, but also mistaken, to couple him with Sedgwick, and to see him, ultimately, as another leading critic of Lyell's uniformitarian synthesis of geology, another critic who responded by joining in developing the case for an alternative to Lyell's

[13] Whewell (1837a), i. 679, 687–8.
[14] Ibid. iii. 594–605.

synthesis; the case, that is, for what has been called the directionalist (or progressionist) synthesis such as one finds it in Sedgwick, or more elaborately, in De la Beche, especially his *Researches in Theoretical Geology* of 1834.[15] That this view of Whewell is mistaken is apparent as soon as we notice that there is no moment in Whewell's writings where he defends, even by implication, several central elements in that progressionist synthesis. Most decisively, there is nowhere in Whewell any defence of the thesis that a progressive decline in the activity of the physical agencies affecting the earth's surface, especially igneous agencies, has made the earth progressively more fit for higher and higher types of life, which have accordingly been introduced as a progressive organic succession adapted to the progressive increase in habitability consequent upon progressive physical cooling and calming.

To be sure, Whewell does accept that a palaeontological research has disclosed a succession, if not a progression, of faunas and floras; he does accept that there are in the fossil record various evidences of climate change, notably a general cooling; and he does accept that there has been a decline in igneous causes of physical change. But he is not seeking to give coherence to his views on geology as a whole by seeing how these various beliefs can be brought together in any integrated synthesis; even less is he knowingly allying himself with one particular synthetic stance.

There is, ultimately, for Whewell, as he explains in his closing palaetiological chapter, 'On the Conception of a First Cause', a synthetic integration of palaetiological science, including geology; but it is not an achievement of any of these sciences themselves, taken singly or even in concert, for it is a theological integration.[16] All the origins that these various sciences point to, but cannot explain by reference to any known course of events, all these different origins are creditable to one God as the First Cause; the origination of the earth, the origination of life, the origination of language may all be presumed due to one and the same originating cause, the same First Cause. This positive outcome, from the palaetiological sciences brought together with theology, depends on the negative outcomes in each of those sciences. What the negative outcome comprises, in any one case, depends on the explanatory successes and failures exhibited by that particular science. There is, then, in Whewell's thinking on palaetiological science, a highly distinctive blend of two ingredients. On the one hand are his quite general and abstract views as to how any such science can be scientific at all; on the other hand are his specific judgements as to what any one such science has or has not done so far.

5. Laws, Causes, Induction and History

Whewell's concern with these issues goes back to the period 1830–2, when he prepared his reviews of Lyell's first two volumes; and when he prepared

---

[15] On the progressionist synthesis, see Rudwick (1971); de la Beche (1834).
[16] Whewell (1847), i. 700–8.

himself to take up where Herschel had left off in the characterization and vindication of the physical sciences as grounded in an inductive philosophy descending, before all other ancestors, from Bacon.

By July, 1831 he was already tabulating '*Types of Progress of Sciences*' in a manuscript notebook.[17] Later, in the same notebook, and in conformity with that tabulation, he took up the question of how to characterize geology as an inductive science that makes inductions of the highest sort, that is, inductions to causes. Within a few sentences the discussion spreads and escalates so as to include a range of very general issues such as would be canvassed in the palaetiological science chapters in the *History* and *Philosophy*.

Whewell begins by insisting that causes are of two kinds: 'the cause of a thing continuing as it is, and the cause of its beginning to be.' There are, that is, 'laws of force' and there are 'accounts of origin'. In other words, there are 'permanent' causes and there are 'progressive' causes; there are 'perpetual' and there are 'primeval' causes. Examples of the first kind are, Whewell says, the force of gravitation, the luminiferous ether, the cooling of bodies as the cause of dew, refraction as the cause of the rainbow. Examples of the second are 'the *true* geological theory of the condition of the earth—Laplace's account *if true* of the formation of the solar system', and 'The true origin of the geographical distribution of plants and animals'.[18]

Now, says Whewell, the second kind appear to be 'as good subjects for science as the other'. And they are 'at least as attractive to the scientific intellect'. However, they do not seem 'to have succeeded at least not on any great scale', there being 'no large instance of an established theory of this kind', the 'parallel roads' of Glen Roy (a geological case) being 'perhaps one of the best'.

In making sense of this very general but not total lack of success, Whewell cites Cuvier's designation of the geologist as an antiquarian, and goes on to relate geology first to our knowledge of man and then to our knowledge of God. When studying the action of physical causes within the human period, as in the case of the ruins of the Temple of Jupiter Serapis, geology may indeed overlap with human history. Here, then, we 'include among our *causes* the acts, habits, institutions, history of man'; these being once again, 'very different . . . conceptions from what we have in permanent causes'. Whereas, at the near end of the past, geology borders on the history of man, at its far reaches it leads to God, to God viewed as an antiquarian views ancient men. In the 'investigation of primeval causes we are led to a nearer view of the operations of the Deity'. It is like 'making out the character of a nation from its architecture', the ancient Egyptian nation for instance.

[17] Whewell Papers, Trinity College Library, Cambridge University, R18 17[15]. All the passages quoted in this section are from pp. 43[r] ff.

[18] Note that Whewell does not identify what the true geological theory is. He is in fact convinced that it does not yet exist. Indeed, he will eventually write (and never retract the judgement) that, in all the palaetiological sciences, there is '*No sound palaetiological theory yet extant*'. See Whewell (1847), i. 664.

Geology is therefore a causal and inductive science; but it is like history in its conceptions of cause. Pursuing this twofold characterization, Whewell reflects that geology, as it has been pursued so far at least, and indeed in regard to the questions the phenomena suggest, 'differs somewhat from the sciences so called'. For the 'object of science is to determine general *laws*', that is, propositions that are 'universally and constantly true, from the nature of their subjects, and independently of any one particular exhibition of them in fact'. The object is not, as it is in history, to ascertain particular events that have happened.

To that extent, Whewell finds he must distinguish 'reasoning' that is 'inductive' from reasoning that is 'historical'. For consider the example he develops, of an eclipse as an object of inductive reasoning by an astronomer and that eclipse as investigated by the historian. An astronomer, says Whewell, can determine that, according to the laws now regulating the motions of the sun and moon, an eclipse must have happened at a certain remote period, and have been visible in certain places. But whether this is the eclipse mentioned by some classical authors and associated with certain ancient traditions, the astronomer does not decide. On such questions, rather, he is examined as a witness by the historian or antiquarian. For he 'gives to the historian or antiquarian the laws which he has determined'.

Now Whewell goes on to emphasize that although the eclipse example is instructive, it is unrepresentative in that astronomers are exceptional in being able to reason back into past time and forward into future time, so as to reach, inductively, events very remote from present ones. To be sure, in a complete science such reasoning can be conducted as exact calculation; and perhaps astronomy can so proceed; but 'in no other case does it belong to the inductive philosopher alone to recall into existence the series of events which form the past'. So, Whewell concludes, when the 'inductive philosopher' goes 'beyond those times in which he can unite his efforts with those of the historical reasoner it is his duty to arrest his narration'. In other words, outside astronomy, the occurrence of past events cannot be inferred in the absence of any records of them, records studied by historians. So where records do not exist, there is usually no ascent to past events and no narrative possible.

Whewell is now in a position to discern the limits of geological knowledge. From this last reflection, it appears, he says, 'that the geologist cannot in accordance with inductive principles' tell us 'what happened when the present or any former condition of the earth began to obtain'. The geologist 'cannot descend from his proper region of necessary and constant connexion', to engage the 'accidents of individual and transient change, or of origin which is the greatest of such changes'.

The ultimate drift of this argument is plain enough, but the rationale of its component steps may be far from obvious. What makes for the difficulty is that Whewell has here a quite different argument from the one that will eventually appear in 1837 and 1840; but it is an argument with much the same conclusion:

the geologist cannot tell us about origins. To see how the present argument works, one does best to go to its closing passage, where Whewell allows that geology can achieve some explanatory causal narrative conclusions, even while being precluded from the beginnings of things.

Whewell closes by insisting that, although 'confining the geologist to his proper field', he is still leaving him able to decide 'many or most of the questions' prompted by 'speculations on the origin of the earth'. For there are cases where the geologist, as inductive reasoner, can move from inductive generalizations that include no causal narrative to confident assertions of that kind. For instance, the geologist may begin with the generalization that 'from trap rocks proceed veins into the adjacent rocks *as if* they had been injected in a state of igneous fusion', and that there is 'a *gradation* without interruption from newest lavas to trap rocks of all kinds'. Then, by coupling this universal appearance of igneous origin, with a refutation of all proposals implying 'aqueous production', one will have established 'the *fact* that trap rocks *were produced* in a state of igneous fusion'. Likewise, to take another example, the 'superior rocks being shown to contain fragments of the inferior, it will be believed that they *were formed* after them'. In this way, says Whewell, 'we may state any of the theories ill or well supported'. For, in such cases—where a theory about the facts as to what has happened in the past is inferred from some generalization about present facts—all that is 'requisite is to arrange the inductions (legitimate or otherwise) which are included in the theory, so that the *narrative* which it includes may be as orderly and clear as it is possible to make it'. For conversely, if the theory is one that is very 'loosely' constructed, there will be in it a noticeable 'hiatus' that will show the 'insufficiency of the structures so elevated'. Exclusion of alternatives and continuity as secured by coherence are, therefore, the grounds of truth in a causal, narrative theory that is constructed when the geologist moves beyond inductive reasoning that is not historical to inductive reasoning that is. Such are the ways whereby geology can be both historical and inductive and so scientific.

Whewell has already set out the limitations that these conditions impose on the geologist's ambitions. If his series of facts takes him back only a finite time into the past, he is 'not at liberty to place a hypothesis in the previous period thus left blank'. Only by including this series in a wider generalization can he pass 'beyond the barriers'. His 'boldness of assertion' is not to increase with the difficulty of investigation, in the 'dim and shadow districts', but is to 'diminish with the scantiness of means of proof and the insecurity of his accumulated pile of inferences'. It being no longer 'in his province to tell us of the beginning of things', he would leave that to those philosophers who have 'imagined that they were authorized' from present facts 'to speak of chaos, and creation, of deluges and conflagrations', to construct cosmogonies and to fix the 'times and circumstances of cataclysms and ecpyroses'. With this modesty about the very remote would go diffidence about the most proximate events. The geologist would leave to others any investigations of the individual events 'preserved by record or tradition, such as have excited the fears and wonders of men or influenced their moral destiny'; for that is a 'province' where the judicious among such men will undertake the interpretation of the evidences by appropriate 'peculiar rules'. Moreover, the 'indistinctness and number of the historical mass of such geological changes in early times is so great', and 'our ignorance' of the details of the 'circumstances' of those that 'geology indicates to be in the order of time' also so 'obvious', that, Whewell concludes, 'an irreconcilable contradiction between history and science is scarcely possible'.

6. The Delimiting of Palaetiological Science

It has been well worth looking at Whewell's arguments and conclusions in this manuscript text, for they can alert us to his most persistent preoccupations as a writer on geology and so on palaetiological science generally. Especially they exhibit his consistent strategy of deploying epistemological explications in the service of delimitational adjudications. The proper duties of those enquirers who work in this or that province are to be explicated, so that the nature of their actual and possible achievements can be identified correctly and so related appropriately to the actual and possible achievements of others. The principle case of such adjudication is that between geology and Scripture.

That the adjudication is an epistemological one may lead us to overlook that it is an adjudication at all. We are used to finding geology and Scripture related to one another on corresponsive grounds, whether particular or general. Buckland upholds the correspondence of a particular geological event to a particular Biblical moment, the Flood; Sedgwick upholds organic progression, over vast eons before man's arrival, as analogous to the progressive revelation in Scripture of the progression from Creation to Judgement. We are then inclined to presume that the rationale for Whewell's philosophy of palaetiological science is not adjudicational, because it is not grounded in a corresponsive thesis, not even a quite general and abstract one. But Whewell's conviction is that there is adjudicational work to be done through epistemological delimitations, even when no corresponsive thesis is involved—much less adopted.

It was reflection on Lyell's geology, especially on his rejection of Lamarck's views on the origin of species in the transmutation of species, that gave Whewell his eventual delimitational strategy as adapted in 1837 and 1840. For that strategy, although quite general in its form—as expressed in the abstract negational maxims, about the undiscoverability of origins and the uninformativeness of science concerning creation—is a generalization of the lessons Whewell drew from Lyell's encounter with the transmutation of species issue. Lyell, as Whewell saw it, had judged advocates of transmutation, such as Lamarck, unsuccessful in their attempt to find, in the changes of character presently wrought in animal and plant species by changed conditions, causes adequate to have brought new species into existence in the past. These changes that species are subject to at present are changes within the course of events as now known to us. So, as Whewell represents it, this failure is a failure to find a beginning for species that is homogeneous with the known course of nature.

Now, as Whewell saw it, in insisting on this failure, Lyell was admitting one major exception to his own claim that all past events recorded in the rocks could be successfully explained as the effect of causes still at work at present, and acting with no lesser or greater intensity than in the past. So Whewell's understanding of the failure of transmutation rests on his understanding of Lyell's most general commitment to the adequacy of existing causes. Whewell understood Lyell's claim as one of causal adequacy on behalf of causes existing today, and presumed to be acting with uniform intensity throughout past, present and future. For anyone committed to this claim, Whewell urged, there were at least three challenges to be met. First, causes must be found for the mechanical effects indicated by such phenomena as tilted and elevated rock strata. Second, past climate changes must be traced to causes now in action. Third, it will have to be shown that changes from one set of animal and plant species to another are explicable in this way.[19]

Now, Whewell welcomes Lyell's comprehensive marshalling of the evidence of causes now in action, and he proclaims that in doing so Lyell has contributed decisively to geological dynamics. However, when it comes to the first of the three challenges in moving beyond dynamics to geological history, Whewell declares himself entirely unconvinced. The extent and form of the elevation of ancient strata, the relations of successive strata to each other and the shapes and connections of river valleys are all beyond the competence of existing causation. In such cases, one has to conclude that the changes belonging to the past stages of the earth's existence, as disclosed by geological inquiry, 'cannot be considered as forming a continuous and homogeneous series' with those 'which are now taking place upon its surface'.[20] As to the second challenge, Whewell hails as ingenious Lyell's novel speculation referring climate changes to the effects of changes in the position of land and sea relative to the poles and the equator; but he declines to accept or reject it. On the third challenge, postponed by Lyell until his second volume, Whewell endorses Lyell's conclusion that transmutation cannot provide a resolution, because the evidence is that species are not now sufficiently mutable for new species to arise from old even in changing conditions over long periods of time. The appearance of new species, at successive epochs, indicated by irresistible 'geological evidence to have repeatedly occurred', is a fact '*not* belonging to the operation of that tendency to change in organized beings which we see still brought into play'. So, Whewell concludes, with 'this striking exception' of the origin of species, he is willing to 'assert', with Lyell and others, 'that all the facts of geological observation are *of the same kind* as those which occur in the common history of

---

[19] Whewell (1831). See especially 193–4.
[20] Ibid. 199–200.

the world'; it being another question whether they are, as Lyell asserts, of the same degree or intensity.[21]

We are now in a position to see how decisive for Whewell's eventual view of the palaetiological sciences was this striking exception. For, in the *History* of 1837, this exception within geology is exhibited as exemplifying the norm for palaetiological science generally. It is in addressing the question of the transmutation of species, as a question within 'organic geological dynamics', that Whewell raises the more general *'Question of Creation as related to Science'*, and so the question of the limitations on all palaetiological speculations that try to move from the present to the remote past.[22]

Whewell here formulates two versions of a dilemma, the first within organic geological dynamics itself, the second within physical geology. The first concerns 'only such causes as we know to be in constant and orderly action'; the second concerns 'the facts which have happened in the history of the world'. The first dilemma is that, if species are not transmutable then 'we must suppose' that their variations, which are 'apparently indefinite' are in truth 'bounded by rigorous limits'; while, if we admit a transmutation of species, 'we must abandon that belief in the adaptation of the structure of every creature to its destined mode of being', a belief many would give up 'with repugnance', and one that the best naturalists have been convinced is true.[23] A decisive element in this version of the dilemma is, then, the matter of the destined mode of being, the way of life of each species. For the transmutation hypothesis, as Whewell views it, supposes that a species may change in taking up a new way of life, so that its eventual structure would not be due to its being fitted to one way of life that it was specially destined for, but due rather to any ways of living that it was forced into or fell into through contingent circumstances.

Consider now the second version of the dilemma, as raised by geology. For this science discloses that 'many groups of species' have 'succeeded each other at vast intervals of time'. The dilemma then arises 'anew': either accept the transmutation of species, and 'suppose that the organized species of one geological epoch were transmuted into those of another by some long-continued agency of natural causes'; or else believe 'in many successive acts of creation and extinction of species, out of the common course of nature; acts which, therefore, we may properly call miraculous'.[24]

Now Whewell postpones discussion of this second version of the dilemma until he has examined the first. His examination not only concludes that species are limitedly mutable, but also that the additional hypotheses are false that transmutationists have thought it necessary to join with unlimited mutability, in order to explain geological and other phenomena; particularly the hypothesis that traces new structures to changes in 'wants', and the hypothesis of a continual spontaneous generation of the simplest organisms.[25]

[21] Id. (1832). See 103–9, 126.
[22] Id. (1837a), iii. 580.
[23] Ibid. 574–5. [24] Ibid. 574. [25] Ibid. 575–80.

272

It is after this examination that Whewell raises the question of creation in relation to science, and asks whether 'since we reject the production of new species by means of external influence', do we 'accept the other side of the dilemma', and 'admit a series of creation of species, by some power beyond that which we trace in the ordinary course of nature'?[26]

It is at this point that Whewell finds himself with a response, if not an answer, that is supplied by his most general conclusions regarding palaetiological science. For he says that 'the history and analogy of science' teach us to 'reply as follows' to this question.

All palaetiological sciences, all speculations which attempt to ascend from the present to the remote past, by the chain of causation, do also, by an inevitable consequence, urge us to look for the beginning of the state of thing which we thus contemplate; but in none of these cases have we been able, by the aid of science, to arrive at a beginning which is homogeneous with the known course of events. The first origin of language, of civilization, of law and government, cannot be clearly made out by reasoning and research; and just as little, we may expect, will a knowledge of the origin of the existing and extinct species of plants and animals, be the result of physiological and geological investigation.[27]

This reply to the question leads, then, to the very issue of science and creation. Although 'philosophers have never yet demonstrated, and perhaps never will be able to demonstrate, what was that primitive state of things in the social and material worlds, from which the progressive state took its first departure', nevertheless, they can 'go very far back' in all these 'lines of research'. They can 'determine many of the remote circumstances of the past sequence of events', they can 'ascend to a point which, from our position at least, seems to be near the origin', and they can 'exclude many suppositions respecting the origin itself'. But it is 'difficult to say' whether 'by the light of reason alone, men will ever be able to do more than this'. It is then 'no irrational opinion, even on grounds of philosophical analogy alone', that in all sciences that seek origins, 'we may be unable to arrive at a consistent and definite belief, without having recourse to other grounds of truth, as well as to historical research and scientific reasoning'. If we are to 'apprehend steadily the creation of things' we will have to 'summon up other ideas than those which regulate the pursuit of scientific truth', and to 'call in other powers than those to which we refer natural events'. Not surprisingly, then, 'in this part of our inquiry, we are compelled to look for other than the ordinary evidence of science'.[28]

Having invoked these general conclusions, Whewell elaborates their implications for geology. This science, together with other palaetiological sciences, leads us to a single 'focus of being', a single source not only of animal and plant life, but also of social and rational life, arts and language, law and order. Proudly, then, Whewell insists, this 'reflection concerning the natural scientific

---

[26] Whewell (1837a), iii. 580–1.   [27] Ibid. 581–2.   [28] Ibid.

view of creation' has not come from 'any wish to arrive at such conclusions'. Rather, 'it has flowed spontaneously from the manner in which we have had to introduce geology into our classification of the sciences'; this classification being 'framed from an unbiassed consideration of the general analogies and guiding ideas of the various portions of our knowledge'. He goes on to warn, however, that although geology, like any other palaetiological science, leads to theology, geology and theology must not be confused or mixed. Theology cannot supply a supernatural geological dynamics to supplement natural dynamics; nor can geology tell us how to interpret any narrative of the providential history of the world and man. Moreover, Whewell insists, we should not expect a 'full insight' into the consistency that we presume exists between 'the results of true geology or astronomy' and the 'statements of true theology'; after all finite minds cannot expect to 'comprehend the infinite'.[29]

7. The Natural and the Supernatural

Whewell saw, therefore, no contradiction or even tension where we may sense one in his thinking about palaetiological science. On the one hand, he seems to consider seriously the possibility of a complete palaetiological scheme, with the origins of the solar system, the earth, life, man, civilization, and the rest all credited to natural causes in a comprehensive, progressive series of changes wrought by a series of successive causal actions over vast stretches of past time. Here the living would be made continuous with the lifeless, the human with the animal, and the civilized with the barbarous. And yet, on the other hand, Whewell is insistent that such natural continuities are not established by any science, and so are not conclusions imposed on any general reflections concerning palaetiological science.

The reason Whewell sees no contradiction here is that he is entirely confident that the more comprehensive in its ambitions is any palaetiological scheme, the more will it be evident that the natural continuities it implies will be mere suppositions, and so entailing no exclusion of God, acting as First Cause, in the successive originations of those quite distinct series, of quite different changes, that are the proper object of the various particular palaetiological sciences.

His confidence in this outcome is most plainly expressed in his treatment of the nebular hypothesis. A chapter is devoted to this hypothesis in the Bridgewater Treatise, where indeed it seems to have been first given that name. Whewell's position there is that the natural theologian need not worry over the prospect of the hypothesis turning out to be true. For God will still be evidenced in two ways. First of all, however far back the hypothesis goes, one can still ask how there originated the state and powers of diffuse matter that it takes as its initial condition; and since such an initial condition is not traced by the hypothesis to

---

[29] Ibid. 583, 586.

any natural causation, it may be evidence for God's wisdom and beneficence. Second, the progression of cause and effect, leading away from that initial condition, and including eventually the formation of a terraqueous globe, its stocking with life and man, will include productions that evidence God rather than the state and powers of the nebular material itself.[30]

For suppose a planet to be produced in accord with the hypothesis; it may be presumed to resemble only 'a large meteoric stone'. How, then, does it come to be 'covered with motion and organization, with life and happiness'? The rhetorical questions and their answer follow easily:

What primitive cause stocked it with plants and animals, and produced all the wonderful and subtle contrivances which we find in their structure, all the wide and profound mutual dependencies which we trace in their economy? Was man, with his thought and feeling, his power and hopes, his will and conscience, also produced as an ultimate result of the condensation of the solar atmosphere? Except we allow a prior purpose and intelligence presiding over this material 'primitive cause', how irreconcilable is it with the evidence which crowds in upon us from every side.[31]

The same line of argument, now quite generalized, can be used by Whewell, in 1840, to support the same conclusion, that in order to form the whole hypothetical series of events—from nebular condensation to human civilization—into a connected chain of causation, supernatural influences may have to be considered as part of the past series of events. In contemplating such a series, we are led, says Whewell, 'by a close and natural connexion, through a series of causes', all the way from those that regulate the remotest heavenly nebulae, to those determining 'the diversities of language, the mutations of art, and even the progress of civilisation, polity and literature'.[32]

However, regarding the earth's formation and stocking with its first and subsequent life, Whewell says he has spoken 'hypothetically' of these events as 'occurring by force of natural causes', only so that 'the true efficacy of such causes might be brought under our consideration and made the subject of scientific examination'; to raise, that is, but not to settle, the question of whether there are natural causes adequate to such events.[33]

It may be found, that such occurrences as these are quite inexplicable by the aid of any natural causes with which we are acquainted; and thus, the result of our investigations, conducted with strict regard to scientific principles, may be, that we must either contemplate supernatural influences as part of the past series of events, or declare ourselves altogether unable to form this series into a connected chain.[34]

We should now be able to appreciate why Whewell felt no tension, much less contradiction, between two themes in his writings on natural science in

---

[30] Whewell (1833c). See Book 2, Chapter 7, 'The Nebular Hypothesis'. On Whewell and the nebular hypothesis, in general, see Schaffer (1989) and the references given there to early historians of the topic, such as Jaki and Merleau-Ponty.

[31] Id. (1833c), 184–5.

[32] Id. (1847), i. 657–8.   [33] Ibid. 658.   [34] Ibid.

relation to natural theology. On the one hand is the theme, dominating much of the Bridgewater Treatise, that God acts in the physical world through law. One enunciation of this theme is especially explicit. In Book 3 ('Religious Views'), chapter 8 ('On the Physical Agency of the Deity') begins with yet another Whewellian negation. 'We are not', he says, 'to expect that physical investigation can enable us to conceive the manner in which God acts upon members of the universe.' Science, even more clearly than 'everyday reason', shows us 'at what an immeasurable distance' we are from being able to conceive '*how* the universe, material and moral, is the work of the Deity'. However, regarding the 'material world'—as distinct, that is, from the moral world—at least we can 'go so far as this;—we can perceive that events are brought about, not by insulated interpositions of divine power exerted in each particular case, but by the establishment of general laws'. This is the 'view of the universe proper to science, whose office it is to search out these laws', Whewell declares.[35]

On the other hand, however, there are Whewell's explicit declarations that among the events inquired into by the palaetiological sciences are some that manifest God, acting as a power beyond the known, lawful course of nature. For instance, he expects Lyell to agree with him, he says, in thinking it undeniable that 'we see' in the transition from one set of species to another, as shown by palaeontological studies, 'a distinct manifestation of creative power, transcending the operation of known laws of nature', geology thereby 'lighting a new lamp along the path to natural theology'. The more closely this transition is examined, indeed, the more it seems to Whewell out of reach 'of any other power than that of which the Creator has confined the regulation and manifestation to the depths of his own bosom'; it is a power that 'appears to belong, not to what we are accustomed to speak of on the laws of nature, but to that Supreme Will, which is their source and foundation'.[36]

Now, there is no inconsistency, for Whewell, in these two lines of thought. For, in judging the origin of some series of events beyond the known laws of nature, he does not need to assert that it has been originated by an act of God that involved breaking or overriding any laws, in the way that the virgin birth of Christ was sometimes represented as brought about by a special, exceptional Divine action in violation or circumvention of a known law. Indeed, for Whewell, the origination of any series of events, such as a new set of species, is a matter of God's introducing new laws rather than either breaking or conforming to laws already in force. He seems nowhere to have said explicitly that these originations are to be construed in that way, as additions to the constitution, additions to the laws of nature; but such would seem very much in accord with his overall stand; for he would then be arguing that from the way

[35] Id. (1833*c*), 356–7.
[36] Id. (1831), 194; id. (1832), 125. In following Whewell's views on these matters, I have benefited from an unpublished paper written some years ago by Michael Ruse: 'William Whewell on the Origin of Species'.

a law is known to be conformed to now in the course of nature, there can be no inferring how it was that the law first came to be instituted. But regardless of whether this was precisely Whewell's position, it is plain that he saw the originations discerned, but left unexplained, in palaetiological science, as moments when additions were made to nature. So his argument is that from the way things go, lawfully, now, after these additions, one cannot arrive at an explanation of how the addition was made in the first place. One may know that it has been made but not how. When Whewell says that God does not interpose his power, on isolated occasions, to produce a particular event, he sees that as consistent with his account of God's originating action in making these additions to the world; for such additions are not particular events, they are initiations of series, lawful series, of events, a set of organic species, for instance, or a new language, or whatever.

8. The Doctrines of Uniformity

It will already be plain that, although an admirer of Lyell's geological dynamics, Whewell's motivational preoccupations as a writer on geology contrast strikingly with Lyell's own. Where Whewell was resigned, even ultimately resolved, to conclude that geology, like any palaetiological science, was inevitably striving after what it could not be expected to achieve, Lyell was wanting to frame the business of geology in such a way that success could be ensured, and the ill reputation of the science—for vague speculation and divisive disagreement—scotched. Whewell is rightly known as the most acute and enlightening critic of Lyell's proposals for the foundations of geology. In reading Whewell on Lyell, we do well to keep in mind this contrast in their goals.

Whewell eventually made his critique and rejection of Lyell's uniformitarianism largely turn on two issues, that may ultimately be seen as one. For the one issue is that of where in geology we can know, and where by contrast we must retreat to suppositions that are more or less arbitrary, and so more or less unsatisfactory supplements to what we do or can know.

Lyell's own position almost comes down to a single twofold thesis: namely, that the only knowable causation is that active in the literate human period, the present in the sense of the last three thousand years through today and tomorrow; and that the present is not special in any other way than in the knowability of its causation, where knowability is pretty much a matter of observability. The agencies active in the prehuman past are the same, and they have worked in similar circumstances, so that the sorts and sizes of their effects have been similar too. Conversely then, a cause that is confined to the prehuman period would be not knowable because not observable, even potentially, by man. The assumption about the uniformity of prehuman and present causation, about the stability and permanence of the system of causation, is then demanded by the requirement that geology attempt always to confine itself to explaining past changes as the effects of causes that are or could be known causes. That requirement derived in turn from the *vera causa* ideal first enunciated in Reid's explication of Newton's first rule of philosophizing. For that ideal specified that one's explanatory causes should be causes that are known to exist—to be real and so not hypothetical—from the evidence of facts other than the facts they were used to explain.

According to Lyell's argument, therefore, that ideal for geology entailed a strong and strict acosmogonical reproductive causal adequacy presumption for geology. For past events are to be presumed to be the effects of causes that are still undiminishedly active today and tomorrow, in that they could, and indeed eventually will, produce, (and so reproduce) in the future effects of the same sort and size as they have in the past. Accordingly, no past effects recorded in the rocks are to be explained as residues from an original state of a nascent earth—an early molten fluid earth or an early landless panoceanic earth for instance—a state that no causes active today could ever produce even in an unlimited stretch of future time. Cosmogony, in the sense of geogony, is therefore excluded from geology. The science has no concern with whether, much less how, the earth did once originate from some preplanetary, prototer-restrial beginnings. The reproductive adequacy thesis, as required by the *vera causa* ideal, requires a presumption of uniformity that precludes cosmogony.[37]

To this Lyellian proposal, Whewell replied with three countering arguments. First, he argued that present causation, in Lyell's sense, is no more knowable than past causation; second that the *vera causa* ideal of explanation by known causes is often too restrictive for any physical science; and third that there is more than one way to subsume the earth's past within a presumption of uniformity in natural causation, and that there are reasons not to prefer Lyell's way, but to prefer another that leads to entirely different expectations as to what geology may or may not establish.

In developing the first argument, Whewell insists that it is arbitrary either to assume a uniformity in the course of nature that precludes catastrophe, or to assume that catastrophic rather than uniform change has occurred. We should look to the rocks themselves to discern whether or not past change has been uniform or catastrophic. This declaration may seem to beg the whole question as Lyell had framed it. For Lyell had been adamant that what is seen in the rocks is always indeterminate, in regard to the issue of whether it has been caused slowly by causes no more violent than today's, or suddenly in catastrophic events. But Whewell thinks that there are plenty of cases where the first option can be confidently eliminated. The physicist in him, especially, would insist

[37] For a more extensive interpretation of Lyell along these lines, see: R. Laudan (1982), and id. (1987), ch. 9. For Whewell's critique of Lyell's appeal to the *vera causa* ideal, see Ruse (1976a). A manuscript entry from the early 1830s reveals, tellingly, that Léonce Elie de Beaumont, whose theories were embraced by Sedgwick and rejected by Lyell, was Whewell's exemplary catastrophist: 'Catastrophists (E. de Beaumont) Uniformitarians (Lyell)', Whewell Papers, R18 17^{12}, p. 22^{r}. The significance of Elie de Beaumont's theorizing, especially for English discussions, is explained in Rudwick (1971), Greene (1983), and R. Laudan (1987).

that certain configurations of effects can only be wrought by certain constellations and intensities of forces. Not that Whewell is less sceptical about our ability to know causes than Lyell is. Indeed, he maintains that causes are never known except from their effects, and that this goes for present causes no less than causes acting in the past. So we know the causes of the elevation of the Alps no less, and no more, than we know the causes of an elevation of land now going on; for we know each only from their effects. Lyell's equation of knowability in causes with action in the human present is, therefore, unacceptable.[38]

Moreover, even if we did agree to try to move from the present, as more knowable, to the past, as less so, we could not know what we would need to know concerning the present. Most decisively, we could not know the extent of any fluctuations in the intensity of causes acting in the present. What we have observed so far, in the way of violence, may be exceeded tomorrow for all we know. So if we cannot set limits to violence in the short run, even less can we for the longer run, including the prehuman past. Nor, if we take in a longer run of the future, can we sustain any presumption of stability in the causal system. Lyell's attempts to justify a perpetual subterranean, combustive motion notwithstanding, the heat of the earth must be presumed to dwindle eventually. There will be an end to the earth, then; just as Whewell had argued, in his Bridgewater Treatise, that there would be an end for the solar system, thanks, eventually, to the resistance of the ether through which the planets make their orbits.[39]

Lyell has, therefore, misunderstood the precedents set by astronomy for geology. A combination of the *vera causa* ideal, with a presumption of stability in the system of causation, is by no means the only proper way to model this young science of geology upon astronomy. Rather, we should 'consider our knowledge of the heavens as a palaetiological science', and, specifically, consider the nebular hypothesis. For there we have 'the beginning of a world'. Whewell stresses that he does not, for his argument, maintain that this hypothesis is 'true'; but if geologists would borrow 'maxims of philosophizing from astronomy', then 'speculations' such as have led to that hypothesis 'must be their model'.[40]

Again, consider other provinces of palaetiological speculation, says Whewell, in 'the history of states, of civilization, of languages'. We may presume 'some *resemblance* or connexion between the principles' determining 'the progress of government, or of society, or of literature, in the earliest ages, and those which now operate', but no one has 'speculated successfully' by 'assuming an *identity*' in these causes.[41]

Where do we now find a language in the process of formation, unfolding itself in inflexions, terminations, changes of vowels by grammatical relations, such as characterize the oldest known languages? Where do we see a nation, by its natural faculties,

[38] Whewell (1837a), iii. 616–17; id. (1847), i. 668–9.
[39] Id. (1847), i. 674; id. (1833c), 207–9.
[40] Id. (1837a), iii. 619.   [41] Ibid. 619–20.

inventing writing, or the arts of life, as we find them in the most ancient civilized nations? We may assume hypothetically that man's faculties develop themselves in these ways; but we see no such effects produced by these faculties, in our own time, and now in progress, without the influence of foreigners.[42]

In the *Philosophy* in 1840, Whewell will give these examples, especially glossology (or philology), more extended treatment. But they never take on a life of their own.[43] Geology continues to dominate the understanding of palaetiological science, and the other instances of the genre never require any revisions in the way that genre is conceived. It is only as instances of a genre that also includes palaetiological astronomy, which is to say hypothetical nebular astronomy, that they can help discredit Lyell's mistaken appeals to unpalaetiological astronomy as a precedent for geology. He continued, in the *History*:

Is it not clear, in all these cases, that history does not exhibit a series of cycles, the aggregate of which may be represented as a uniform state, without indication of origin or termination? Does it not rather seem evident that, in reality, the whole course of the world, from the earliest to the present times, is but *one* cycle, yet unfinished;—offering, indeed, no clear evidence of the mode of its beginning; but still less entitling us to consider it as a repetition or series of repetitions of what had gone before?

Thus we find in the analogy of the sciences, no confirmation of the doctrine of uniformity, as it has been maintained in geology.[44]

This argument against Lyell is so typical of Whewell that we may overlook how characteristic it was of his time. There were in his day not only more distinct sciences than ever before, but more self-conscious efforts than ever before to distinguish and advance novel sciences. Whewell is the child of his time in taking this kind of effort as exemplary for the history of science as a whole, making his history a history of sciences, plural but comparable. Arguments from the analogy of science in this sense, arguments from comparing comparable, but only comparable, sciences, are what his writing on the palaetiological sciences is all about.

9. Whewell, Chambers, and Darwin

Whewell may well have sensed, fearfully enough, that sooner or later someone would have a go at putting the nebular hypothesis together with spontaneous generation and the transmutation of species, to construct just the kind of complete palaetiological scheme that he had insisted was only to be considered in order to see that the gaps in it could not be closed by natural causes. Whether or not he did sense this, when the time came he was indeed exasperated, the more so as the job was done by an anonymous author (it was Robert Chambers)

[42] Ibid. 620.   [43] Id. (1847), i. 649–51; 661–3.
[44] Id. (1837*a*), 620.

in a book obviously designed for the general reader (*Vestiges of the Natural History of Creation*, 1844) so taking the issues it raised to a wide public, in circumvention of the natural guardians of epistemological rectitude in such matters, Whewell himself and his Christian, savant friends. Whewell responded publicly by having printed a collection of extracts of his writings on physiology, comparative anatomy, and geology, under the title *Indications of the Creator* (1845). Within a year, he was writing a preface for a second edition of this anthology. In that preface he engaged Chambers's own criticisms of his writing, criticisms made, anonymously again, in Chambers's *Explanations: A Sequel to Vestiges of the Natural History of Creation* (1845).

As with his treatment of the transmutation issue in the *History*, Whewell's response was twofold. On the specific question of whether *Vestiges* made its cases empirically, for spontaneous generation and transmutation, Whewell argued that the facts, as assessed by the best authorities, were not in favour. Here, Whewell had to take up, especially, the state of discussion surrounding various generalizations in comparative embryology that were decisive for the *Vestiges'* attempt to establish that the development of life on earth proceeded, naturally, in accord with powers and tendencies exhibited in higher animal embryos.

However, Whewell was no less concerned to counter *Vestiges* by arguing that it presumed to have established what we had no reason to expect could be had, and that is a palaetiological scheme that finds natural origins for such series of causes and effects as are presented by the history of life and man. In arguing this theme, Whewell had to insist, against Chambers's accusations, that he, Whewell, was not saying that we should not try to seek scientifically after such a scheme; rather, he was saying that we should try, but without expecting to succeed.[45] This general rebuttal of Chambers is combined by Whewell with the specific criticisms, to provide his answer to a direct challenge: 'It may be said', Whewell acknowledges, 'if the hypothesis of creation' in the *Vestiges* is 'not the true one, what then is?' To this question, says Whewell, 'men of real science do not venture to return an answer'. Moreover, as he eventually conceded, 'I must necessarily confess that we cannot obtain from science a complete view of the history of the universe.' There is 'no surprise or humiliation' in our inability to 'fathom and comprehend the acts of the Creator and Governor of the World'. He closes his second edition preface by recognizing that he represents 'the results of the History of Science as being in a great measure negative' regarding all such questions. He quotes Chambers's plea that science should explain to us the great ends of the Author of Nature and our relations to him, to good and evil, to life and eternity; and Chambers's disappointment at the emptiness of any answer from science when it tells only

---

[45] Id. (1846), 9. For the immediate context of Whewell's response to the *Vestiges*, see Brooke (1977a), and Yeo (1984). See further Brooke (1977); Raub (1988). I cannot join Raub in identifying Whewell as the reviewer of Chambers in *Fraser's Magazine*: see n. 9 of her article.

of classes of animals, of electrical and chemical agencies, and so on.[46] To this Whewell retorts as his parting shot:

Strange caprice of the human mind! that one who feels such answers to be worthless, should think that he has remedied all their deficiencies, and supplied all that the soul can desire, when he has told us that *life grows out of dead matter, the higher animals out of the lower, and man out of brutes!*[47]

To Whewell it was both reprehensible and incomprehensible that anyone could want to conduct his theodicy publicly through a palaetiological scheme that included such theses.

As viewed by Whewell, Darwin's *Origin* was the transmutation of species coming round for the third time. But Darwin himself, an eminent scientist, a friend, and the author of a book that made no appeal to popular favour, could not be countered with such hauteur as one would indulge in countering the anonymous composer of a book like *Vestiges*. For us, as Whewell's and Darwin's historians, the matter is further complicated because there were interactions between the two men that can instruct us about both.

The subject of Darwin's relations with Whewell, relations of many kinds, personal, institutional, ideological, and so on, is such a complicated one that it cannot be resolved by any sweeping generalization on the one hand, nor, on the other, by fastening on to any particular text, moment, or transaction as a single, privileged key to the whole story.

A comprehensive and critical account of the two men's relations with one another must wait for another occasion. However, it may be appropriate to indicate here how the subject could be advanced beyond where the literature now leaves it. Several points, of rather disparate sorts, may be worth making in particular. First, there are motivations from social historiography and from intellectual historiography that have conspired, mostly unwittingly, to make many people wish that Whewell will turn out to be the brains behind Darwin. On the social side there is the neo-Namierite presumption that individuals are best understood as products and protégés of some informal, invisible school, network, or college that fostered and raised them. Cannon's Darwin, as child of a Cambridge network having Whewell as a dominant presence, is very much in this tradition.[48] On the intellectualist side, there is a neo-Kantian presumption that behind many a scientist is a philosopher who has taught him how to think about the universe as a whole and about science as the way to explain its contents. Ruse's Darwin, with his indispensable debts to Herschel and Whewell, as teachers of philosophy, is very much in this tradition.[49] Even more generally there are, perhaps, hopes in many a scholarly heart, that two giants, Whewell and Darwin, will turn out to be, ultimately—and if it is an irony, then all the better—collaborators in a single revolution of thought. So,

---

[46] Whewell (1846), 21–3, 29–30.   [47] Ibid. 30.
[48] W. Cannon (1961); S. Cannon (1978).
[49] Ruse (1975a). For a more recent paper in the same vein, see Recker (1987).

in a word, the challenge to critical scholarship is not to get carried away by a tide of assumption that would tend a priori to greatly overrate the importance of Whewell for Darwin.

Secondly, the truths about Whewell and Darwin, it is now becoming plain, are to be found in the details. Thanks to recent editions of Darwin's early notebooks and correspondence, especially, we can reconstruct with some confidence what was going on and when, between the two men, at least in the decisive years from 1836 to 1839, in letters, in conversation, and in reading each other's published work.[50] When this evidence is put together with other relevant documentation, notably Darwin's marks and annotations in his copy of Whewell's *History*, then it is apparent that Darwin learned much from Whewell, had many debts to him, was much influenced by him, if one prefers; but that this indebtedness was far more plural and fragmentary and much less coherent, integral, and synoptic than either the Cannonical or the Rusean presumptions would suggest. For an itemizing of Whewell's contribution to Darwin's life and work would form a long list of things that Darwin either accepted, was challenged to reject, or was concerned to modify and reinterpret and subsume within his own thinking. Thus, to cite only some instances, there is Whewell's version of Kant's teleology that Darwin rejected;[51] there is Whewell as aid in understanding the physical geology of wave motions in rock material;[52] there is Whewell on necessary principles as elements in all scientific knowledge, to be reinterpreted as instinctive, ancestral, animal legacies for humans as capable of success in pursuit of scientific knowledge;[53] there is Whewell, very acceptably to Darwin, distinguishing formal from physical laws;[54] there is Whewell on the nature of a natural system in systematic botany and zoology, to be partly accepted, partly rejected;[55] there is Whewell on the teleology of sleeping and daylight, to be scorned;[56] and so on and so forth. Later, in writing the *Origin*, there is Whewell on God working through law to be used as an epigram at the front of the book; and later still, there is Whewell on hypotheses, to be cited in legitimation of a 'provisional hypothesis' of pangenesis, a hypothesis about generation or reproduction and so inheritance and variation; and so on and so forth.[57]

[50] Barrett (1987). Darwin's annotations on the *History* are discussed, for instance, in Manier (1978) and in the essay review of this book by Schweber (1979). The most recent, enthusiastic advocate of Darwin's debts to Whewell is Curtis. See his two articles cited above in n. 8. It will be apparent that I can accept some but not all of Curtis's proposals.

[51] Cornell (1986).

[52] Darwin to Whewell, 18 June 1837, in Stair-Douglas (1881), ii. 24–5. Whewell's two presidential addresses to the Geological Society show him to be knowledgeable concerning Darwin's work at the time. They were printed, respectively, in Whewell (1838a), 624–47, and (1839).

[53] Curtis (1987).

[54] Ruse (1975a).

[55] See Darwin's 'Notebook D', p. 26 in Barrett (1987).

[56] See 'Notebook D', p. 49, ibid.

[57] Thagard (1977), 353–6. See the introductory section of Chapter 27 of Darwin's treatise on *The Variation of Plants and Animals under Domestication*.

What all the evidence, taken together, makes quite implausible is that Darwin owed to Whewell either his most general convictions concerning the universe or his most general convictions concerning science. Central to Darwin's understanding of the universe is his view of the relationship between the physical world—of the stars, planets, earth, ocean, climate, and so on—and the organic world of plants and animals, including man. On the physical side, Darwin never dissented in any fundamental respect from Lyell's assumptions concerning the stability of the physical system of terrestrial causation, with its corollary the exclusion of cosmogonical theorizing from geology. His view of the organic world, directly and knowingly in disagreement with Lyell's view of the organic world, was nonetheless accommodated to Lyell's view of the physical. A principal challenge for Darwin was to explain how progress, although not invariable, would be inevitable in some lines of descent at least, in the adaptation of life to a terrestrial surface that was changing only unprogressively à la Lyell. In his understanding of this challenge and how it could be met, Darwin shows no signs of indebtedness to Whewell's views on the physical world and the relation to it of the organic. In working out his earliest conclusions concerning organic progress Darwin drew on Lamarck (as expounded and criticized by Lyell) on Owen and on Carus, among others, but neither then nor later did Whewell contribute. Nor is that surprising; Whewell's writings do not convey a consistent, positive, detailed teaching on such matters.

As for Whewell's teachings on the evidential and explanatory ideals appropriate to inductive science, there is no reason to think that Darwin's early work was at all directed by them. Most especially is this true of Whewell's leading proposal, the consilience of inductions. No one had access to this proposal in printed form until Whewell published it in 1840 in the *Philosophy*. Contrary to legend, there is no exposition of it in the *History*; indeed, it may well be that Whewell had not worked it out when he was writing that book. Nor is there any sign that Whewell was likely to have communicated the consilience doctrine to Darwin in conversation at the Geological Society, or indeed in any other setting. What is more, when Whewell did defend consilience in 1840, he offered it as a replacement for the *vera causa* ideal, which, he insisted (as he had in 1837), was too restrictive.

Now Darwin's argument for his theory of natural selection was conformed to that ideal and not to Whewell's consilience as a replacement for it. Certainly there is a part of Darwin's argument that would have been appropriate had he been presenting his theory as consilient. But that part is as much required by the *vera causa* ideal.[58] So its presence provides no ground whatever for reading that part as a trace of Whewell's influence; the place of that part in the whole of that argument is the place it should have, given that Darwin is conforming to the ideal Whewell had sought to discredit and replace. What is more, in July 1837, nearly a year and a half before Darwin had first come to natural selection,

[58] Hodge (1977), 237–45; id. (1989).

he was arguing—for an earlier theory of adaptive species formation in branching descent—along very much the same lines that he would give in his argument for natural selection in 1839, 1842, 1844, and 1859.[59] So there is simply no ground at all for finding any moment in Darwin's life as a moment when, from reading in Whewell's writings, receiving letters from him, or talking with him, he learns to embrace the consilience of inductions as a decisive new ideal for any theory of the origin of species.

Perhaps, then, if Darwin's idea of the universe and his ideal of science are not derived from Whewell, may his concern with man, mind, and morals not be traced to that source? Again, the evidence all suggests otherwise. That any theory of transmutation should include man, his mind and morals, was a conviction Darwin clearly had in July, 1837, when he opened his 'Notebook B'. There is every sign that Darwin's reading in Lyell's account of Lamarck, and in his grandfather's books, contributed to this conviction. There is no sign that Whewell had anything to do with it. A year later, Darwin opened a 'Notebook M', devoted to 'metaphysics', that is to say, mind and morals in men and animals. It has been suggested that in opening this notebook, Darwin was acting on a resolution to counter Whewell's views on such subjects.[60] Again, the trouble with this suggestion is that there is no sign that Whewell played this role in Darwin's life. His earlier 'Notebook C' and his contemporary discussions with his family show that he had many good reasons, at that time, for paying even more attention than he had before to metaphysical matters. Nothing suggests that Whewell provided Darwin with those reasons.

But if Whewell is not a decisive resource on broad issues for Darwin in the younger man's London years, was he not such a resource in the Cambridge undergraduate years before the *Beagle* voyage? One should doubt it, and for two reasons. First, a great deal that Whewell could have taught later and Darwin could have learned later was not available to them at that time. The Whewell of the *History*, let alone the *Philosophy*, was largely a child of the years when Darwin was away (1831–6); and Darwin himself at Cambridge had no specific ambitions, let alone achievements, as a scientific theorist, that Whewell could have helped him pursue. As to general educational culture, Whewell may well have been an inspiring and instructive presence at soirées. However, to go beyond this conclusion, to make Darwin primarily a product of informal Cambridge culture as exemplified by Whewell, is to risk, with Cannon, continuing the tendency that traces to Darwin's own century (and especially to the biographical efforts of his son Francis, a don at Cambridge), to claim Darwin for Cambridge and England rather than Edinburgh and Scotland. That this tendency now needs critical reappraisal has become apparent with enhanced understanding of the influence on the early Darwin of two Scottish and Edinburgh legacies: his Lyellian geology and his Grantian zoology.[61] To

[59] Hodge (1985); Hodge and Kohn (1985).
[60] W. Cannon (1976), 377–84; Curtis (1987); Barrett (1987).
[61] Hodge (1985); Sloan (1985); Hodge (1990).

persist in making Darwin descend from Whewell may be to persist, in effect, in English and Cambridge efforts to make Scotland and Edinburgh peripheral to the nation's understanding of its own heroes and history.

10. Whewell and Darwin's Theory

Moving on from the question of Darwin's debts to Whewell, to the older man's response to the younger, one notices right away that although the Whewell who writes about Darwin's *Origin* in the 1860s is that much older than the author of the *History* and the *Philosophy*, the view taken of palaetiological science remains the same. One might expect it to be so from the discussion of geology and of the nebular hypothesis in Whewell's writings on the plurality of worlds in the 1850s. For, despite geology being deployed in a new argument there, against the plurality of worlds inhabited by man, the account of geology itself contains nothing not in the *History* and *Philosophy*.[62]

Eventually, in a new preface to a seventh edition of his Bridgewater Treatise, Whewell explained, in 1864, why he found Darwin's *Origin*—he left book and author unnamed even when quoting them—unpersuasive.[63]

Whewell takes Darwin to be engaging the old choice between chance and design, as that choice was first formulated by the Greek atomists and those, like Cicero, who opposed their claim that the world has been produced by a 'fortuitous concourse of atoms'. There is, says Whewell, a 'new approach' to the fortuitous concourse doctrine 'in those who maintain, in its extreme form, the Nebular Hypothesis'. And he directs his readers to his old arguments against anyone who puts that hypothesis in opposition to the belief in an Intelligent Creator. The same arguments, he proposes, can be brought to bear on a current version of a proposal concerning animal structure. 'In opposition to the view of the Design shown in the structure of animals, as manifested by the adaptation of their parts each to its purpose', that structure has been credited to the 'operation of external circumstances and internal appetencies'. This is Whewell harking back to Lamarck as encountered in Lyell; but he goes on to say that 'very recently' the argument has been stressed that 'all those structures of animals which are less adapted to their mode of life, cease to be continued in the species, precisely because of this want of adaptation', so that 'no structures remain for us to consider' except those where 'the organisation answers the purpose of life perfectly'.[64]

'This doctrine', says Whewell, 'again taking the extreme form of it', makes the universe the 'result of an infinite number of trials, the failures being constantly rejected by the fact of their being failures'. So, in its extreme form, it says 'the world' is the 'work of chance'. But the same arguments that refute the Nebular hypothesis refute also 'this hypothesis in this form'. The arguments

---

[62] Whewell (1854a).   [63] Id. (1864).   [64] Ibid. pp. xv–xvi.

are that, if any animal structure has been produced by 'necessary causation' from some earlier state of the animal world, then 'there must have been in that previous state the germs of the adaptations' later produced; and 'these germs and possibilities of adaptation must, in reality, contain evidence of design'. The argument holds, moreover, no matter how far we go back.[65]

Whewell does not think, however, that this argument is needed. For he thinks one faces 'a mere work of fancy' in attempts recently made 'to show that the organs of animals, as they now exist, with at least a seeming adaptation to their purpose, may be derived by necessary causation from a state of things in which there was no such adaptation'. Darwin, he argues, in confronting cases such as the structure of the mammalian eye, has to resort to two assumptions that deprive his reasoning of all force. The first is that the mere possibility of imagining the required transitional steps is a reason for thinking them to have taken place; and the second is the assumption that to confirm that they have done so, one needs only to suppose that there has been an unlimited number of generations over an indefinite time. Beyond his resorting to these two assumptions, Whewell urges, Darwin has not given us any new insight to replace the teleology offered us by Cuvier and other authorities earlier in the century. There is then, in Darwin's speculations, nothing that calls for any revision of the natural theology of the first edition of his own Bridgewater Treatise.[66]

Whewell had already explained, in writing to an Aberdonian Professor of Theology in 1863, that recent discussions in 'geology and zoology' did not seem to him to have 'materially affected the force of the arguments' given in his *Indications of the Creator*. In 'tracing the history of the world backwards, so far as the palaetiological sciences enable us to do', it seemed that 'all the lines of connexion stop short of a beginning explicable by natural causes; and the absence of any conceivable natural beginning leaves room for, and requires, a supernatural origin'. Darwin's 'speculations' do not 'alter this result'. He still has 'an inexplicable gap at the beginning of his series'. Whewell's God of the gaps is, therefore, still defensible. Furthermore, most of Darwin's hypotheses 'are quite unproved by fact', says Whewell. 'We can no more adduce an example of a new species, generated in the way his hypotheses suppose, than Cuvier could.' Darwin still has to admit that the 'existing species of domestic animals are the same as they were at the time of man's earliest history'. Nor can unlimited geological time rescue Darwin. The 'advocates of uniformitarian doctrines in geology' persist in helping themselves to 'additional myriads of ages', but the 'best and most temperate geologists' still accept 'great catastrophes'; so the 'state of the controversy on that subject' is not 'really affected permanently'. What Whewell has written in the 1830s is, still, then 'a just representation of the question between the two doctrines'.[67]

A few months later, he wrote to his friend James Forbes about the antiquity

---

[65] Whewell (1864) pp. xvi–xvii.
[66] Ibid. pp. xvii–xviii.
[67] Todhunter (1876), ii. 433–4.

of man. The 'course of speculation' on this subject has 'somewhat troubled me', he admits.[68]

I cannot see without some regrets the clear definite line, which used to mark the commencement of the human period of the earth's history, made obscure and doubtful. There was something in the aspect of the subject, as Cuvier left it, which was very satisfactory to those who wished to reconcile the providential with the scientific history of the world, and this aspect is now no longer so universally acknowledged. It is true that a reconciliation of the scientific with the religious views is still possible, but it is not so clear and striking as it was. But it is still a weakness to regret this; and no doubt another generation will find some way of looking at the matter which will satisfy religious men. I should be glad to see my way to this view, and am hoping to do so soon.[69]

It was not to be. Whewell never, it seems, fulfilled this hope, at least not in print. Elderly men's retrospects can be misleading, but Whewell's recurrent invocation of Cuvier in these years is not a false note. Cuvier always represented many things for Whewell. Most obviously, he was a leading authority on natural history and on palaetiological science in the years when Whewell was forming his own views on these subjects. But Cuvier was also the enduring, magisterial, autocratic, Protestant careerist who had survived to maintain, throughout successive French political upheavals, a stance that the Tory, Anglican Whewell could find reassuring even if he did not identify himself quite unequivocally with the Frenchman. Whewell had been largely content in the 1820s with Cuvierian science, including its successful resistance to the challenges from Lamarck and Geoffroy St Hilaire. It would have been easier for Whewell if the later challenges, from Lyell, Darwin, and others, had been taken care of in advance. Whewell often lived his life, as an analyst of palaetiological science, as if they could have been; as it was, it often fell to him to find a level of generalization about this genre of science from whence those challenges could be represented as leaving the original stance, when abstracted from dispensable details, essentially inviolate.

11. Conclusion

In his reflections and writings on palaetiological science, Whewell lived and died by history. His views constitute a highly instructive response to the state of the art at one moment, the early 1830s. After that, Whewell was defending conclusions drawn from that momentary encounter. There may be a certain hubris in his apparent presumption that he could make a foray into this area, and leave it again having derived lessons that he would be unlikely to have to

[68] Ibid. 435.

[69] Ibid. 435–6. See also three letters to Lyell (pp. 429–32), concerning languages and the antiquity of man, that also show Whewell resting his case on the conclusions he had drawn nearly three decades before.

revise. His influence in geology, among the palaetiological sciences, seems to be very limited. To be sure, he contributed through the BAAS and the Geological Society to the high standing the science then enjoyed. But that standing was an achievement already secured by others before he came on the scene in the 1830s. On the inorganic side, he did join Sedgwick and others in promoting the pursuit of a distinctive Cambridge commitment to physical geology, as exemplified by the work that William Hopkins began to publish in the middle of that decade.[70] On the organic side, especially in the 1840s, his support probably advanced the cause of morphology as developed by his friend from Lancaster, Richard Owen, but only after he had convinced himself that this morphology did not displace teleology. Beyond such particular influences, there remains the possibility that he may have provoked Chambers into pursuing the hope of a comprehensive palaetiological scheme that relied only on natural causes.

In his encounter with palaetiological science, Whewell shows himself to be the judicious critic and cautious reformer that he is usually made out to be. Never subversive, never reactionary, he is not a radical or a visionary either, not a Bentham or a Coleridge. The reasons for his particular blend of judiciousness and caution do not have their sources in anything peculiar to geology or palaetiological science generally. That blend is characteristic of the kind of national institutional success story that is Whewell's career, in his triumphant early maturity in the 1830s, a career so exemplary of his generation and so exemplary of such motivations as the English nationalist Tory containment of the disconcerting consequences of the French and Industrial Revolutions. Whewell on palaetiological science is, therefore, a case where context has not been conditioning text topic by topic or argument by argument, but rather in the overall outlook and attitude. Nothing that is taken seriously by prominent respectable savants is to be lightly dismissed, and nothing that is cherished by moderate, intellectual divines is to be directly challenged. The authority of the formal and informal institutions committed to the congruence of these is to be cultivated, as a condition for the well-being of the nation and its contribution to civilization.

[70] For Sedgwick, Whewell, and Hopkins, see the full account in Smith (1985).

BIBLIOGRAPHY CHAPTER VI

Barrett, P.H., Gautrey, P.J., Herbert, S., Kohn, D. and Smith, S., eds. 1987. *Charles Darwin's Notebooks, 1836–1844: Geology, Transmutation of Species, Metaphysical Enquiries*. London: British Museum (Natural History) and Cambridge University Press.

Brooke, John H. 1977. 'Natural Theology and the Plurality of Worlds: Observations on the Brewster-Whewell Debate.' *Annals of Science* 34: 221–86.

—. 1977a. 'Richard Owen, William Whewell and the *Vestiges.*' *British Journal for the History of Science* 10: 132–45.

Butts, Robert E. 1968. *William Whewell's Theory of Scientific Method.* Pittsburgh: University of Pittsburgh.

Bynum, William F., Browne, E. Janet, and Porter, Roy, eds. 1981. *Dictionary of the History of Science*. London: Macmillan Press.

Cannon, S.F. (formerly W.F.) 1978. *Science in Culture: The Early Victorian Period*. New York: Science History Publications.

Cannon, W.F. 1960. 'The Problem of Miracles in the 1830s.' *Victorian Studies* 4: 5–32.

—. 1960a. 'The Uniformitarian-Catastrophist Debate.' *Isis* 51: 38–55.

—. 1961. 'The Impact of Uniformitarianism: Two Letters from John Herschel to Charles Lyell 1836–1837.' *Proceedings of the American Philosophical Society* 105: 301–14.

—. 1961a. 'The Bases of Darwin's Achievement: A Revaluation.' *Victorian Studies* 5: 109–34.

—. 1976. 'The Whewell-Darwin Controversy.' *Journal of the Geological Society of London* 132: 377–84.

Cornell, J.F. 1986. 'Newton of the Grassblade? Darwin and the Problems of Organic Teleology.' *Isis* 77: 405–21.

Curtis, R.C. 1986. 'Are Methodologies Theories of Scientific Rationality?' *British Journal for the Philosophy of Science* 37: 135–61.

—. 1987. 'Darwin as an Epistemologist.' *Annals of Science* 44: 379–408.

de la Beche, Henry Thomas. 1834. *Researches in Theoretical Geology*. London: Charles Knight.

VI

Greene, Mott T. 1983. *Geology in the Nineteenth Century*. Ithaca: Cornell University Press.

Hodge, M.J. [Jonathan] S. 1977. 'The Structure and Strategy of Darwin's "Long Argument".' *British Journal for the History of Science* 10: 237–45.

—. 1985. 'Darwin as a Life-Long Generation Theorist.' In David Kohn, ed., *The Darwinian Heritage*, pp. 207–43. Princeton: Princeton University Press.

—. 1989. 'Darwin's Theory and Darwin's Argument.' In Michael Ruse, ed., *What The Philosophy of Biology Is: Essays Dedicated to David Hull*, pp. 163–82. Dordrecht: Kluwer.

—. 1990. 'Darwin Studies at Work: A Re-examination of Three Decisive Years (1835–37).' In T.H. Levere and W.R. Shea, eds., *Nature, Experiment and the Sciences: Essays on Galileo and the History of Science in Honor of Stillman Drake*, pp. 249–73. Dordrecht: Kluwer.

Hodge, M.J.S and Kohn D. 1985. 'The Immediate Origins of Natural Selection.' In David Kohn, ed., *The Darwinian Heritage*, pp. 185–206. Princeton: Princeton University Press.

Hooykaas, Reyer. 1959. *Natural Law and Divine Miracle: A Historical-Critical Study of the Principle of Uniformity in Geology, Biology and Theology*. Leiden: E.J. Brill.

Huxley, Thomas Henry. 1887. 'On the Reception of *The Origin of Species*.' In Francis Darwin, ed., *The Life and Letters of Charles Darwin*. 3 vols., pp. 179–204. London: John Murray.

Laudan, Rachel. 1982. 'The Role of Methodology in Lyell's Science.' *Studies in History and Philosophy of Science* 13: 215–49.

—. 1987. *From Mineralogy to Geology: the Foundations of a Science 1650–1830*. Chicago: University of Chicago Press.

Lyell, Katherine Murray, ed. 1881. *Life, Letters and Journals of Sir Charles Lyell*. 2 vols. London: John Murray.

Manier, Edward. 1978. *The Young Darwin and His Cultural Circle: A Study of Influences which Helped Shape the Language and Logic of the First Drafts of the Theory of Natural Selection*. Dordrecht: Reidel.

Morrell, J.B. and Thackray, A. 1981. *Gentlemen of Science: Early Years of the British Association for the Advancement of Science*. Oxford: Oxford University Press.

Raub, C.Q. 1988. 'Robert Chambers and William Whewell: A Nineteenth-Century Debate over the Origin of Language.' *Journal of the History of Ideas* 49: 287–300.

Recker, D. 1987. 'Causal Efficacy: The Structure of Darwin's Argument Strategy in the *Origin of Species*.' *Philosophy of Science* 54: 147–75.

Rudwick, M.J.S. 1971. 'Uniformity and Progression: Reflections on the Structure of Geological Theory in the Age of Lyell.' In Duane H.D. Roller, ed.,

Perspectives in the History of Science and Technology, pp. 209–77. Norman: University of Oklahoma.

Rupke, Nicolaas A. 1983. *The Great Chain of History: William Buckland and the English School of Geology (1814–1849).* Oxford: Clarendon.

Ruse, Michael. 1975a. 'Darwin's Debt to Philosophy: An Examination of the Influence of the Philosophical Ideas of John F.W. Herschel and William Whewell on the Development of Charles Darwin's Theory of Evolution.' *Studies in History and Philosophy of Science* 6: 159–81.

—. 1976a. 'Charles Lyell and the Philosophy of Science.' *British Journal for the History of Science* 9: 121–31.

Schweber, Silvan S. 1979. 'The Young Darwin.' *Journal for the History of Biology* 12: 175–92.

Sloan, Phillip R. 1985. 'Darwin's Invertebrate Program, 1826–1836: Preconditions of Transformism.' In David Kohn, ed., *The Darwinian Heritage*, pp. 71–120. Princeton: Princeton University Press.

Smith, C. 1985. 'Geologists and Mathematicians: The Rise of Physical Geology.' In P.M. Harman, ed., *Wranglers and Physicists: Studies on Cambridge Mathematical Physics in the Nineteenth Century*, pp. 49–83. Manchester: Manchester University Press.

Stair-Douglas, Janet. 1881. *The Life and Selections from the Correspondence of William Whewell D.D.* London: C. Kegan Paul & Co.

Thagard, P.R. 1977. 'Discussion: Darwin and Whewell.' *Studies in History and Philosophy of Science* 8: 353–6.

Todhunter, Isaac. 1876. *William Whewell, D.D. Master of Trinity College Cambridge: An Account of his Writings with Selections from his Literary and Scientific Correspondence.* 2 vols. London: Macmillan.

Whewell, William. 1831. 'Lyell's *Principles of Geology* volume 1.' *British Critic* 9: 180–206.

—. 1832. 'Lyell's *Principles of Geology*, volume 2.' *Quarterly Review* 93: 103–32.

—. 1833c. *Astronomy and General Physics Considered with Reference to Natural Theology.* London: William Pickering.

—. 1837a. *The History of the Inductive Sciences, from the Earliest to the Present Time.* 3 vols. London: J. W. Parker.

—. 1838a. *Address Delivered at the Anniversary Meeting of the Geological Society of London, on the 16 of February 1838; and the Announcement of the Wollaston Medal and the Donation Fund for the same year.* London: R. and J.E. Taylor.

—. 1839. 'Presidential Address.' *Proceedings of the Geological Society of London* 3: 61–93.

—. 1846. *Indications of the Creator*. 2nd edn. London and Philadelphia: J.W. Parker.

—. 1847. *Philosophy of the Inductive Sciences: Founded Upon Their History*. 2nd edn. 2 vols. London: John W. Parker.

—. 1854a. *Of the Plurality of Worlds: An Essay*. 3rd edn. London: John W. Parker and Son.

—. 1864. *Astronomy and General Physics Considered with Reference to Natural Theology*. 7th edn. London: Pickering.

Yeo, Richard R. 1984. 'Science and Intellectual Authority in mid-Nineteenth Century Britain: Robert Chambers and *Vestiges of the Natural History of Creation.*' *Victorian Studies* 28: 5–31.

VII

The Universal Gestation of Nature: Chambers' *Vestiges* and *Explanations*

INTERPRETIVE ISSUES

Sometimes a book proves hard to interpret correctly not because of what is in the book, but because of what is in another book. Darwin's *Origin of Species* has made it very difficult indeed to give a correct reading of Robert Chambers' *Vestiges of the Natural History of Creation* (1844) and *Explanations: A Sequel to the Vestiges* (1845). In fact, I know of no such reading, and a main aim of this paper is to attempt one, at least for the sections where Chambers deals with organic diversity.[1] But let me dwell a moment on the difficulty involved. For there are some important historiographical issues at stake.

The biology students I have been introducing to the history and philosophy of biology during the last two years come to the subject with several predictable preconceptions, mostly derived from scientific textbooks. A parody of one preconception might go like this: Darwin synthesized a molecule with two atoms—the ideas of evolution and natural selection. The second was new, the first was not. Darwin did a couple of key things, then. He produced more "evidence for evolution" than anyone before, and he provided a novel "mechanism" for it. Conversely, his "precursors" who "believed in evolution" lacked Darwin's weight of evidence and his mechanism. For the biology student there is an easy assumption, as natural as it is

1. I am grateful to the valuable comments made by the Editor and two referees on an earlier version of this paper. That version in turn was based on part of my thesis, "Origins and Species: A Study of the Historical Sources of Darwinism and of the Contexts of Some other Accounts of Organic Diversity from Plato and Aristotle on," unpub. diss., Harvard University, Cambridge, Mass. 1970.

false, that the familiar "forerunners of Darwin" were using their "insufficient evidence" to support what Darwin was to "prove" with his "sufficient evidence."

Consider next a typical graduate student in the history of science who has been doing the usual reading on what he would probably describe as "the backround to Darwin" or "the history of evolutionary biology." His lines, I find, can be not unfairly summarized as follows: Many people prior to Darwin believed in evolution. But, like the belief in God or progress or democracy, this belief took widely varying forms. The "idea of evolution" has assumed many guises. Darwin had one idea of evolution, Lamarck another, Chambers a third, and so on. In the customary phrases, these authors "looked at evolution in fundamentally different ways"; they proposed "thoroughly different accounts of evolution." For, it is emphasized, not only have several easily distinguishable "mechanisms for evolution" been proposed—Lamarck's Lamarckian inheritance, Darwin's survival of the fittest, and so on—"evolution itself" has been understood in no less disparate ways. Chambers saw it as a planned linear developmental escalation, Darwin as a tree branching without goal or guidance. On what grounds, then, are these all called theories of evolution? Well, all recognize— or better, perhaps, postulate—that species are mutable. An evolutionist is simply someone who explains the origin of species in this characteristic way.

Now the graduate student can hardly be accused of fallacy because, it would seem, there are no inferences here, only classification and definition. And the definition—of an evolutionist—squaring well with present usage, as it does, must be correct. But there is an error; the error lies in what the whole line of thought leads us to overlook. "The evolutionists," it is insisted, have had different answers, but the label is useful, for, the presumption is, all had a common problem. However various their responses may have been, their goals, and so their intentions, were in a single crucial respect the same. Challenged on this, our candidate can quote from the fullest discussion of the *Vestiges* to date. "His aim," wrote A. O. Lovejoy of Chambers, "was to prove, against the special-creationists, the doctrine of the gradual production of new species by natural causes operating through ordinary processes of reproduction—to vindicate the doctrine of organic evolution, not to propound a universal theory of the nature or *modus operandi* of its causes." What then was "the question which to Chambers seemed primary"? He "held that the main

The Universal Gestation of Nature

matter was to prove that the theory of natural descent of many diverse species from common ancestors was a *fact,* whatever the specific process of the production and perpetuation of variations." [2]

EXPLANATORY INTENTIONS

But there is trouble here; as a description of Chambers' intentions this will not do at all. Quite apart from the highly infelicitous talk of proving a theory to be a fact, what is included is partly false and what is omitted no less misleading. We can easily guess how the error could have arisen. Lovejoy, with the usual textbook analysis of the Darwinian molecule in mind, has simply assumed that if you subtract from the *Origin* Darwin's intentions as a natural selectionist, you are left with his intentions as an evolutionist. Now *qua* evolutionist, *qua* propounder of "the doctrine of organic evolution," Chambers must have had the same aims.

This conclusion too is as mistaken as it is natural. For if, as we shall do shortly, we forget "evolution" entirely—either as "doctrine" or "fact"—and examine the *Vestiges* afresh, what do we find? We find that, quite inconsistently with Lovejoy's statement, Chambers was (a) necessarily just as centrally concerned with the origins of life—in spontaneous generations —as with those of species; (b) was not arguing for—in fact explicitly rejected—a *common* ancestry for diverse species; and (c) in one respect, at least, was not proposing a gradual production of new species from old.

But why make such an uncharitable fuss about the slips in a brief description of Chambers' intentions? Because, of course, nothing is more important than intentions. A person's scientific theorizing, his explaining, is something he does, an act he performs. And common sense, ordinary language, and the many philosophers who have lately looked long and hard into these matters are all agreed: intentions do make people's actions intelligible, their behavior explicable. Analysis of an author's intentions is, then, something we need to undertake with the greatest care. As one ideal, of course, we should strive to reconstruct an account of his intentions that he himself would have accepted as accurate and insightful at the time he was carrying them out. Here, needless to say, any uncritical use of anachron-

2. A. O. Lovejoy, "The Argument for Organic Evolution before the *Origin of Species,* 1830–1858," in B. Glass, O. Temkin, and W. L. Straus (eds.) *Forerunners of Darwin: 1745–1859* (Baltimore, 1959), pp. 356–414.

istic terms or assumptions will frustrate the whole purpose of the historical exercise.[3]

This suggests a point of permanent policy. In our learning —and even more in our teaching—about books like *Vestiges*, we need a new place to start. Rather than seeking a definitive decision on the "ideas" present, we should begin by attempting to understand the ends pursued in the book's argumentation, the explanatory goals the author intended to achieve with and by his assertions and inferences. This has the advantage of focusing our attention from the start on the arguments and on the particular use they have for the data brought into them. Chambers, for instance, draws, in *Explanations*, on the Galapagos data given in Darwin's *Voyage of the Beagle*. No one, as far as I know, has remarked on this but, as we shall see, Chambers' arguments had a very distinctive use for these data. It was quite different from the use to which Darwin and Wallace, wholly independently of one another, put them in their argumentation.

Now, if we concentrate in this way on what an author is trying to explain and on how his arguments are designed to explain it, the old analysis in terms of ideas becomes totally redundant. There is simply nothing to be gained from deciding whether or not the "idea of evolution," however defined, is truly in the *Vestiges* or was clearly in Chambers' mind when he wrote it. The graduate student, in my experience, finds this last result very hard to assimilate. He craves an answer to his query whether Lamarck, or Buffon, or Chambers "really had the idea," was "actually talking about evolution" and so on. That this whole family of questions is not merely unhelpful and potentially misleading but completely *unnecessary* is a disconcerting suggestion to anyone raised on the standard reading requirements. The best way to convince him that the suggestion is worth accepting is, of course, to show by doing it, that, first, we can talk more accurately and comprehendingly about the *Vestiges*, say, without discussing "evolution" at all;

3. Intentions, actions, and associated topics have received much expert attention of late. Two recent representative publications with useful bibliographies are A. R. White, *The Philosophy of Mind* (New York, 1967) and A. R. White, ed., *The Philosophy of Action* (London, 1968). For an insightful employment of current analytic conclusions respecting the explanation of actions, in a powerful critique of the "historiography of ideas," and an extensive plea for the reconstruction of intentions as a prime duty of the intellectual historian, see Q. Skinner, "Meaning and Understanding in the History of Ideas," *History and Theory*, 8 (1969), 3–53, an article all the more valuable for the historian of science because Skinner is mainly discussing examples from the history of political theory.

The Universal Gestation of Nature

second, that the deep logical differences between the *Vestiges* and *Origin* raise for us, as historians, far more instructive queries about the two books than do their superficial resemblances; and, finally, that in this task the *argument* proves a much more discriminating unit of analysis than the "idea." Later in this paper I try to follow my own advice. If I have failed, that does not, of course, mean that the advice must have been faulty. It may still be worth pondering in its own right.

II

AN ANONYMOUS AMATEUR'S THEODICY

So much for general historiographical issues. There remain some special difficulties with the *Vestiges,* due to Chambers' anonymity and amateur standing.[4] From the outset he observed the utmost secrecy concerning the book. As a result we are not likely ever to know in detail either what prompted him, apparently around 1836–37, to begin the project, or what reading, observations, and experiments may have influenced him subsequently. He decided on anonymity mainly to safeguard his publishing ventures and to get a fairer hearing for his views. In Chambers were combined, of course, the concern with literacy, hard work, and commercial competence for which Scotsmen were famous in the golden age of self-help. A cheerful, confident soul, he felt he could tackle almost anything to do with books. Mechanics Institutes were everywhere teaching the sciences to people like himself. But Chambers, if some later remarks are any guide, was always dissatisfied with the narrow-mindedness of established "men of science." He asks his reader, a fellow layman presumably, to recall the impression made on him by the transactions of learned societies and the pursuits of individual scientists. "Did he not always feel that while there were laudable industry and zeal, there was also an intellectual timidity rendering all the results philosophically barren!" Picking on Sir John Herschel's cautious vindication of science, as saving men from attempts at

4. For Chambers' career, see M. Millhauser, *Just Before Darwin: Robert Chambers and 'Vestiges'* (Middletown, Conn., 1959); F. Egerton, "Refutation and Conjecture: Darwin's Response to Sedgwick's Attack on Chambers," *Studies in the History and Philosophy of Science,* 1 (1970), 176–183; the long introduction by Alexander Ireland to the twelfth edition of the *Vestiges* (Edinburgh, 1884) and W. Chambers, *Memoir of Robert Chambers with Autobiographic Reminiscences* (Edinburgh, 1872).

VII

the impossible and as showing the quickest, most economical
means to practical ends, Chambers—in a passage of rare and
revealing vehemence—retorts that some results may well keep
the odd "country gentleman from a hopeless ruining speculation" or aid the "powers and profits of an iron-foundry or
cotton-mill; but nothing more." The "awakened and craving
mind," he cries out, wants to know "what science can do for
us in explaining the great ends of the Author of nature, and
our relations to Him, to good and evil, to life and eternity."
But the "man of science" turns to his "collection of shells"
and his "electric machine." Inevitably, then, "the natural sense
of men," with no taste for these esoteric pursuits, "revolts" at
their "sterility." No less inevitably, we are told, a bold synthesis like the *Vestiges* can never get a just appraisal from
the specialist.[5] As a common man, Chambers had sought the
widest possible account of God's world. Having had to supply
it himself, he must take his case back to the ordinary people.
Bearing the good news that progressive development permeates
the heavens, earth, and all mankind, he insisted that not only
does science reveal God's goal to be happiness in creatures
ruled under invariable law; in discovering those laws, science
enables us to spread happiness and reduce misery among
masses and squires alike.[6]

We can not be certain, but some such attitude—a striving
for a comprehensive, populist theodicy—must surely have
inspired the first Chambers' work on the *Vestiges*. At any rate,
we do know that his successive achievements as an Edinburgh
man of letters brought, along with membership in learned
societies and access to pertinent publications, acquaintance
with scientists themselves. Dr. Neill Arnott, Sir Charles Bell,
Edward Forbes, and Dr. Samuel Brown could be tapped, without being let in on the secret, for advice on physics, anatomy,
physiology, geology, natural history, and chemistry. Further
out, nearer the edges of scientific respectability, Chambers fed

5. [R. Chambers], *Explanations: A Sequel to the "Vestiges of the Natural
History of Creation." By the Author of that Work,* 2nd ed. (London, 1846),
pp. 177–178.

6. For what I have called Chambers' theodicy, see especially his final
chapter in all editions of *Vestiges:* "Purpose and General Condition of the
Animated Creation." The editions referred to here are as follows. They
will be identified hereafter by date alone. Robert Chambers, *Vestiges of the
Natural History of Creation. With an Introduction by Gavin De Beer*
(Leicester and New York, 1969), a facsimile reprint of the first (London,
1844) edition; *Vestiges . . . ,* 3rd ed. (London, 1845); *Vestiges,* 10th ed.,
with Extensive Additions and Emendations, (London, 1853) and *Vestiges
. . . ,* 11th ed. (London, 1860).

The Universal Gestation of Nature

eagerly on recent psychology and the new Continental social statics, these proving to him respectively, that the minds of humans, like those of animals, fall fully under physiological rule because their brains do, and that all our actions, even suicides, conform, on average at least, no less strictly, to predictable regularities.

Chambers could draw, therefore, at first or second hand, on almost all sciences save mathematics, and within each on many of the theoretical traditions and controversial proposals then current. He would, for instance, have early been familiar with Lyell's geology and the contemporary objections to it. But the *Vestiges* adopts only one of the many possible permutations of positions in these various fields. Why, then, it may be asked, did Chambers select what he did? We can only guess. The documents permit nothing else. In a psychological-historical mood, we might speculate that, for a self-made man with a career going daily from strength to strength, there would be a special attraction in any theories of progress and law-abiding development. In a direct projection, Chambers could have been composing his cosmogony as an unconsciously autobiographical success story. I offer the possibility not to urge its probability, but more to show what we are reduced to when the vital evidence is lacking.

CRITICS AND COMMENTATORS

The *Vestiges'* immediate reception hardly enlightens us either. Most critics were upset by the book; those responding in print harped on empirical "blunders" and railed at "atheistical tendencies." All the furor did not spring solely from religious sensitivities; the hostility of several scientists, like young T. H. Huxley and Asa Gray, for instance, was largely a defensive reaction from a new breed of natural history professionals combatting an anonymous but suspiciously amateur attempt to take various issues over their heads to the recently expanded reading public. True, several eminent "men of science," notably Richard Owen and Baden Powell, wrote favorably of the work in correspondence; and another, W. B. Carpenter, still uncertain of the author's identity but probably suspecting Chambers, supplied some unsigned notes—while disclaiming any endorsement—for its later editions, editions which were, as even the harshest reviewers acknowledged, impressive improvements on the original version. Granted, too, that more than one profoundly religious mind welcomed the main drift of Chambers' case for the universal reign of law throughout the inorganic, organic, and "moral" worlds; still, none of these commentators, hostile or friendly, public or private, undertook a detached or detailed dissection of Chambers' theses.[7]

Darwin's and Wallace's reactions are of special interest, obviously. But they offer no analytical guide to Chambers' arguments. The youthful Wallace lauded the *Vestiges* in his well-known letter to an apparently unimpressed Bates.[8] Darwin, of course, was already a Darwinian, and in writing to friends characteristically praised the author for bringing so many things together, while calling his conclusions hopelessly wide of the mark.[9] Later, both Darwin and Wallace routinely recognized Chambers along with other "forerunners," and Wallace cited the *Vestiges* as a major personal influence. Both, however, were keen to stress that, in their eyes, he had made no start on the problem of how new and distinct species arise, each closely adapted to a peculiar place in the natural economy. Equally unsurprisingly, and for our purposes here equally unhelpfully, Chambers insisted that the *Origin* contained very little of major importance that was not already in his book. He had not overlooked adaptation, he countered. Furthermore, the *Origin*, "in no essential respect," contradicted the *Vestiges*. Quite the reverse; "while adding to its explanations of nature, it expresses substantially the same general ideas." [10] Many people familiar with the *Vestiges* were making this mistake about the *Origin*. It would have been very remarkable had Chambers not been among them.

We are thrown back, then, on the book itself and on one precious, strangely neglected commentary. The commentary is Chambers' preface to his tenth edition (1853), a preface unique to this edition and written well before the *Origin* had

7. See, in general, Millhauser, *Just Before Darwin;* Owen's letter to "Author of 'Vestiges'" in R. Owen, *The Life of Richard Owen by His Grandson*, 2 vols. (London, 1894), I, 249–252; Gray's remarks quoted in G. H. Daniels, *American Science in the Age of Jackson* (London and New York, 1968), pp. 57–58; and the introduction to the *Vestiges* (Edinburgh, 1884), where Powell's letter to the anonymous author was published.

8. See H. L. McKinney, "Wallace's Earliest Observations on Evolution: 28 December 1845," *Isis*, 60 (1969), 370–373. The fullest account of Wallace's relation to Chambers is McKinney's Ph.D. thesis, "Alfred Russel Wallace and the Discovery of Natural Selection (Unpub. diss., Cornell University, 1967). See also his article with the same title, *J. Hist. Med.*, 21 (1966), 333–357.

9. See the Darwin letters quoted by Egerton in the review essay cited in note 4.

10. *Vestiges* (1860), pp. lxiii–lxiv, in the appended "Proofs, Illustrations, Authorities, etc."

The Universal Gestation of Nature

led him to say misleading things about his own book. He recounts in this his route to the *Vestiges'* treatment of organic diversity. Now prefaces, especially autobiographical ones, are notoriously suspect sources. When, therefore, lacking other material, we have to take such a document seriously, we need to proceed very warily.

INTENTIONS AND MOTIVES

Completely general criteria for distinguishing sharply between intentions and motives may prove impossible to find. Nevertheless, the distinction remains an indispensable one. If someone is firing a gun, his intention may be murder, his motive money. Likewise, a young man in a hurry might deliberately investigate DNA to win wealth, fame, and women. These would be his motives. But he also has intentions; he intends to discover and explain various things. He would have one intention, at least, in seeking the structure of DNA. This would be to solve certain problems, answer certain questions about inheritance. His intentions and his motives are, of course, related in instructive ways. It tells us much about the social context of empirical inquiry in our own day, as contrasted, say, with the high Middle Ages, that an able youth with these motives would find them far from frustrated by a career spent studying heredity. But such intentions and such motives need not always go together. The apprentice geneticist might well work with a rich, retiring, and celibate associate whose explanatory intentions were the same but whose motives were entirely different.

Scientists by and large find it easier to be candid and explicit about their intentions than about their motives. This can lead the historian into error. He is inclined to think that the only real and enlightening story lies hidden in the masked motives. But intentions, though not so hard to discern, are just as real and in their own way equally rich sources of historical understanding. Now certainly, in one sense, a theorist can leave us a quite false account of his intentions. His autobiography may claim that his theory was originally constructed and designed as an answer to some specific question. But we find on examining notebooks he wrote at the time that in fact he was then grappling with a related but by no means equivalent problem. Likewise, though a scientist may at any time recommend his theory as the one best fitted to solve a certain mystery, that mystery may have been far from his mind when he first arrived at the theory. Nevertheless, there remains a situation, and a sense, in which a theorist's account of his own intentions can hardly fail to be true. A man, wheeling concrete in a wheelbarrow he has made, might elucidate its design by showing how it meets the present purpose better than any alternative conceivable to him. This elucidation would be tantamount to claiming that, even had carrying concrete been nowhere in his thoughts when he first designed it, if he were now to make one solely for that purpose he would adopt the same design.

I would not want to push the parallels very far, but the parable's general point is valid. However sceptical we must be about a scientist's retrospective indications of his earlier intentions, we can still gain important insights when at any time he deliberately recommends his theory as still being the best able to do some rather specific explanatory job. For that job is what the theory is intended to do, now. This is what he intends by continuing to expound it.

III

EXPLANATORY NEEDS

As Chambers tells it, the job his theory had to do arose as follows.[11] God, he was already convinced, conducts all the

11. All quotations in the next four paragraphs are from *Vestiges* (1853), pp. v–viii. We can of course accept Chambers' claim to have hit on this analogy by himself, without denying that this kind of comparison goes way back. In ancient times, the world's origin was often likened to a growing organism. Christ's more philosophical contemporary, Philo, commenting on *Genesis* I, remarks that just as any grown man has started life as a humble seed, so it is with the world, when the Creator formed living creatures, "those first in order were inferior, if we may so speak, namely fishes, while those that came last in order were best, namely men," with middling creatures like birds coming in between. In Chambers' own generation, such analogies were especially familiar in the speculative cosmogonies of German *Naturphilosophen*. In a passage Chambers might well have seen, Lyell wrote of a recent anatomical discovery that had appeared to some persons "to afford a distant analogy, at least, to that progressive development by which some of the inferior animals may have been perfected into those of more complex organization." In mammals, the brain of the foetus "assumes, in succession, forms analogous to those which belong to fishes, birds, and reptiles before it acquires the additions and modifications peculiar to the mammiferous tribe." Thus in the movement from the embryo to the adult mammal, "there is a typical representation, as it were, of all those transformations which the primitive species are supposed to have undergone during a long series of generations" since the remote geological past. But Lyell, of course, rejected this conclusion. Chambers may have been aware of all this, but it seems doubtful; that he originally set out to vindicate what he knew was a venerable tradition is

The Universal Gestation of Nature

world's "passing affairs" strictly and without exception according to fixed laws. So, on happening to learn about Laplace's nebular hypothesis, he was greatly impressed. The "physical arrangements of the universe had been originated" no less lawfully; God's action displayed "a perfect unity." In this way, then, he says, was his attention first attracted "to the early history of animated nature." But as explanations for "the commencement of life and organisation," he was soon dissatisfied with the "'fiats,' special miracles,' 'interferences,'" and the like then "in vogue" with geologists. For law must prevail here, too. The problem was thus obvious: Given a lawfully formed earth, to get all its life onto it, lawfully. Two comments are called for. First, the fundamental contrast with the problem the Darwinians tackled is plain. Second, though Chambers later gave up the general nebular hypothesis—which had planetary systems continually forming from nebulae throughout the heavens—he never abandoned the special Laplacian speculation as to the solar system. Always, and as his main rhetorical strategy, he argued from a natural, lawful formation of the earth and the other planets around the sun to a like origin for their inhabitants.

Light did not break in immediately. Lamarck's proposal seemed circular and "wholly inadequate" to account for living species, while Geoffroy Saint-Hilaire's writings were then unknown to him. However, he did know at second hand "some of the transcendental views" found among French and English "physiologists." These he joined, significantly, with "some knowledge" of the geological succession of life, and so pursued that "Great Mystery," the "genesis of species."

We notice right away an essential consequence of Chambers' explanatory intentions. His task always included both the origin of life and the origin of species. "The first fact"—that is, the first event to be explained—was "the passage from the inorganic to the organic." This he found less baffling, illustrated as it was by "organic chemistry." What really held him up, Chambers recalls, was deciding how animals above the very simplest had originated, since their complexity, and with creatures like mammals their need for internal gestation, made a spontaneous generation impossible.

A distinction between the simple and the complex among organisms was thus necessarily built into Chambers' inquiry

even less likely. See Philo, *De Opificio Mundi*, Trans. F. H. Colson and G. H. Whitaker in Loeb Classical Library; C. Lyell, *Principles of Geology*, 5th ed. 4 vols. (London, 1837), II, 439.

from the start. At his very point of departure he was remote, indeed, from the Darwinians. For consider the typically Darwinian question: Given this group of indigenous finch species on these islands now, why have they all originated here and not in any of the other equally suitable sites around the world? While manifestly involving quite general principles applicable to the entire history of life on earth, such a question can be posed without any decision on the relative "complexity" of the species involved. Most valuably, the *Vestiges* shows us what momentous considerations flow directly from this apparently minor matter.

"Long cogitation," Chambers' preface continues, yielded the insight, owed to no one else, "that the ordinary phenomenon of reproduction was the key" to the origin of species too complex to be traceable to spontaneous generation. What was so enlightening in reproduction? The "gradual evolution of higher from lower, of complicated from simple, of special from general, all in unvarying order, and therefore all natural, although all of divine ordination." Any presentation of Chambers' treatment of organic diversity must be, in effect, a commentary on the telling phrases of this succinct sentence. For with that insight, everything fell into place; in an analogical scaling-up of the orderly development displayed by the individual embryo, Chambers had the foundation for a theory. "Might there not be in those *secula seculorum*" of geological history "a similar or analogous evolution of being, throwing off, as it were, the various species as it proceeded, until it rested (if it does rest) with humanity itself?"

His essential analogical insight had for Chambers two compelling advantages. It implied "a slow and gradual process" requiring "no new power" beyond the usual ability of organisms to reproduce. It also plainly harmonized with four familiar sets of facts: the "unity of organisation" among diverse types; "the affinities seen in lines of species"; the existence of "rudimentary organs" and "above all" the historical course of life disclosed by paleontology. Geography—and the significance of the omission will appear later—he was not taking into account at this time.

That the leading clue to the origin of species was, for Chambers, an analogy between the germinal and geological histories of life, would of course by itself have stopped us reading the *Vestiges* and the *Origin* into one another. For there is simply no place for this analogy anywhere in the Darwinian case for common descent. Mammal embryos may show traces of their fish ancestry, but Darwinian theory allows no sense, analogical

The Universal Gestation of Nature

or otherwise, in which those fish ancestors were "immature" mammals whose "adult" future was thus already determined. A good way to read the *Vestiges* is, then, to explore the manifold manifestations of this difference. On the one hand, we have a case made for the developmental analogy and, on the other, a case for the common descent of allied species. The argumentation appropriate to the one must often run directly counter to what the other requires. We can see this best by briefly following Chambers as he uses his central analogy to get from distant nebulae to the Galapagos Islands. Though the route changes slightly in later editions, the navigational principles remain the same; they can be perceived most clearly by concentrating on their original exposition.

NEBULAR HYPOTHESES

By embracing the nebular theory, Chambers allied himself —in one important respect—with William Whewell against Charles Lyell in geology. In a passage almost certainly known to Chambers, Whewell, in 1837, had discussed three ways men might take the explanatory achievements of astronomy as a model for geological theorizing. Lyell's and Playfair's followers adopt the first two, Whewell tells us, but he can accept neither and proposes instead a third.

First, geologists attempting "to refer the whole train of facts to known causes" cite as a precedent Newtonian astronomy. This successfully "seeks no hidden virtues" but explains "all by the force of gravitation, which we witness operating at every moment." Second, geologists invoke astronomy to confirm "the assumption of perpetual uniformity." For, they claim, "the analysis of the heavenly motions" supplies "no trace of a beginning, no promise of an end." But, says Whewell, rejecting both analogies, look where astronomy has "analogy with geology." Consider it "as a palaetiological science; as the study of a past condition, from which the present is derived by causes acting in time." Surely we then have "evidence of a beginning" and "of a progress." For consider what "the Nebular Hypothesis" implies. "A luminous matter is condensing, solid bodies are forming, are arranging themselves into systems of cyclical motion;" we have, in short, "the beginning of a world." Without positively endorsing the nebular hypothesis, Whewell concludes nevertheless that "if geologists wish to borrow maxims of philosophising from astronomy, such speculations as have led to that hypothesis must be their model." [12]

12. W. Whewell, *History of the Inductive Sciences*, 3rd ed., with addi-

And that is exactly what Chambers did (to Whewell's ample chagrin as it turned out; he was to be a vigorous opponent of the *Vestiges*). In the early editions, Chambers' first two chapters introduce us to the observations of the two Herschels, which reveal not only countless nebulae but "nebulous stars" in "every stage of concentration" right down to that of "a common star with a slight *bur* around it." All these, Chambers observes, "are but stages in a progress, just as if, seeing a child, a boy, a youth, a middle-aged and an old man together," we may presume "that the whole were only variations of one being." For this repeated celestial progression, Chambers adduces as the cause the combined effect of gravitational and centrifugal forces in a rotating mass of matter. All is in accord, then, with the laws of gravity and inertia.[13]

BEGINNING AND PROGRESS

In every edition, Chambers applies these two laws to the earth's formation from diffuse rotating matter; the scene is thereby set for life's start on the new planet.[14] This very strategy determines his appeal to the fossil records. He must begin by taking issue—in direct conflict here with Darwin's unpublished *Essay* of the same year—with those "few geologists" who have tried to show that the absence of organic remains in the primary rocks "is no proof of the globe having been then unfruitful or uninhabited," on the grounds, as these geologists argue, that the heating of those rocks when solidified could have obliterated any such remains.[15]

---

tions. 2 vols. (New York, 1890) II, 594. The text of the first (1837) edition remained unaltered in subsequent editions. Whewell simply added new sections, which were indicated as such.

13. *Vestiges* (1844), pp. 7–8, 19–21; cf. *Vestiges* (1845), pp. 7–8, 19–21.

14. *Vestiges* (1844), pp. 23–26; (1845), pp. 23–26; (1853), pp. 12–17; (1860), pp. 12–15.

15. *Vestiges* (1844), pp. 53; cf. *Vestiges* (1845), pp. 51–53; (1853), pp. 28–31; and (1860), pp. 24–26. Cf. Darwin's statement: "If the Palaeozoic system is really contemporaneous with the first appearance of life, my theory must be abandoned, both inasmuch as it limits from *shortness of time* the total number of forms which can have existed on this world, and because the organisms, as fish, mollusca and starfish found in all its lower beds, cannot be considered as the parent forms of all the successive species in these classes. But no one has yet overturned the arguments of Hutton and Lyell, that the lowest formations known to us are only those which have escaped being metamorphosed"—Darwin's *Essay* of 1844 in G. De Beer, ed., *Charles Darwin and Alfred Russel Wallace: Evolution by Natural Selection* (Cambridge, 1958), p. 158.

The Universal Gestation of Nature

Chambers has to disagree, because for his thesis the rocks are useless as evidence unless they show vestiges of both a first beginning and a subsequent progressive development of life. Any heat sufficient to solidify rocks and eliminate fossil traces, he answers, would never have let organisms come into existence and survive. Further, animal and plant fossils show, he claims, "an advance in both cases, along the line—or, it may be lines—leading to the higher forms of organization." Among the animals "we see zoophytes, radiata, mollusca, articulata, existing for ages before there are any other forms"; then the various vertebrates appear in progressive succession.[16]

But Chambers was never content to confine himself to this planet. Countless others are inhabited or fit and destined to be so. And "one set of laws overspread them all with life."[17] His attempt to establish these universal laws provides the context for explaining the origin of species. Chambers had learned from the so-called transcendental anatomists—including perhaps his countryman Robert Knox—that a unity of type runs through all the diversity of living species. His next step, accordingly, is to identify a "fundamental form of organic being." Finding "a perfect resemblance" between the mammalian ovum and "the young of infusory animalcules," he concludes that the form, from which all else takes its "origin" is a *globule, having a new globule forming within itself.* Electricity seems able to produce such globules from inorganic matter.[18] Now, Chambers cited numerous recent observations in support of spontaneous generation, just as he did in support of species transmutations. But in both cases, he insisted, his argument did not require that these events still be going on around us. His aim was only to make it credible that in the planet's past the simplest and more complex species had originated naturally, by spontaneous generation and transmutation respectively. We see today, perhaps, only "vestiges" of God's lawful workings in bringing about these earlier events—events which may well have culminated with the crowning achievement, man.[19] Chambers had a definite and instructive purpose in entitling his tract as he did. Again, we note the contrast with the Darwinians' common descent which, as they both stressed,

16. *Vestiges* (1844), pp. 53, 148; cf. *Vestiges* (1845), pp. 150–151; (1853), pp. 117–118: (1860), pp. 106–107.
17. *Vestiges* (1844), p. 164; cf. *Vestiges* (1845), pp. 165–166; (1853), p. 122; (1860), p. 110.
18. *Vestiges* (1844), pp. 171–173; cf. *Vestiges* (1845), pp. 173–176; (1853), pp. 125–127; (1860), pp. 113–115.
19. *Vestiges* (1853), p. 130; (1860), p. 117. Cf. *Explanations* (1846), p. 29.

entails that many existing varieties and subspecies are right now on the point of becoming new and distinct species.

How then are we to get from this universal origin of a universal form to the distinctive forms of diverse adult organisms? Clearly, just as all celestial bodies are more or less perfected and developed nebulae, so all animals and plants are more or less developed globules. And so, if laws of matter regulating gravitational and centrifugal forces universally ensure advances toward astral perfection, then there must also be laws of development responsible for all advances in animate progress. That is the original argument. Lord Rosse's and other big telescopes, by resolving many Herschellian nebulae into star clusters, eventually led Chambers to modify it. But gravity, he continued to urge, is the universal means of inorganic change, and development is the equivalent for the organic realm. Gravity has its laws; so too then does development.

As a provisional law Chambers accepts that "each animal passes, in the course of its germinal history, through a series of changes resembling the *permanent* forms," which is to say, adult forms, "first of the various orders" below it in the "entire scale" and next "in its own order." But how can this law provide the clue to the production of new species, in "the gestation (so to speak) of a whole creation?" For, surely, each species invariably produces its like, and throughout observed history "the limits of species" have been rigidly adhered to. There would seem to be no grounds for supposing that any species can advance "by generation to a higher type of being." [20]

REPEATED ROUTES TO MATURITY

The obvious difficulty is confronted head-on. Chambers argues for "a higher generative law," to which "like-production is subordinate." The rule that offspring be specifically identical with their parents does not hold in certain conditions. So offspring can sometimes develop beyond the rank and species of their parents, to have as adults a different and higher form.[21]

---

20. *Vestiges* (1844), pp. 198, 210–211. This law is reasonably called the Meckel and Serres law of parallelism. In itself it makes no mention of or assumption about ancestries and so is quite wrongly called a law of recapitulation. For the early nineteenth-century formulations of this law by J. F. Meckel and A. E. R. A. Serres, see E. S. Russell, *Form and Function: A Contribution to the History of Animal Morphology* (London, 1916). Cf. *Vestiges* (1845), pp. 200 ff; (1853), pp. 147 ff; and (1860), pp. 130 ff.

21. *Vestiges* (1844), pp. 211, 231.

The Universal Gestation of Nature

Chambers can now exhibit the essential link between the germinal and geological histories. On close inspection, we see any developing mammal passing through "stages in which it is successively fish-like and reptile-like." But, he says, anticipating from critics of this law the objection that no mammal is ever a fish, "the resemblance is not to the adult fish or the adult reptile, but to the fish and reptile at a certain point in their foetal progress." A fish, reptile, bird, and mammal foetus all proceed initially through similar stages to the same point. Here, the fish embryo "diverges and passes along a line apart, and peculiar to itself, to its mature state." A branched diagram shows how the reptile and bird successively part company from the mammal which alone "goes forward in a straight line to the highest point of organisation." The diagram only charts the comparative embryology of the four main branches of the vertebrates, but additional minor ramifications could represent "the subordinate differences" of orders, tribes, families, genera and species.[22]

The pedagogical stage is finally ready for a new species. To advance from one class to another, an embryo, instead of diverging from the straight line at the usual point, merely continues to the next so that the progeny will be not a fish, say, but a reptile. But, of course, for the slighter shift from one species to another within the genus, a protraction of "the straightforward part of the gestation over a small space" is quite sufficient. And this could result from the "force of certain external conditions operating upon the parturient system."[23]

The great value of comparative embryology for Chambers' nomological alternative to the miraculous creation of species was always, then, that it exhibits a standard sequence of forms rehearsed by every individual making its way to maturity. And so, to support protracted gestation in unusual conditions, Chambers always sought cases where repeated exposure to the same conditions invariably yields the same advance from one species to another. This is why he continued to prize most highly a report that "whenever oats sown at the usual time are kept cropped down during the summer and autumn and allowed to remain over the winter, a thin crop of rye is the harvest at the close of the ensuing summer." Here, he suggests, the cropping, by prolonging "the generative process," invariably allows "the type" to advance so that "what was oats becomes rye."[24] Needless to say, no such datum could have a use in the Darwinians' arguments for common descent, which all

22. Ibid., p. 212.   23. Ibid., p. 213.   24. Ibid., p. 221.

assumed the origin of each species to be not only a geographically confined but also an unrepeatable event.

Later editions introduced a sharp distinction between the advances from one major type to another, and the adaptive proliferation of different species within that type. On the new account, the rather few major stages of developmental advance are achieved in one generation, an embryo born of a fish growing straight into reptilian form. But with this advance achieved, the environment can act much more gradually on successive individuals. Various species are slowly produced and adapted over many generations, partly in animals, as Lamarck had suggested, through the cumulative effects of changed habits. Some modifications may even be regressive, higher reptile species occasionally yielding lower ones. But how this works and how the successive individuals achieve their new adaptations, Chambers does not claim to know.[25]

LINES OF DEVELOPMENT AND REGULAR AFFINITIES

Chambers never departed, therefore, from his original proposal; there are "lines of development" in which any gaps must soon be filled by the continuous elevation of the lowest types, themselves supplied in turn by some "new germinal vesicle" formed "out of inorganic matter."[26] The initial introduction of any major type he was bound to explain by describing the conditions for its reappearance: the progress of progeny from parents immediately inferior in the scale of organization.

When Chambers brought taxonomy and geography into his argument, his inferences had, then, to be at every turn the exact reverse of those on which Darwin and Wallace converged. For the fundamental geographical question raised by Chambers' principles is one that never arises for someone seeking to demonstrate how the present distribution of organic types has been determined by the *common* descent of allied species. Because, on Chambers' developmental assumptions, the most pressing problem posed by geographical facts is to decide how many *independent* lines of *advance* there have been, and how closely their corresponding stages match one another. Do the same families, or even genera, appear in separate lines, or is there an identity only in the orders or classes through which they progress?

25. *Vestiges* (1853), pp. 139–184, and (1860), pp. 123–163.
26. *Vestiges* (1844), p. 222.

The Universal Gestation of Nature

To understand such a question, one must grasp the logical forces put into play by this kind of talk—talk of "advance," "stages," "development," "growth," and "maturity." For there are definite presuppositions implied by any claim that something is a less developed or immature version or an earlier stage of something else. We only identify a thing as a puppy when we are satisfied that in any conditions sufficient for its survival it will become a dog. That assumption is clearly implicit in our identification of something as a puppy rather than a kitten or a colt or, for that matter, a stone or a haystack. Likewise, to assert that an embryo is a mammal embryo is necessarily to assert that if it lives long enough it will become a mammalian adult. For we mean by "embryo" something that will grow if it lives, and will *grow up* if it grows.

Chambers has to establish that any spontaneously generated infusorial globule is, in effect, an embryonic, an immature, mammal. For this globular form is the initial stage of all geological as well as germinal growth. Under all conditions in which a line of geological development advances beyond that initial stage, it will go on eventually to reach mammal form. A line of developmental advance—what Chambers calls a *stirp*—is simply the maturation of an infusorial globule. He therefore rejects what he takes to be Lamarck's version of geological history with its proposal of "form going onto form merely as the needs and wishes in the animals themselves dictated." We have in germinal growth a regularly repeated succession of forms. And so, according to Chambers' analogy, there must have been in geological development nothing "irregular" or "arbitrary." [27]

The analogy thus demands regularity throughout the natural arrangement of animals and plants by affinities. And, indeed, Chambers declares, this arrangement is "as symmetrical as the plan of a house, or the laying out of an old fashioned garden." [28] For its account of this symmetry, the *Vestiges* remained loyal to William Macleay's taxonomic principles as elaborated by William Swainson. They had been applied most fully to the birds. Within the class Aves, five orders were identified, and in each of these orders five families and so on. Each order has affinities with two others; the five thus make a circle when arranged by affinities. The same is true within all further subdivisions of the class. Furthermore, one order of the five exhibits most perfectly the class character and is therefore "the best representative of that class." And, what is "most remark-

27. Ibid., pp. 230–232.   28. Ibid.

able," the five families of the most perfect order have representative characters matching and analogous to each of the five orders of the whole class. Such at least are the principles; while admitting that naturalists have yet to identify these circular quinarian representatives throughout the animal and plant kingdoms, Chambers insists that we can see portions of a regular plan. Every species has its place in this "system of both affinities and analogies." Lawfulness in the progression of life is proven, then, for Chambers, by the consistent production of these regular affinities within and analogies between distinct lines of development.[29]

The lawfulness Darwin and Wallace found in common descent demanded, we recall, quite *irregularly* branching lines of affinity. They would never have reached their position had they not first rejected what Chambers' analogy required him to defend. This point of contrast is part of a larger one, of course, as Chambers' next move very vividly demonstrates. For having defended a single, symmetrical pattern in natural classification, he asks next whether it agrees with the geographical and historical distribution of species.

THE UNDERDEVELOPED COUNTRIES
AND THE SEA BETWEEN

Chambers introduces the standard geographical generalization. The earth has several isolated regions which, although their climatic conditions are the same, often differ in their species and even families of animals and plants. But a developmental theory always correlates differences with inequalities of advancement. Chambers accordingly ranks the chief isolated regions in an order—Africa, Asia, America, and Australia—indicated by the luxuriance of their native vegetation and the swiftness and agility of their indigenous animals.

Four general conclusions of historical geography thus appear permissible. First, life started independently at several places, "distinct foci of organic production." Second, in the sub-

29. *Vestiges* (1844), pp. 240–241; cf. *Vestiges* (1860), pp. 164–229. It would be a mistake to think that Macleay's and similar regular systems of taxonomy were only an aberration of cranky amateurs during this period. Many of the most respected specialists, including the young Huxley, sought for this kind of regularity in the affinities and analogies between groups and subgroups of species. I owe confirmation of this conclusion to conversations with Miss M. P. Winsor, whose unpublished doctoral dissertation discusses early nineteenth-century taxonomy in detail: "Issues in the Classification of Radiates, 1830–1860," unpub. diss., Yale University, (New Haven, Conn.; 1971).

The Universal Gestation of Nature

sequent advances in isolated regions, the same classes and orders have originated wherever conditions permitted. Third, there are, however, native species, genera, and families peculiar to certain areas. With "no physical or geographical reason" for this "diversity" it is presumably owing to "minute and inappreciable" causes diverting the direction of organic development in these, the smaller, divisions. Fourth, due probably to "the comparative antiquity of various regions," development has not gone equally far in all of them, "being most advanced in the eastern continent, next in the Western and least in Australia." [30]

For Chambers the continental representation of an order by families found nowhere else is, therefore, both a disproof of any common ancestry for the order and a striking confirmation of regular development. What better proof of regularity in quite independent advances than finding some families of an order "confined to one continent and some to another, without a conceivable possibility of one having been connected with the other in the way of ancestry?" Consider, says Chambers, the two great families of quadrumana; the Cebidae are "exclusively American," while the Simiadae belong "entirely to the Old World." [31]

The distance from Darwinian presuppositions hardly needs remarking. The great virtue in a common ancestry for allied species was, they argued, not only its explaining why there is a natural classification but one with no symmetrical regularity; common descent explains also precisely what Chambers must refute—why there are real affinities between all the various species of Cebidae and Simidae; and why their common characters cannot be explained as common adaptations to common conditions but can be traced to a single quadrumanal ancestry for the two groups.

Consistently enough, Chambers explicitly denies that these similarities between the families of an order are, properly speaking, natural affinities at all. "In reality," he says in his *Explanations*, "they are only identical characters demanded by common conditions, or resulting from equality of grade in the scale." The only "true affinities are the affinities of genealogy," which is to say, of course, the affinities linking "an order in one class to the corresponding order in the class next higher," as, for example, the cetacean mammals and marine saurian reptiles.[32]

30. *Vestiges* (1844), p. 259; cf. *Vestiges* (1860), pp. 224 ff.
31. *Vestiges* (1844), pp. 259–260; cf. *Vestiges* (1860), p. 228.
32. *Explanations* (1846), p. 73.

VII

Given this restriction on affinities, Chambers can now exhibit any historical and geographical data as support for progressive development. If two allied groups differing in rank have appeared simultaneously in the fossil record, there is no difficulty. On the contrary, the seeming contemporaneity of the "carnivora and monkeys" confirms a "true development theory" recognizing "distinct lines of development, which might well advance to a certain stage (namely that of terrestrial mammal) about the same time." [33]

Like geology, geography offers "vestiges" of what we cannot observe directly. For "absolute proof" or developmental transitions between species may not be possible, since we can never be sure we have witnessed one. But present geographical situations, being differently advanced progressions, can often reveal "a fraction of the entire phenomena, in conformity with the hypothesis as to the whole." Look at certain "isolated parts of the earth which we know to have become dry land more recently than others." Such is the Galapagos archipelago. Quoting Darwin's new 1845 edition of the *Voyage of the Beagle*, Chambers notes first that the islands contain many indigenous species "though with an affinity to those of America," some being "even peculiar to particular islands." But, Chambers announces, "the most remarkable fact bearing on the present inquiry" is that, barring a rat and mouse brought in by foreign vessels, *"there are no mammals on the Galapagos."* Their absence splendidly vindicates his theory. Time "is necessary for the completion of the animal series in any scene of its development." On the Galapagos there has not yet passed "the full time required" for life to reach the mammal stage. Now, "had there been mammals and no reptiles," then on the contrary there would have been "one decided fact against the development theory." [34]

Here, of course, we can make a direct comparison. Wallace had been reading Darwin's *Voyage* too. In 1855 he was as yet completely in the dark as to what causes a new, distinct species to arise by modification of an earlier one. His working out of natural selection was still three years away. But he was ready to discuss "the law which has regulated the introduction of new species." Phenomena such as those in the Galapagos Islands, "which contain little groups of plants and animals peculiar to themselves, but most nearly allied to those of

33. Ibid., p. 94.
34. Ibid., pp. 161–163; cf. *Vestiges* (1853), p. 258 and *Vestiges* (1860), p. 229.

148

The Universal Gestation of Nature

South America, have not hitherto received any, even a conjectural explanation," he claims. The Galapagos, he continues —with his chronological point in perfect opposition to Chambers'—are "a volcanic group of high antiquity," probably never more closely connected with the mainland than they are now. "They must have been first peopled, like other newly-formed islands, by the action of winds and currents," and long enough ago to have had "the original species die out, and the modified prototypes only remain." Likewise we can explain "the separate islands having each their peculiar species," either by supposing "that the same original emigration peopled the whole of the islands with the same species from which differently modified prototypes were created," or, alternatively, that "the islands were successively peopled from each other, but that new species have been created in each on the plan of the pre-existing ones." In a word, Chambers' theory requires a relatively young Galapagos to explain an incomplete advance. Wallace's theory requires a relatively ancient Galapagos, ensuring a long isolation, explaining the wide gap—comprising generic, not merely specific differences—between mainland and island indigenes. We hear nothing in Wallace's paper about the absence of native mammals there, nor any talk of "higher" and "lower" forms among the Galapagos species. His reasons for introducing the Galapagos data make any such discussion beside the point. (Had he applied his principles to the Galapagos' lack of mammals, he would have argued, as Darwin had already done, that all mammals have descended from a common stock, none of whose descendant species has ever migrated successfully to the islands and served as a source for indigenous mammal species there.) [35]

35. A. R. Wallace, "On the Law which has Regulated the Introduction of New Species," *Annals and Magazine of Natural History*, 16 (1855), 184–196. Wallace's law reads: "Every species has come into existence coincident in both space and time with a pre-existing closely allied species." This law says nothing of "complexity" or "advance"; and in explaining geographical and geological facts Wallace always couples it with "the analogy of a branching tree, as the best mode of representing the natural arrangement of species and their successive creation," insistence on the "complicated branching of the lines of affinity," and rejection of "all those systems of classification which arrange species or groups in circles, as well as those which fix a definite number for the divisions of each group." It seems misleading, then, for McKinney to quote Chambers' law of developmental advance and proceed to say of Wallace's law that it is "in a similar vein." We know the *Vestiges* made a powerful impression on Wallace (whether he read the *Explanations* is unclear), but in analyzing the actual influence of the book we must understand how Wallace, thanks to other influences, reached a position differing in many and fundamental

Equally, Wallace's assumptions in reconstructing Galapagos history are completely incompatible with those Chambers must make. For how does the *Explanations* explain the frequent resemblance between island and nearby continental species? The entire terrestrial faunas and floras have developed quite independently in the two bits of land; but the two developments, though independent, took their start from a "common source" in the sea between. For animal and plant progress alike begins with marine species as "the progenitors of terrestrial species" and proceeds with life creeping, as it were, "out of the sea upon the land." Accordingly, Chambers, citing the Galapagos and other equivalent data, explicitly rejects the possibility of migratory colonization from the mainland, with or without any earlier land bridges, and concludes that a "community of forms" in the neighboring lands "merely indicates a distinct marine creation" in the intervening oceanic area, which "would naturally advance into the lands nearest to it as far as circumstances of soil and climate were found agreeable." And so ends his treatment of plant and animal geography.[36]

CONCLUSION

Nothing brings out better the logical distance between two theories than finding them accepting the same observational reports as data but explaining them in quite different ways. The discrepancy in the present case can obviously tell us as historians some important things—but only, I submit, if we handle it aright. Many students, I find, tend to respond in the manner of a Chambers rather than a Wallace. They accept that there is a difference, but seek always to interpret it as one between two stages of development in a single entity— the "idea of evolution." Darwin's and Wallace's thinking is simply seen as a more advanced point and Chambers' a more immature stage in the upward life this idea was then enjoying. The dangers in this analysis need, I trust, no further stressing here. It suggests an inevitable growth which required little more than time for its original completion and which today calls for no hard explanatory work from the historian. The alternative response requires that we begin by taking the gap

---

respects from Chambers'. For McKinney's remark, see his "Alfred Russel Wallace and the Discovery of Natural Selection," *J. Hist. Med.*, 21 (1966), 333–357.

36. *Explanations* (1846), pp. 166–168.

The Universal Gestation of Nature

between Chambers' developmental and the Darwinians' common descent argumentation seriously in its own right, and then inquire how that gap could and did arise. This is not the place to pursue that inquiry, but I hope I have shown something of what such an inquiry must involve.

VIII

The Structure and Strategy of Darwin's 'Long Argument'

Charles Darwin's Natural Selection, Being the Second Part of his Big Species Book Written from 1856 to 1858. Edited from manuscript by R. C. Stauffer. London and New York: Cambridge University Press, 1975. Pp. xii + 692. £20.00

As Robert Stauffer explains at the very beginning, *On the origin of species* was largely written as an abstract of what Darwin called his 'big book'. This treatise was well over half finished and running to more than a quarter of a million words when Wallace's famous letter precipitated the Linnaean Society presentation of the theory of natural selection. But although Darwin had the *Origin* ready in only nine months after that meeting he never went on with the parent work. He did not abandon the manuscript altogether for he later developed the first two of its ten and a half chapters into *The variation of animals and plants under domestication* (2 vols., London, 1868), and allowed portions of the tenth chapter, on instinct, to be included in a book by George Romanes, *Mental evolution in animals* (London, 1883), that appeared shortly after Darwin's death. What is edited and printed here for the first time are all of those ten and a half chapters except for the first two (apparently no longer extant) on variation under domestication.

Having had the author's own long abstract for over a century, we expect no great surprises or revelations in the present text. And so we find it. As Dr Stauffer says himself, what we now have that we formerly lacked is mainly more abundant and detailed examples supporting and illustrating the arguments of the *Origin*, together with the comprehensive citation of sources that is so impressive in *Variation*.

Anyone who has worked with the manuscript can confirm what is manifest from this exemplary edition of it: the task of transcribing Darwin's messy text and tracking down his myriad references must have been dreadful to contemplate in advance and exasperating to execute once undertaken. Wallace put us all in his debt by moving Darwin to produce a reasonably readable exposition of his theory; Dr Stauffer has put us no less in his by giving us this careful, judicious, and informative edition of Darwin's definitive account of the origin of species by means of natural selection.

Introducing Darwin's own words are two chapters by Dr. Stauffer that explain fully how the 'big book' was written, starting in 1856, and the editorial decisions taken in preparing this printing of it. In addition, each chapter is preceded by a discussion of its particular place in Darwin's schedule of work, especially as shown by his correspondence and diary entries from that time. Then at the end there are printed as appendices various 'fragments of the manuscript, letters and related materials' that supplement the main text. Finally, to follow the extensive bibliography, Dr Sydney Smith and three collaborators have prepared a valuable collation of the *Origin* and *Natural selection* and a truly excellent index to the big book. This collation and index will, needless to say, provide great help for the many detailed exegeses of particular passages that we can expect Darwinian scholars to present in the years to come.

However, *Natural selection*—this book seems destined to be known as that— helps us not only in understanding particular passages in the *Origin*, but also in seeing how that sometimes seemingly formless work is organized as a whole. For pretty certainly we now have in print almost all the surviving pages of every version Darwin ever wrote of the *Origin*, so that if we are to improve our grasp of the strategy and structure of its argumentation it must be largely by comparing the recently published 'outline and draft of 1839', the *Sketch* of 1842, the *Essay* of 1844, *Natural selection* (1856–1858) and the six editions of the *Origin* itself, starting with the first in 1859 and ending with the sixth of 1872.[1]

Stauffer has identified *Natural selection* as the 'second part' of Darwin's big treatise, and he explains that it is the second part of a three-part work. With a couple of qualifications this identification is correct as far as it goes; the two qualifications, to be developed further below, are that the first of the eight and a half chapters printed here is really from the first part of the work; while the half chapter at the end belongs to the third part.

It might be thought that nothing very much depends on how Darwin divided his case for the origin of species by natural selection, whether in his own mind or in his published writing. But in fact we can learn a great deal that is of interest by trying to clarify this point as thoroughly as we can. In this task, I want to suggest, there are two things which are of decisive importance: (i) the realization that there is a sense in which the *Sketch*, and every subsequent version of it, is a two-part work, and a sense in which it is a three-part work; and (ii) the realization that Darwin's commitment to the *vera causa*—or 'true cause'—principle shows us that the bi-partite division is more fundamental than the tri-partite one.

Now before turning to the *Sketch*, *Origin*, and lastly *Natural selection* to confirm these points, it will be as well to take care right away of a pair of objections that are likely to be made. For after all the suggestion that the doctrine of *verae causae* is relevant here may seem implausible: first, because there is a historical gap of nearly two hundred years between Darwin's *Origin* and the 'rule of philosophizing' in Newton's *Principia* which was the *locus classicus* for all later discussion of 'true' or 'known' causes; and second, because we usually expect a vast gap of credibility between scientists' aphoristic professions of methodological principle and their actual performances in expository practice, let alone their private thoughts as recorded in research notebooks and diaries.

But as it turns out, both of these gaps can be closed without difficulty or dispute in the present case. We may take the historical one first. Vincent Kavaloski, in a penetrating and comprehensive study of the history of the *vera causa* principle (hereafter VCP) from Newton to Darwin, has no trouble in documenting the explicit and far from inconsequential acceptance of various versions of the principle shared by numerous scientists and philosophers (Whewell was a notable dissenter) in the two centuries after 1687.[2] Indeed the principle, as he shows, was eventually invoked by people on both sides of the controversy over Darwin's theory—Huxley and Sedgwick, for example. Even more instructively for our business here, Kavalowski shows that the two people Darwin most hoped to impress with the arguments of the *Origin*, John Herschel and Charles Lyell, were at one in seeing the principle as a crucial contraint and indispensable standard for theories not only in physics but also in geology and biology. In offering the principle as a main clue to the argumentative ideals, and so also to the organization of the *Origin*, I am going beyond Kavalowski's own account of the VCP's history, but in ways that are only possible after his pioneering study.

Appropriately enough, one of the best introductions to the VCP is provided by Herschel's discussion of it when comparing Lyell's theory of geological climate change with his own. But since Herschel begins with a nod towards Newton, we too had better have the *Principia* before us. In the original edition, the four rules of philosophizing called in the second and third editions 'Regulae philosophandi' were actually included in a list of nine 'Hypotheses'.3 But in the case of the first rule, it was only this denomination and some of the supporting comments which changed, not the crucial opening sentence expressing the rule itself:

> Hyp. I Causas rerum naturalium non plures admitti debere, quam quae et vera sunt et earum Phenomenis explicandis sufficiunt.

> Hyp. I We ought to admit no more causes of natural things, than such as are both true and sufficient to explain their appearances.4

With Reid and other eighteenth-century commentators it became customary to read this rule as laying down, in Reid's words, 'two conditions' to be met by explanatory causes: first they must be 'true', that is, *known* to have a 'real' existence and not to be conjectured so as to have a merely 'hypothetical' existence; second they must be adequate, must suffice to 'produce the effect'.5 To avoid confusion, then, we will take a *vera causa*, a true cause, a real cause, a known cause, to be one meeting the first condition, and a VCP cause to be one meeting both conditions.

As a rough approximation, then, we may take the whole rule or principle to specify the following: in explaining any phenomenon, one should invoke only causes whose *existence* and *competence* to produce such an effect can be *known* independently of their putative *responsibility* for that phenomenon. As an elementary but not too misleading example, consider the explanatory challenge presented by a dead rabbit in the garden; the neighbour's cat or lightening bolts seen the night before would both be true and sufficient causes for this, for both are already known to exist and to be adequate for this sort of effect, so that their providing a possible explanation for this particular phenomenon would not be the sole evidence for their existence and competence to cause it. By contrast, to ascribe the rabbit's death to a burst of cosmic radiation conjectured to accompany every sunrise would be to dodge the requirement of independent evidence for the existence of explanatory causes; while to blame the neighbour's hamster would be to violate the requirement of independent evidence of competence.

Herschel's discussion brings out well the expository corollary of the separateness of these two evidential requirements of the VCP from one another, and, no less, the separateness of both from the requirement of independent evidence for causal responsibility. To anticipate, we can see in Herschel's and Lyell's upholding of the VCP the source for Darwin's taking up, in the *Sketch*, the following in turn: (i) the case for the *existence* of natural selection; (ii) the case for its *competence* to produce new species; and (iii) the case for its having been *responsible* for the production of extant and extinct species. And to anticipate further, we can see here the source for Darwin's presentation, there and thereafter, being fundamentally bipartite. For, with the existence case (i) and competence case (ii) both required to establish natural selection as a VCP cause for species production, these cases contribute a first VCP half of the whole exposition, which must come before the transition to the responsibility case (iii) that makes up the second half. Looking still further ahead, we can see one obvious potential source of confusion over Darwin's equally explicit tripartite division of his exposition. For Darwin himself sometimes distinguishes in that VCP half of his exposition two parts, but not an existence case (i) part and a competence case (ii) part. Rather he distinguishes a part I on variation under domestication from a part II on variation under nature, a division that does not coincide with that between case (i) and case (ii). Indeed, one of the keys to understanding the structure of the *Sketch* and so the *Origin* is to see how this part I/part II division is related to, or rather unrelated to, the existence case (i)/competence case (ii) division.

To illustrate what Newton and others understood by the term *vera causa*, Herschel explained how both Lyell's and his own very different causal candidates for the cooling of the earth since the geologist's secondary eras were true causes. Lyell's—the changing distribution of land and sea relative to the poles and equator—is a real cause because that distribution is known now to be affecting climate; and Herschel's candidate—the slow diminution of the eccentricity of the earth's orbit around the sun—is similarly real. But Herschel stresses that whether either of these two causes is adequate, whether it could ever cause a change of the full amount required, is 'another consideration'. Moreover it goes without saying that which of them is more likely to have been responsible for the climate change since the secondary eras—if indeed it was either of them at all—would be yet another consideration.[6]

But is there any evidence from the *Sketch* and the descendant texts that Darwin really did adapt his exposition to the demands of this principle? There is; and what is more there is a postscript to a letter (22 May 1863) of Darwin's to the botanist and erstwhile logician George Bentham that shows us how to find that evidence.

> P.S.—In fact the belief in Natural Selection must at present be grounded on general considerations. (1) On its being a *vera causa*, from the struggle for existence; and the certain geological fact that species do somehow change. (2) From the analogy of change under domestication by man's selection. (3) And chiefly from this view connecting under an intelligible point of view a host of facts. When we descend to details, we can prove that no one species has changed [i.e. we cannot prove of any particular species that it has changed]; nor can we prove that the supposed changes are beneficial, which is the groundwork of the theory. Nor can we explain why some species have changed and others have not.[7]

At first sight this sequence of three considerations may seem out of order. Surely the second should proceed the first, seeing that variation under domestication is always taken up before variation in nature? But a telling text in the *Sketch* itself shows that this query is mistaken. In a passage headed 'summing up this division', Darwin reviews the whole first half of his entire exposition in the *Sketch* as follows:

> If variation be admitted to occur occasionally in some wild animals, and how can we doubt it, when we see thousands of organisms, for whatever use taken by man, do vary. If we admit such variations tend to be hereditary . . . If we admit selection is steadily at work, and who will doubt it, when he considers amount of food on an average fixed and reproductive powers act in geometrical ratio. If we admit that external conditions vary, as all geology proclaims they have done and are now doing—then . . . there must occasionally be formed races . . . differing from the parent races . . . Take *Dahlia* and potato, who will pretend in 5000 years that great changes might not be effected: perfectly adapted to conditions and then brought again into varying conditions. Think what has been done in few last years, look at pigeons, and cattle . . . And therefore with the [adapting] selecting power of nature, infinitely wise compared to those of man, I conclude that it is impossible to say we know the limit of races, which would be true to their kind; if of different constitutions would probably be infertile one with another, and . . . adapted in the most singular and admirable manner . . .—such races would be species.

The existence and competence of this cause for species are then established. But what of its responsibility? Darwin continues by assuring us that that is just what he is going to take up next, as an independent issue:

> But is there any evidence that species have been thus produced, this is a question wholly independent of all previous points, and which on examination of the kingdom of nature we ought to answer one way or another.[8]

If we turn to the opening half of the *Sketch* that this summary recapitulates, we find a sequence of topics as follows: I(a) ability of new conditions to cause hereditary variants in domestic species (b) tendency of free crossing to swamp such variation (c) ability of artificial isolation and selection to counteract that swamping tendency; II(a) geology shows wild species often get exposed to changed conditions that would therefore cause hereditary variations (b) so these are variations that could be accumulated by any selective breeding (c) the struggle for existence entails that just such a selective breeding exists (d) this natural selection has longer to work and is more discriminating than artificial selection, so (e) if artificial selection has been able to produce races, this natural selection will have been able to produce races which are not merely races, but which meet the criteria for specific distinction. So extant and extinct species *could* have originated as races produced by natural selection.

So far we have passages I (a) to II (c) arguing for the existence of natural selection. These passages correspond, then, to consideration (1) in the letter to Bentham; for (especially *via* the struggle for existence and geological causes of hereditary variation) they establish natural selection as a *vera causa*, as a true, not a fictional, cause, analogous to artificial selection. Equally clearly, passages II(d) and II(e), arguing for the sufficiency, the competence, of this existing cause to produce species, correspond to consideration (2) in the letter to Bentham. Appropriately then, Darwin follows II(e) with passages taking care of two 'Difficulties on theory of selection', two *prima facie* objections to the case for its *competence*: (f) the difficulty of 'perfect organs' useless in their early stages, and (g) the difficulty of species differing in instincts and mental powers (there being no question from the argument and phrasing of (g) that Darwin really sees it as on the same footing as (f), although he inadvertently obscures this equivalence of the two difficulties by giving the topics in (g) a 'separate section'). And it is, of course, after (g) that we have the summing up prior to the transition to the second half of the book, the half devoted to consideration (3) in the postscript to Bentham.

Why then do we need to distinguish, in the topics prior to the 'summing of this division', between those we have numbered here I(a)—I(c) and those numbered II(a) to II(g)? We need to, simply to mark Darwin's own implicit designation of I(a)—I(c) as constituting a part I on variation under domestication, and II(a)—II(g) a part II on variation under nature. But if Darwin does not actually indicate it in his manuscript, what is the evidence for this implicit designation? As Francis Darwin records, the new beginning that follows the 'Summing up [of] this division' just quoted is headed in the manuscript 'Part III'.[9] And this heading, as he acknowledged, accords well with an outline plan in his father's hand, a plan possibly for the *Sketch* iteslf but in any case for some such exposition of the theory of natural selection, and very likely dating from about 1842:

I The Principles of Var. in domestic organism.

II The possible and probable application of these same principles to wild animals and consequently the possible and probable production of wild races, analogous to the domestic ones of plants and animals.

III The reasons for and against believing that such races really have been produced, forming what are called species.[10]

But Francis Darwin was being reasonable enough in dividing the *Sketch* not according to this tripartite plan and the 'Part III' heading that follows the 'Summing up [of] this division', but according to the more fundamental bipartite structure of its argumentation. For this bipartite structure was explicitly recognized by its own author—as his filial editor understood correctly—when he promised that a discussion of 'whether the characters and relations of animated beings are such as favour the idea of wild species being races descended from a common stock' would form 'the second part of this sketch'.[11] For the reference here is clearly to the chapters (beginning with the geological ones) concerned with the responsibility case that was to follow the cases for the existence and competence of natural selection as a VCP cause of species. Equally correctly, Francis Darwin realized that this bipartite division did not replace and supersede that tripartite plan. Indeed, it was to that concerto-like plan that Darwin returned in the 1860's, in distributing his writing efforts on behalf of natural selection into three distinct books. As he explained in introducing *Variation under domestication* (1868), he was in that work treating, 'as fully as my materials permit, the whole subject of variation under domestication'; then in a 'second work' he would present variability 'in a state of nature', including, of course, chapters on the struggle for existence and natural selection itself; before, finally, in 'a third work' going on to 'try the principle of natural selection by seeing how far it will give a fair explanation' of 'several classes of facts', from geology, geography, morphology, embryology and the like.[12]

But it should now be clear that this tripartite scheme always represented a convenient division of Darwin's personal labour, rather than a natural articulation in his public argument. For of course variation under domestication, no less than variation under nature, was appealed to twice over in the VCP argumentation making up the opening half of his whole exposition: first of all, in establishing that a natural selection of hereditary variation *existed* in the wild, he appealed both to the known tendency of new conditions to cause such variation indirectly in domestic species, apparently by affecting the reproductive system itself, and to the known fact that such variation can be accumulated by selective breeding; then secondly, in establishing the *competence* of natural selection to produce *species*, he pointed to the known power of artificial selection to produce distinct *varieties*, and to the reasons for thinking that the natural selective breeding entailed by the struggle for existence would have the greater effect—*species*—within its power because it is more precise and prolonged than any practised by man.

A tabular view of the structure of the *Sketch* will provide the best way to summarize the various suggestions just offered. It will also allow us to see directly that, in writing the *Essay*, *Natural selection* and the *Origin*, Darwin made successive departures from this early version of the structure, and that, though really minor and superficial, these departures were eventually enough to render the strategy and organization of his most famous book unhelpfully and quite unnecessarily obscure. (See Table I, page 243).

Moving now to the *Essay* of 1844, we can see this orderly three-part, two-division and three-case scheme beginning to be violated. For the first thing we notice is that the difficulties for the *competence* case posed by 'perfect organs' and by instincts have been moved, so that they now come after the principal summary of that case and not before it as in the *Sketch*; while the second thing to stand out is that Division Two now opens with a topic not recognized separately as such in the *Sketch*, namely the difficulty for the *responsibility* case posed by the lack of intermediate stages of morphological change among fossil organisms.

Moreover when we reach the *Origin* itself we find further such changes in its first edition, with the result that that book appears, quite misleadingly, to have the following three stages in its exposition: A. Four chapters developing

TABLE 1. The structure of the *Sketch* of 1842

Part I	Topic a	Chapter I	Consideration 1		Division One
Principles of variation and selection under domestication	b c	Variation under domestication	Existence case		Natural selection established as VCP cause for species
Part II	a	Chapter II			
Application of those principles to species under nature	b c	Variation in nature, and natural selection			
	d		Cons. 2	The case made	
	e		Compe- tence case		
	f			Diffi- culties	
	g	Ch. III Instinct etc.		consi- dered	
Part III		Ch. IV Geology	Consideration 3		Division Two
Trial of theory of natural selection as explanatory of species production		Ch. V Geology Ch. VI Geo- graphy Ch. VII Classifi- cation Ch. VIII Mor- phology Ch. IX Mor- phology	Responsibility case		Natural selection as, on balance, probably responsible for species
		Ch. X Recapi- tulation			

the arguments for natural selection; B. A rather miscellaneous collection of chapters devoted to sundry digressions and difficulties—i.e. Ch. V on laws of variation; Ch. VI difficulties posed by, *inter alia*, perfected organs; Ch. VII difficulties of instincts and the like, Ch. VIII difficulties with hybrids; Ch. IX difficulties with missing intermediates in fossils; and then C. Four chapters triumphantly returning to favourable facts from geology, geography and the rest. But, in fact, the two-division and three-case structure originally required by the VCP has not been replaced by anything so casual and arbitrary as this apparent three-stage sequence might suggest. For Ch. V is really a supplement to Ch. IV ('Natural selection') which makes the competence case for natural selection; while Chs. VI, VII and VIII all take up difficulties for the competence case as made in Ch. IV. By contrast Ch. IX opens what is in effect the *Origin's* Division Two, and opens it just as it was opened in the *Essay*, by taking on right away the obvious paleontological objection to this Division's responsibility case. So in outline we have a structure as in Table 2, page 244.

Returning, finally, to what we have of *Natural selection*, we can see from

TABLE 2. The structure of the *Origin* (1859)

Part	Chapter	Consideration	Case	Division
Part I — Variation and selection under domestication	Chapter I	Consideration 1 — Existence case		Division One — Natural selection established as VCP cause for species
Part II — Variation and selection under nature	II			
	III			
	IV	Consideration 2 — Competence case	The case	
	V			
	VI		Difficulties considered	
	VII			
	VIII			
Part III — Trial of theory of natural selection as explanatory of species production	IX	Consideration 3 — Responsibility case	Geological difficulty	Division Two — Natural selection as probably responsible for species production
	X		Evidence favouring responsibility	
	XI			
	XII			
	XIII			
Recapitulation	XIV			

the text itself, and even more from Darwin's own detailed table of contents for it, that it should be mapped as in Table 3, page 245.

From this table together with the other two, we can see that Chapters III and XI in the big treatise would have belonged in the first and third, respectively, of the three works Darwin hoped to complete in the 1860s and 1870s. In the case of Ch. XI this is obvious from its subject, geographical distribution; and in the case of Ch. III, we have Dr Stauffer's own report that Dr Alice Guimond found material from that chapter worked into *Variation* while helping with his edition of *Natural selection*, which accords well with the place of its topic—crossing—in the part of the *Essay* devoted to domestic variation.

It is surely unfortunate that Darwin did not stick more closely, in *Natural selection* and in the *Origin*, to the format of the *Sketch* or at least to that of the *Essay*. For even in the *Essay* there is manifest the relation of the two fundamental divisions—their chapters numbered separately as Francis Darwin emphasized— and there is no mistaking there the instructive adaptation of the whole to the VCP, and thereby to the precedent set by the structure and strategy of Lyell's *Principles of geology*. For of course Lyell's Books II (on the physical world) and III (on the organic) had explicitly sought (after Book I had put the general case for adapting the Huttonian system to the current state of the science) to establish various 'known' agencies as true and sufficient causes—the struggle for existence as such a *vera causa* for species extinctions for example. While his Book IV (later split off to form the *Elements of geology*) argued for these causes probably having been responsible for various past effects recorded in the rocks on

TABLE 3. The structure of *Natural selection* (1856–8)

Part I Domestic variation and artificial selection	[Chapter I]	[Variation under dom.]		Division one Natural selection as VCP cause
	[II]	[Variation under dom.]		
	III	Crossing etc.	Existence and competence cases	
Part II Natural variation and selection	IV	Natural variation		
	V	Struggle for existence		
	VI	Natural selection		
	VII	Laws of variation		
	VIII	Perfect organs etc.	Difficulties for competence case	
	IX	Hybrids		
	X	Instinct		
Part III The theory tested	XI	Geographical distribution	Responsibility case	Division Two Natural selection as responsible

the grounds that they offered at least as comprehensive and coherent explanations for these records as did the causes invoked by those geologists who denied the adequacy of any present causes acting with their current intensity to bring about certain past effects.[13]

But Darwin, it hardly needs remarking, had little of that love of schematic clarity and consistency which kept Lyell to the same format through no less than twelve editions of the *Principles* and more than forty years of revisions, some of them substantial—even radical. Nor was this lack a failing or a handicap—not at the time anyway, for the cognoscenti would have had little trouble recognizing the basic organization of the *Origin*, even obscured as it was by Darwin's mingling of arguments and shuffling of topics that he had once had explicitly, though privately, arranged in their natural order. It does, however, make it easy for his readers today to miss the wood for the trees. Now, with *Natural selection* apparently completing the publication of all the versions of the book still extant, we must take full advantage of Darwin's monumental researches and Dr Stauffer's meticulous scholarship in seeking to improve our mapping of that wood even as we discern ever more of the detail in the twigs on the trees.

---

[1] For the first of these, see P. J. Vorzimmer, 'An early Darwin manuscript: the "Outline and draft of 1839"', *Journal of the history of biology* viii (1975), 191–217. Vorzimmer discusses in detail the dating of the outline and of the draft, admitting that while some evidence favours 1839, the two documents may be of different dates and both later than 1839. The thirteen-page draft deals

only with domestic variation. The outline, published earlier by Francis Darwin in his introduction to the *Sketch* and *Origin*, identifies the three parts of a projected work on the origin of species by natural selection, as I have explained later in this review. Apart from Stauffer's edition of *Natural selection*, I refer here to the edition of the *Sketch* and of the *Essay* in C. Darwin and A. R. Wallace, *Evolution by natural selection* (Cambridge, 1958. Reprinted, New York and London, 1971) which includes Francis Darwin's 1909 introduction to those writings, and to the facsimile of the first (1859) edition of the *Origin of species* (Cambridge, Mass., 1964).

2 V. C. Kavaloski, *The vera causa principle: a historico-philosophical study of a metatheoretical concept from Newton through Darwin*. Ph.D. dissertation, University of Chicago, 1974.

3 A. Koyré, 'Newton's "Regulae philosophandi"', in Koyré, *Newtonian studies* (Cambridge, Mass., 1965), pp. 261–72.

4 Text and translation in Koyré, op. cit., p. 265. For a fuller citation of the texts in Newton's three editions, see A. Koyré and I. B. Cohen (eds.), *Isaac Newton's Philosophiae naturalis principia mathematica. The third edition with variant readings* (Cambridge, 1972), ii, 550.

5 Thomas Reid, *Essays on the intellectual powers of man* (Edinburgh, 1785) as quoted in Kavaloski, op. cit., p.6.

6 J. F. W. Herschel, *Preliminary discourse on the study of natural philosophy* (London, 1830), p. 148.

7 F. Darwin (ed.), *The life and letters of Charles Darwin* (3 vols., London, 1888) iii. 24–5.

8 Darwin, *Sketch* in *Evolution by natural selection*, pp. 57–8.

9 Ibid., p.59.

10 As in F. Darwin, 'Introduction', op. cit., p. 29. Cf. Vorzimmer's discussion of this text in his article cited in note 1.

11 *Sketch*, loc. cit., p. 46. Cf. F. Darwin, 'Introduction', pp. 29–30.

12 *Variation of animals and plants under domestication* (2 vols., New York, 1868), i. 13, 15 and 21.

13 For this division of the *Principles* into four books—not explicit in the first edition—see the table of contents in *Principles of geology* (5th edn., 4 vols., London, 1837).

IX

Darwin's Theory and Darwin's Argument

1. Introduction: Darwin in New Jersey and elsewhere

One admirable characteristic of David Hull's work is that it is both genuinely interdisciplinary and consistently disciplined. His work presents therefore a splendid counterinstance to the usual tendency for discipline – in the sense of boundaries – hopping to be accompanied by discipline – in the sense of standards – dropping. In moving back and forth between history, philosophy, sociology and biology, David has always wanted to get things right by meeting, not dodging, the demands set for any good work on the subject in hand.

For this reason, he has been second to no one in our time in raising the level of interdisciplinary discussion of Darwin and his theory of natural selection. He has moreover contributed decisively to one special disciplinary element in this interdisciplinary achievement. Darwin and his theory have now been brought, in North America at least, within the domain of high, academic, professional philosophy of science, as can be seen by scanning recent *PSA* or *Philosophy of Science* volumes, or indeed by checking the index of such a prominent monograph on general philosophy of science as van Fraassen's *The Scientific Image* (1980). Others who have played major roles in advancing and securing this disciplinary development – Michael Ruse, Mary Williams, John Beatty, Philip Kitcher, Elliott Sober and Elizabeth Lloyd come at once to mind – owe a great deal to an author who had published both *Darwin and His Critics* (1973b) and *The Philosophy of Biological Science* (1974) before Nixon had had his comeuppance from Congress and the people.

My paper here offers to further this philosophical rehabilitation of Darwin's theorising; and it offers to do this in a way that is directly indebted to David's work. In his paper, 'Charles Darwin and Nineteenth-Century Philosophies of Science' (1973a), and in *Darwin and his critics*, David brought out for the first time how complex were the relations between Darwin's argumentation on behalf of natural selection, and diverse ideals of evidence and explanation then articulated by writers on philosophy of science such as Herschel, Whewell and Mill. Shortly afterwards, understanding of this topic was also enhanced by a dissertation by

Vince Kavaloski (1974), then a University of Chicago graduate student who had benefited from David's encouragement and counsel. What Kavaloski did was to show that one methodological ingredient in Darwin's theorising is indispensable to any attempt to grasp how that theorising stands in relation to the ideals of good science upheld by the philosophers of the time. This ingredient is Darwin's view of natural selection as a true cause, a *vera causa*, in the sense first adumbrated in Newton's first rule of philosophising as explicated by Thomas Reid in the eighteenth century.

Since Kavaloski made this contribution to the subject, others, including Michael Ruse (1976), Rachel Laudan (1982, 1987) and myself (1977, 1987), have extended his insight by applying it in more detail to the structuring and strategies of Lyell's *Principles of Geology* (1830–33) and of Darwin's *Origin of Species* (1859) itself. However, even more recently the subject has undergone a further transformation. For a succession of people – notably Paul Thagard (1978), Elizabeth Lloyd (1983), Philip Kitcher (1985a) and Doren Recker (1987) – have sought to relate the structure and strategies of the *Origin*'s argument to philosophical views unheard of in Victorian England; and unheard of there for the very good reason that they mostly trace to doctrines constructed in the last three or four decades, and prominently represented in what was in Darwin's day the College of New Jersey but is now Princeton University. For, in particular, Darwin's arguments on behalf of natural selection have been called in to illuminate, vindicate, and indeed sometimes to deprecate, Harmanian inference to best explanation, Hempelian explanatory unification and the Fraassenating semantics of empirical adequacy.

In these higher order, philosophical endeavors there has been recurrent if not invariable reference to Darwin's commitment to the *vera causa* ideal. So the new, often Princetonian, issues have been thoroughly entangled with a specific matter of historical scholarship. Not surprisingly, therefore, coming from the historical side as I do, I have been prompted to ask the two inescapable, Hullian questions: Are they getting the history right? And, if not, how can they get the philosophy right? By the end of this paper, I will, I trust, have shown two things. First, that often they have not been getting the history right and, second, that unless and until they do, they can not get the philosophy right either. Less belligerently, then, this paper seeks to advance the state of the philosophical arts as practised on Darwin's theorising, and to do so by establishing some minimal historical conditions for not going astray philosophically in such practice. More positively, this paper suggests some general considerations that we all, philosophers and historians alike, can do well to keep in mind whenever a particular scientific theory, such as Darwin's, is invoked in the evaluation of some general philosophical theory of theories.

The need to raise such general considerations is shown by a significant development concerning the philosophical study of the Darwin case. For a long time, there was a standard exercise involving this case. The logical empiricist tradition dominant in mid-century had elaborated what came to be called the 'received view' of scientific theories, the one familiar from the books of Carnap, Nagel, Braithwaite and others and examined so instructively by Suppe in the long introduction he wrote for *The Structure of Scientific Theories* (1974). The standard exercise was, therefore, to ask how well the received view fitted the Darwin case; or rather, perhaps, it should be how well the Darwin case fitted that view. The most sustained and informed answer to this question was given by our editor, Michael Ruse (1975). Michael knew his logical empiricists well, especially Braithwaite, and he knew his Darwin well; and he was indeed writing about Darwin's theory as set out in the *Origin*, not about any subsequent twentieth-century version of the theory. He concluded, did Michael, rather against his initial hopes, one suspects, that Darwin's theorising fits the received view only to a limited extent and that some of the fitting requires a fair bit of fudging. For a judicious critique of this fitting and fudging, one can now read Recker (1987).

However, the business has moved into a new phase since the days when comparing or contrasting Darwin's theorising with the received view was a dominant concern. For Kitcher (1985a) and Lloyd (1983) have broken with that concern. It is not merely that neither of them favors the received view. It is rather that they are claiming a much more positive outcome from their encounter with the *Origin*. Each is claiming that when the theory of theories he or she favors is brought to bear on the *Origin*'s theorising, then both the favored theory of theories and Darwin's theorising come out looking better than before. There is one obvious snag. Although Kitcher and Lloyd are both from the same small place in New Jersey and have been colleagues in San Diego, they disagree over what is the best theory of theories, the best metatheory, if this use of the word is allowable. Lloyd accepts that version of the so-called semantic theory of theories developed by van Fraassen. Kitcher, however, is developing his own theory of theories. It draws on Bromberger (the Why? man) and on Hempel; but not the Hempel of the covering law analysis of explanation so integral to the old received view, rather the Hempel of explanatory unification. To my knowledge, Kitcher has never elaborated his view of theories at any great length. But it is clear, from his books on mathematics, evolution and sociobiology and from related papers (1981, 1982, 1983, 1985a, 1985b) that his metatheoretical view has been designed from the start to fit two historical cases he has studied closely. One is the differential calculus as Leibniz and Newton launched it, the other is Darwin on natural selection. So, with Kitcher, we have what we have not had before, I think; and that is a philosophical theory of theories that has been deliberately constructed, not merely to illuminate, but actually to legitimate Darwin's theorising. For those of us working in the Hullian, hybrid zone of hypenated historico-philosophical Darwin studies, Kitcher's efforts on behalf of his theories must hold a special significance as a historical landmark if not a philosophical triumph.

2. Theories and metatheories in philosophy and in history

The need for a greater sense of the history of all such metatheoretical endeavors will form one conclusion to be drawn from the present paper. For we need a sense of history concerning what is a peculiarly twentieth century quest – the quest for a philosophical theory of theories, a general characterisation of how the structure distinctive of natural scientific theories allows them to function as they do. For, make no mistake, this quest does not go further back than the received view itself. Ask what it was the received view replaced, and one sees that it did not replace any comparable view or views. It was not a new set of answers to an old set of questions, for it addressed a need that had not previously been felt in that form. Even Duhem, who wrote on physical theory, its aim and structure, before the received view was invented, was not concerned with the common problems the received view and the semantic view have been designed to solve in divergent ways. For Duhem's quest (as Larry Laudan has pointed out to me) is to characterize any physical theory by appropriate contrast with metaphysical theorising, but he does not do this by offering a general account of the relationship, in scientific theories, between logical form and empirical content.

So what we have to keep in mind is that, in the eighteenth and nineteenth centuries, all sorts of things in all sorts of scientific domains were called theories, while no one thought to enunciate some general account of the structure and function of scientific theories such as the received and semantic views have offered to supply in our day. But surely all that we now know about people such as Lyell, Darwin, Herschel, Mill and Whewell, not to mention Comte and Mach, shows that they had some quite general views about theories as such. Indeed, they did. But it is just a mistake to presume that because those views were general they must have constituted, as the received and semantic views do, a general answer to the question, what is a scientific theory, in the relevant structural and functional senses.

This warning against this presumption is appropriate here, because we have to appreciate that in approaching Darwin's theorising through its place in the *vera causa* tradition, we are not placing it in relation to anything truly equivalent to the structural and functional views of scientific theory familiar in this century. This lack of equivalence will then prove to be one reason why fitting a twentieth-century style theory of theories to the *Origin* is a difficult exercise both to conduct and to evaluate. This difficulty and its historical source should be borne in mind, therefore, as we move from Darwin's understanding of his theory as a *vera causa* theory, to the issues raised by the various Princetonian explications of the *Origin*.

Before actually turning to the argument of the *Origin* itself, however, there are some preliminary matters to be mentioned. First, it will already be plain that I am focusing here only one among many clusters of philosophical issues surrounding Darwin's theory. In doing so, I do not mean to suggest that there is no philosophical interest in relating Darwin's theory to teleology, essentialism, creationism, probabilism, naturalism or capitalism and so on. Second, for our purposes here it is appropriate to focus on natural selection among Darwin's many theoretical proposals. Despite the claims of Michael Ghiselin (1969) and others, Darwin was no methodological unitarian. The structure and strategies of argumentation in play in the *Origin* are one thing, those in play in pangenesis (Darwin's theory of reproduction) another, those in play in his geological paper on Glen Roy another again. Third, our concern here is not with Darwin's original arrival at the theory of natural selection, but with his public case for it. There are instructive questions to be asked about the relation between the private inception in 1838–9 of Darwin's theory and its later presentation, but those relations do not form our principal business here, although some attention to them will be needed in sorting out Darwin's relations with Herschel and Whewell. Fourth, sometimes in historical case study work there is an explicit denial of any concern with the theorist's own intentions regarding his theorising. Indeed, it is sometimes implied that our interpretation of any theory should be kept strictly apart from the interpretation given it by its original author. This separatism may well be possible, even desirable, in other cases. But in the present instance it would be misleading. The most recent philosophical explicators of Darwin, notably Lloyd, have been insistent in claiming that their explications make sense not only of Darwin's theory but also of Darwin's own interpretation of its epistemic and systemic character. Any subsequent integration of historical scholarship and philosophical doctrines is thus invited to follow this precedent, and the present paper does so.

3. Darwin's one long argument and the *vera causa* ideal

Turning now to the structure and strategy of the 'one long argument' that the *Origin* presents, let me stress that the textual evidence for what follows is, much of it, available in an earlier piece of my own (1977); so, I will take various documentary details as given on the present occasion.

What can be shown, mainly by comparing the *Origin* with its two earlier unpublished, draft versions, the *Sketch* of 1842 and *Essay* of 1844, is that Darwin's text, in all three versions, was structured by him in accord with two distinct divisions or partitions, as the diagram for the *Origin* (Figure 1) shows.

The easiest partition to discern is a threefold one: an opening Part I concerns principles of variation and selection under domestication; a middle Part II concerns the application of those principles in the wild; and a final Part III exhibits the explanatory virtues of the theory of natural selection, in relation to various classes of facts about species. We may refer to this partitioning as the *three part* structuring of the argument.

Less easy to discern in the *Origin*, but not in the *Sketch* and *Essay*, is another threefold articulation that does not coincide totally with the *three part* one. For

Part I	Chapter I	Consideration 1		Division One
Variation and selection under domestication		Existence case		Natural selection established as VCP cause for species
Part II	II			
Variation and selection under nature	III			
	IV	Consideration 2	The case	
	V	Competence case		
	VI		Difficulties considered	
	VII			
	VIII			
Part III	IX	Consideration 3	Geological difficulty	Division Two
Trial of theory of natural selection as explanatory of species production	X	Responsibility case		Natural selection as probably responsible for species production
	XI		Evidence favouring responsibility	
	XII			
	XIII			
Recapitulation	XIV			

Fig. 1. The structure of the *Origin* (1859).

Darwin's argument comprises, successively, a case (1) for the *existence* of natural selection; a case (2) for the *competence* or *adequacy* of natural selection to produce new species from old, and diversify them adaptively in the formation of new genera, orders and so on; and a case (3) for the *responsibility* of natural selection for the production of those species now extant and those extinct species commemorated in the rocks. This trichotomy we may call the *three case* structuring.

So, when we see that Part III in the *three part* structuring is the same as case (3) in the *three case* structuring, we see that this coincidence allows us to identify another articulation, this time a dichotomous one. For everything prior to case (3) – and so prior to Part III – constitutes a Division I, which sets out the theory of natural selection, so as to establish its *vera causa* credentials; then, in Division II, there is the verifying of the theory as an explanation for the origins of extant and extinct species.

Since the rationale for the *three case* and the *two division* articulations is the *vera causa* ideal, we need to get a clear view of that ideal without delay. Here it should be emphasized that the *vera causa* ideal is no esoteric historians' rarity. Anyone who has read in the philosophical literature surrounding science in the century from

Hume to Mill will have encountered the ideal in many more or less explicit invocations. Equally, specialist historians of the philosophy of science, such as Larry Laudan (1981) and Bob Butts (1970, 1973) – who have given us sophisticated exegeses of authors such as Thomas Reid and William Whewell on *verae causae* – have removed any excuse that the rest of us might have had for going astray on this subject.

In its original form, the *vera causa* ideal was given canonical explication by Reid in his interpretation of one of Newton's rules of philosophising. For Reid, the decisive element in the ideal was the contrast it drew between true, known, real or existing causes, on the one hand, and hypothetical, imaginary, unknown, conjectural or supposed causes, on the other. For the exemplary contrast was between Newton's gravitational force as a true cause, and Descartes' vortices of subtle matter as a hypothetical cause. Both causes, it was argued by Reid, are adequate causes. That is to say both would be sufficient or adequate to cause the effects, the orbiting of the planets around the sun, that are to be explained. As to their sufficiency, adequacy or competence, they stand equal. Where they differ is over the evidence for their existence. For the gravitational force there is independent evidence (from tides, for instance, and falling bodies on earth) for its existence; it is evidenced by facts other than the planetary orbits it is to explain; whereas for the existence of the Cartesian vortices there is no evidence except those orbits. The gravitational force, unlike the vortices, is therefore both a true, known, real or existing cause as well as a sufficient, adequate or competent cause. It meets, then, the requirement that one is not allowed to suppose, or infer, that one's explanatory cause exists merely because it is adequate to explain what it is to explain. Truth, existence and reality, on the one hand, and sufficiency, adequacy or competence, on the other, are independent considerations and their evidential demands must both be met for a cause to meet the contraints of the *vera causa* ideal.

So far, in this thumbnail epitome, the *vera causa* ideal may seem straightforward enough, especially as it has many echoes in twentieth-century discussions of, for example, what it is for a hypothesis to be *ad hoc*. But it should be plain, too, that there was always plenty of scope for complications and sophistication. Especially was this so concerning the kinds of evidencing that were thought appropriate in establishing either the existence of a cause or, conversely, its adequacy. Reid, for one, seems – sometimes at least – to insist on direct experiential acquaintance, generalisable by enumerative induction, as the only acceptable form of evidence for the known truth, reality or existence of a cause (L. Laudan 1981). More, much more, could be said, here, about such issues, and about the development of the *vera causa* ideal in the centuries since Newton, but our thumbnail epitome will allow us to bring out what we need to recognise in Darwin's argument.

Darwin's case for the existence of natural selection is more difficult to represent precisely than is often appreciated, as Ruse (1971) showed once and for all. But the gist of it is plain enough. Species in the wild are subject to changes in conditions of

170

life; and domesticated species show that any animals and plants exposed to changed conditions vary and heritably so. There is hereditary variation in the wild, then. There is also a struggle for existence, for there is superfecundity and this entails a struggle for food, space and other limited requisites for life. In this struggle for existence, there is a differential survival and reproduction of hereditary variants, for some hereditary differences affect chances of survival and reproduction. There exists in nature, therefore, a process of selective breeding, a process analogous to the selective breeding practised by farmers and gardeners.

Darwin does not claim to establish the existence of natural selection by mere inspection of nature; that this process is going on in the wild is not given from experiental acquaintance with its presence there. However, his view is that its existence there is a reasonable conclusion from various premises, concerning heredity, geological change and superfecundity, that are either already accepted or are evidenced observationally.

Again, Darwin's main case for the adequacy of natural selection is far from straightforward in all steps, but its main outline – lucidly explicated recently by Kenneth Waters (1986) – is unmistakeable. Artificial selective breeding is known to be sufficient, adequate, competent to produce, within a species, distinct races – of dogs, for instance – adapted to distinct human ends. These races do not count, according to customary criteria, as distinct species. But natural selection has vastly longer to work and is much more comprehensive and discriminating. It can then produce races that do count as species, for they would be more permanently and more perfectly adapted and divergent in their organisation, and hence infertile with one another, true breeding and without intermediate varieties.

Finally, the case Darwin makes for the responsibility of natural selection for species formation, although tortuous in places, pursues one line throughout: namely, that the theory that natural selection did it is more probable, and so to be preferred over any rival, because it is better than any other at explaining facts of several general kinds or classes: biogeographical facts, embryological facts and so on.

4. Natural selection, true causes and Whewellian consilience of induction

Even this very brief consideration of how the *Origin* was structured so as to conform to the *vera causa* ideal, the old Reidian ideal, can put us right on many of the issues that have to be understood correctly, if we are not to go astray in attempts to derive lessons from this case for the philosophical theory of theories. One cluster of such issues concerns William Whewell and the wave theory of light. For Whewell and the wave theory clashed with that old ideal. What is more, after the *Origin* was published and in defending it, especially in correspondence, Darwin sometimes invoked the wave theory and Whewell. So, it may seem that we have to conclude that, after all, the old Reidian ideal is not as relevant as we thought, and that Darwin and natural selection really belong among the influences subverting that ideal. Indeed, this line has recently been taken in two valuable papers by Ronald Curtis (1986, 1987).

This is not the place for a full treatment of this line of historical interpretation (Hodge forthcoming). However it is necessary to insist here that the view Darwin took when constructing his theory in the 1830s, and when writing the *Sketch*, the *Essay* and the *Origin* in the 1840s and 50s, was one thing, and that the position he adopted in defending it later against objections was sometimes another.[1]

So the decisive precedent for Darwin's argument in the *Origin* was the old *vera causa* legitimation of gravitational astronomy, rather than the new epistemological rationale being given, in the 1840s, for the wave theory of light. Generally speaking, the wave theory was not being defended as *vera causa* legitimate (L. Laudan 1981). On the contrary, it was, especially by Whewell, being cited as not conforming to that ideal and, therefore, because it was good science, discrediting that ideal as a constraint on physical theory. Now, after he had finally published his theory of natural selection in the *Origin*, Darwin sometimes retreated to a similar style of defence, citing the wave theory in doing so. That is, he declared that even if the *vera causa* credentials of natural selection were disputable, it was not to be dismissed totally on that ground, because there were precedents, notably the wave theory, for taking seriously, even accepting as probably true, theories that did not meet the evidential requirements of that ideal. More particularly, it was agreed, Darwin reminded people, that although the existence of light waves was not evidenced independently of the many classes of optical facts the theory was able to explain, nonetheless that explanatory adequacy and achievement were grounds for thinking the theory at least probable rather than hopelessly conjectural.

However, although Darwin's move to this form of defence is apparent, it should not distract us from the fact that the *Origin* had not been written that way and that it was never rewritten that way either, and that the five revised editions of it that Darwin prepared from 1860 to 1872 kept the same structure and strategy of argument inherited from the *Sketch* of 1842. Moreover, if one reads carefully in those texts where the wave theory precedent is cited, one can see that that abandonment of the old ideal is often more apparent than real, and less than wholehearted.[2] For he will still insist that there is, after all, evidence for the existence and adequacy of natural selection independent of its explanatory virtues as exhibited in Division II of his book. But, be that as it may, the argument of the *Origin* in all its editions is a Reidian *vera causa* argument.

We can not, therefore, read the argument of the *Origin* as presenting a Whewellian consilience of inductions. The reason for this is quite simply that Whewell not only rejected the old Reidian *vera causa* ideal, he also offered his consilience ideal as an alternative to that older ideal. Most particularly and explicitly, he disputed the leading assumption made by the old ideal, the assumption that the truth, the existence of an explanatory cause is a separable issue from its adequacy (Whewell 1840, vol. 2, pp. 441–445). Often, the best we can do evidentially for an explanatory cause, Whewell argued, is establish its probable existence by establishing its explanatory adequacy for many distinct classes of facts, in a consilience of inductions as he dubbed it. For often we will not have independent – much less direct, independent – evidence of its existence. So one main rationale for Whewell's consilience doctrine was that the inductive credentials of a theory can be established without measuring it against that impossibly restrictive ideal of independent, direct evidencing of the existence of any causes it introduces.

So, to be sure, Darwin in his Division II, is doing very much the kind of thing that Whewell called consilience. But the old *vera causa* ideal also required one to do that, too, as can be seen by reading Herschel (1830) on Newton's gravitational theory (Good 1987). The decisive difference was that the old ideal laid down that this is not the only kind of evidencing one has to produce. One has also to produce independent argumentation for the existence of one's cause. So, the argument of the *Origin* taken as a whole is conformed to the ideal that Whewell sought to replace, not to the consilience of inductions that he offered as its replacement. It is, then, not correct to do as Ruse (1979) has done and propose that Darwin is being Herschelian (and so Reidian) in the first part of his argument and Whewellian, in the second. The whole argument is Reidian, and so, as a whole, un-Whewellian; and, as in the whole, so in the parts.

The attractive, persistent, but unsustainable, notion that Whewell's consilience doctrine is the inspiration for the *Origin*'s argument structure has no support from what we know of Darwin's reading of Whewell and other contacts with him. The consilience view is not presented explicitly until Whewell's *Philosophy of the Inductive Sciences* of 1840, and there is no sign that Darwin had read that book when he composed his *Sketch* in 1842. He did read Whewell's *History of the Inductive Sciences* of 1837, in the Autumn of 1838, and he may have read in it the year before. Also, the book emphasises the virtues of the wave theory as an explanation of many, diverse kinds of facts, but it does not set those virtues explicitly against the *vera causa* ideal. The book includes also a rejection of the old *vera causa* ideal as appropriate to geological science (Ruse 1976). All the signs are that this rejection left Darwin entirely unmoved. On the contrary, he shows every sign of remaining loyal to Lyell and Herschel's teaching that the old ideal is indispensable to the foundation of geology as an inductive science. Geology for Darwin, following Lyell, was the science that included theories about the origins of species.

Avoiding the Whewellian consilience reading of the *Origin* is very worthwhile, if only because we can then avoid mistakenly embracing the Princetonian proposal made by Thagard (1978). He holds that Whewellian consilience is tantamount to inference to the best explanation, as defended by Gilbert Harman, as the form of inference proper and typical for much natural science. And Thagard offers us Darwin in general, and the *Origin* in particular, as Whewellian and so, on his account, Harmanian. To this proposal, one must respond that even if the assimilating of Whewell to Harman is conceded, and there are reasons for resisting it, it does not make Darwin's book Harmanian because the overall argument of the *Origin* is not Whewellian.

Recker (1987) has rightly seen that Thagard's view of the *Origin* does not capture the way the whole argument is constructed; and he brings out well that there are difficulties for Thagard's further proposals, concerning simplicity, for instance. However, Recker himself has not succeeded fully in discerning how the whole argument proceeds. For Recker divides the *Origin*'s chapters into three blocks. The third and last block coincides, almost exactly, with Division II on the mapping I have given. But his first two blocks do not correspond either to the distinction I have given between Parts I and II or the distinction I have given of case (1) and (2). For Recker takes his first block of chapters to set out a general, positive case for the 'causal efficacy' of natural selection in relation to evolution; the second block to answer possible objections to the causal efficacy thesis; and the third to support that thesis by exhibiting its explanatory virtues regarding biogeography and so on. On Recker's account, then, it is a causal efficacy thesis, concerning natural selection, that is being argued for right through the whole book. Now, Recker is well aware that the positions of the Reidian Herschel and of the anti-Reidian Whewell on *verae causae* are indispensable in making sense of Darwin's argument. But because Recker sees a single thesis being argued for throughout the book, he has to see Darwin as deploying different *vera causa* strategies in different blocks of chapters, on behalf of this single thesis. Accordingly, he follows Ruse in seeing a Herschelian 'empiricist' *vera causa* strategy in play in the early chapters making up the first block, and a different, Whewellian 'rationalist' strategy in play in the third block. But this single thesis, three block, two strategy analysis, although it allows Recker to contribute many insights concerning Darwin and the *Origin*, rests ultimately on a mistake: the mistake of not seeing that the old Reidian *vera causa* ideal, as upheld by Lyell and Herschel, and learned from them by Darwin, required three evidential cases to be made on behalf of natural selection: an existence case, an adequacy case and a responsibility case. Recker holds that the unity of the *Origin* lies in its pursuing throughout a single causal efficacy thesis; but this is not so, rather it lies in Darwin's making three cases on behalf of his one cause. It is the cause, natural selection, itself, that provides the unity, together with the requirement that the case for the responsibility of natural selection for species origins requires that prior cases also be made for its existence and for its adequacy for such effects.

Although much more might be said, on the way the *Origin* reveals its descent from Lyell's and Herschel's reaffirmation of the Reidian tradition, enough has been done to show that correctly representing the structure and strategy of the book's argument is not always a straightforward task; but that it is an indispensable task if we are to correctly locate Darwin's work in relation to the epistemic and systemic ideals of evidence and explanation under consideration at the time.

5. Darwin and a semantic view of theories

It is a no less indispensable task if we are to assess fairly and fruitfully the suggestions, very different suggestions, made by Lloyd (1983) and by Kitcher (1985a), in their respective efforts to clarify 'the nature of Darwin's support for his theory of natural selection' and the nature of 'Darwin's achievement'.

Following van Fraassen, Lloyd adopts a semantic view of theories. She does not set out explicitly what this view is, but Ronald Giere (1983), another semanticist, proves a very good guide on this as on many other matters. Giere explains that on the van Fraassen version of the semantic view, that he shares with Lloyd, a theory is what is defined by a definition. The definition defines a kind of system: for instance, a Newtonian gravitational system is defined as a system with two or more masses moving in accord with Newton's three laws of motion and one law of gravitation. On one analysis of this view, moreover, the definition is taken to specify a state space in defining a kind of system. Thus for the Newtonian system, six state space variables, three for position, three for momentum, are assigned to each particle. So, a theory is what is defined by a definition of a kind of system, as specified in a specification of a state space. There is a sense, then, Giere explains, in which a theory is a model. A model, he says, is a set of objects satisfying a linguistic structure. The linguistic structure that a theory satisfies is the definition of the kind of system. Giere does not say so, but it would seem that a model, in this sense, need not be existentially instantiated; possible objects can satisfy linguistic structures.

Lloyd, although she does not say so, seems to accept all of this, as is shown by at least one of her other recent papers (1984). She is certainly in accord with the semantic view, as Giere expounds it, when she maintains that the theory of natural selection, in Darwin as elsewhere, is a group or set of model types. For a model type, she explains, is what one has if one takes a model that has values for some parameters and one leaves out the specification of those values. A model type has, then, parameters with unspecified values. Natural selection is a group of model types, rather than a single model type, because, Lloyd holds, the natural selection of bodily structures, say, is represented in one or more model types, while the natural selection of instincts, say, is represented by other, distinct model types. A particular model, of some particular instinct being modified, for instance, would have particular values specified for selective advantages, for variation and so on.

Now Lloyd's main proposals concern the support Darwin is giving his theory. Support, for Lloyd, following van Fraassen, is a semantic relation. And, semantic relations are to be distinguished from pragmatic relations, according to Lloyd, for pragmatic relations involve the uses of a definition or a model or whatever, by some user, for some purpose; as, for instance, when a theory is used by someone to explain some fact.

This invocation of the standard distinction, between what is semantic and what pragmatic, is decisive for Lloyd's analysis of Darwin's support for his theory, as Recker's (1987) appreciative critique of Lloyd brings out well. For Lloyd insists that Darwin's argument involves establishing semantic relations, more particularly semantic relations of empirical adequacy, relations that are independent of pragmatic relations of explanatory use. To see how Lloyd defends this conclusion concerning support and semantic relations of empirical adequacy, we have to see, therefore, how Lloyd understands the distinction between what I have designated Division I and Division II of Darwin's argument. For the business of Division I is, according to Lloyd, to support the group of model types that constitutes the theory of natural selection; while in Division II, by contrast, a host of particular models is being supported.

The support for the theory, in Division I, is provided, Lloyd holds, by showing how the assumptions of the theory, as a group of model types – assumptions about heredity, variation and differential survival and reproduction – are empirically adequate. Empirical adequacy is a semantic relation, an assumption being empirically adequate, on Lloyd's account, if its deductive consequences are instantiated by facts. In Division II, what is going on, then, according to Lloyd, is the construction of many particular models and an exhibition of their empirical adequacy. Thus a model for the origin of the Galapagos fauna is constructed and its empirical adequacy exhibited.

Now, for Lloyd, this exhibition of the theory is supportive because it is not a pragmatic achievement. Darwin, she insists, is not explaining how the Galapagos species come to be as they are. Rather, he is showing that such an explanation is possible. To show this is not to use the theory in explaining, it is to exhibit its empirical adequacy, a semantic not a pragmatic achievement. For what is shown is that the group of model types, that is the theory, includes a model type that can be specified in its parameters, so as to yield a particular model that is empirically adequate to the Galapagos facts.

There is more to Lloyd's suggestive and sophisticated discussion than the sampling given here presents, but we now have enough before us to begin assessing her main proposals. On the positive, favorable side, there are three manifest virtues in her account. First, the articulation of Darwin's long argument into two divisions is taken seriously, and the contrast between the generality of Division I and the particularity of Division II is appreciated. Second, the lack of direct confirmation for the conclusions of Division II, the lack that arises because there is no record – such as a vast film would provide – of the past history of life on earth, is very properly emphasised. Third, there is full appreciation of the way Darwin often provides sketches for possible explanations, sketches of possible sequences of events leading, say, to the present Galapagos species.

On the negative side, however, there are several fundamental difficulties. First, Division I of the *Origin* is not merely an attempt to support a definition of a kind of system. Lloyd seems to accept that a definition as such, apart from its applications and instantiations, has no empirical content. For she represents Darwin, in Division I, as providing empirical support not for the definition itself, but for the assumptions that must be made if the definition is to have any applications or instantiations. But, against this view, we must insist that Darwin is not concerned merely with empirical assumptions, in the sense of presuppositions, made in any application or instantiation of natural selection as a theory. Rather, he is concerned to show that natural selection, as he defines it, is existentially instantiated, in nature, in order that he can go on to show what kinds of consequences it is having and so what it is sufficient to produce. Lloyd's semantic view of the theory as a group of model types does not lead us to see, therefore, how the distinct existence and adequacy cases form the business of Division I.

Second, in insisting that Darwin's explanation sketches are exhibitions of semantic rather than pragmatic relations of support, Lloyd has to distinguish between Darwin's arguments in using his theory and his arguments in support of his theory in Division II. But this distinction has no foundation in Darwin's text as he himself understood it. It is a distinction required by the presuppositions of Lloyd's exegetical enterprise, not by the content of the case study itself. Lloyd's metatheory is itself not shown to be empirically adequate here, in that its consequences are not instantiated by the textual facts.

Third, Lloyd omits all consideration, even mention, of causation in general, and of natural selection as a causal process in particular. This omission is striking, to put it mildly, for in many of the texts Lloyd is quoting from there is abundant causal talk, including, in at least one case, explicit identification of natural selection as a *vera causa*.[3] But this omission is not surprising. We know that van Fraassen (1980) thinks that causation, like explanation, involves pragmatic not semantic considerations. For causation has to do, it is alleged, with a context of human actions and interests, and is not therefore to be confused with context independent semantic relations between statements, or with what makes statements empirically adequate. Lloyd is, then, being consistent in including in her entire paper not a phrase about causation or even any allusion to it. However, the consequence is that none of the features of Darwin's argument that have their rationale in his commitment to causal explanation, and to the *vera causa* ideal, are done proper justice in Lloyd's semantic explications.

The empiricism of van Fraassen's philosophy of empirical adequacy is an explicitly anti-realist and, especially, anti-causal-realist empiricism.[4] There are, then, deep reasons for Lloyd's failing in her attempt to exhibit Darwin as a fellow agnostic concerning causality and reality. For consider two traditions in the contrasting of mathematics and physics. One tradition – the one van Fraassen belongs to – says that mathematics and physics (including mathematical physics) differ ultimately in that mathematics, as such, has no empirical suport, while physics, even mathematical physics has some. Another tradition, however, holds that what marks off mathematics from physics is that the first has no concern for causal explanation. Darwin was obviously an heir to this tradition in its eighteenth-century forms. A historian has to read Lloyd's own paper as being, therefore, an incongruous conjunction of disparate traditions, the one Darwin belonged to and the one Lloyd herself descends from through van Fraassen. Lloyd's attempt to shift Darwin from his tradition to her own is, unknowingly, an attempt in effect to change the past, something even God is traditionally unable to do. Lloyd cannot be blamed, therefore, for failing.

6. Darwin and explanatory unification

Kitcher (1985a) starts from a view not of what theories consist of, but of what they do for us. A theory is a good one, he urges, if it unifies our beliefs by providing a few basic patterns of argument that can be used in the derivation of the many sentences that we accept. Accordingly, he holds that Darwin's theory is an explanatory device, a collection of problem-solving patterns, a collection of schemata, aimed at answering families of questions about organisms by describing the histories of those organisms.

For Kitcher, explanatory promise is not purely pragmatic because it is not irreducibly context dependent. Whether or not arguments can be applied to provide explanations, by providing answers to explanation-seeking questions, can be established independently of context, he says. The issue is whether those arguments can be part of a store of explanatory resources that provide explanations, by conforming to certain argument patterns. Explanatory promise is, then, a matter of potential unification.

Darwin's irregular branching tree of descent, and his process of natural selection, are therefore analysed by Kitcher as a very generalised explanatory resource. Darwin's exhibition of the resourcefulness of that resource store is achieved by using it in providing explanation sketches for a wide range of phenomena, sketches that exemplify a few patterns of argument. Most especially is this done in Division II of the *Origin*, where the common characters distinguishing some supraspecific taxonomic group of species – or their common confinement to some geographical region – are traced to a common ancestry for the group, and a subsequent history of limited adaptive divergence and migratory dispersal.

Kitcher urges that the genuine and controversial innovation in Darwin's *Origin* was not the quasi-deductive derivation of natural selection from heredity, variation and superfecundity, but nothing less than bringing historical questions about species origins within science for the first time, questions previously the subject of theological speculation. Darwin's main claim is thus, on Kitcher's view, that we can understand many biological phenomena in terms of what Kitcher calls Darwinian histories, with these histories conforming to a few basic patterns. The theory is simply the assertion that these phenomena are understandable in that way, for the theory consists of the demonstration that such unifying histories can be constructed.

On four counts, I submit, Kitcher's analysis is to be welcomed. First, he does not force on Darwin an inappropriate distinction between semantics and pragmatics. Second, Kitcher rightly refuses to identify Darwin's theory with the quasi-deductive existence case. Third, Kitcher is illuminating on the relation between Division I and II, and on the domination of Division II by a few recurrent lines of argument. Fourth, Kitcher brings out, better than anyone before him, the full force of those early criticisms of Darwin over the difficulty of getting testable consequences out of the theory and checking them against ascertainable facts. For, as Kitcher emphasises, there was considerable looseness over what were acceptable auxiliary assumptions to incorporate into Darwinian histories; and there was looseness over what were or were not relevant, independently-established facts, available for deciding whether the implications of those histories were confirmed by empirical findings.

Against these four welcome elements in Kitcher's achievement, there are some three less favorable considerations to be weighed. First, Kitcher underestimates Darwin's commitment to a positive causal adequacy case for natural selection. Before Darwin takes up difficulties, such as complex adaptations, which he says could possibly be due to natural selection, he works to establish under what conditions species will probably, indeed invariably, be formed by natural selection. Kitcher's theory of theories has inclined him, too much, to see Division I as merely preparatory to Division II, this last constituting Darwin's main achievement. He has not seen that Division II is often invoking the positive causal adequacy thesis of Division I. Second, although no aetiophobe, Kitcher never quite acknowledges Darwin's view that selection is explanatory because it is causal. Kitcher seems here still under Hempel's positivistic influence, in his unwillingness to make aetiological, rather than or as well as nomological, unification central to his account of explanatory unification. Third, Kitcher is largely mistaken in his claim that Darwin's achievement included – perhaps, for Kitcher, above all else – bringing the problem of the origin of species for the first time into science from theology. For a start, one needs to consider Lyell's explicit stand against those like Humboldt (no strong theist, by the way) who held that the origins of new species presented a mystery that lay beyond science (Hodge 1982). More generally, one needs to look at the discussions given of these issues by a writer such as Baden Powell (1855) writing in the wake of Robert Chambers' *Vestiges of the natural history of creation* (1844). Beyond that one needs to consider writers such as Buffon, in the previous century, who were not obviously excluding the explanation of species origins from science.

It is true that the problem of species origins, as Darwin found and engaged it in Lyell, was not seen as separable from theology. But then it is arguable that Darwin himself never separated it entirely from theology. For atheists there is no theology, granted. But for many theistic scientists – Darwin himself included, in the 1830s especially, but also still in the 1860s – the topic has remained partly a theological one. Indeed, the whole question of when and how theology and science became separated, institutionally as well as intellectually, is far from easy to assess and resolve, as a question of history. Fortunately, a critique of Kitcher's theory of scientific theories, as illuminating the Darwin case, does not need to take up this question, for Kitcher's unsatisfactory approach to the question is independent of his metatheoretical program. It stems more, one suspects from his campaigns (1982) against scientific creationism in our time, and from time spent at Harvard rather than at Princeton.

7. Epilogue: structure and function in scientific theories

One of Kitcher's most general metatheoretical theses (1985a) is a thesis he finds supported by two very dissimilar case studies: the development of the Leibniz and Newton differential calculus, and Darwin's natural selection. The thesis is that theories can be strong in two ways that may not covary. The calculus was, from the start, strong in axiomatisability; it could be given from birth an axiomatic structuring. It was also strong as a source of arguments applicable to geometrical, kinetic and dynamic problem solving. By contrast, Darwin's theory had low axiomatisability, but high explanatory promise, in that many unifying explanatory arguments could be drawn from it.

A historian cannot supply a verdict on the philosophical correctness and value of this thesis of Kitcher's. However, a historian may judge of its historiographical fruitfulness. Ever since Aristotle's *Posterior Analytics*, of course, if not before, there have been discussions of how the structural characteristics of explanatory arguments are related to their explanatory functioning. However, such concerns have not always been in the ascendant. As the Reidian *vera causa* tradition illustrates, in the eighteenth century, epistemological concerns about scientific theories were pursued often without sustained concern with structural issues. Familiarly enough, structural issues have come back to take the center of the philosophical stage in our own century, in the wake most obviously of the revival of formal logic late in the last century and the rise of logicism in the philosophy of mathematics.

It should not surprise us, then, when we look over the long run of history, to find that a theory, such as Darwin's, that draws its epistemic and systemic ideals from the eighteenth century, does not have a structure that is easily explicable by comparison and contrast with twentieth-century structural ideals. Moreover, and here is a stronger suggestion to end with, it may be worth giving a history a further role beyond that of merely taking the surprise out of such a finding. For, after all, Darwin's theory has a direct neo-Darwinian descendant in the synthetic theory that came of age in the middle of the twentieth century. The epistemic and systemic ideals of that theory may, then, be appropriately interpreted as continuous with the old eighteenth-century ideals (Hodge 1987). Perhaps, therefore, in so far as the synthetic theory is a canonical example of twentieth-century science, we should demand of a twentieth-century philosophical theory of scientific theories that it find a place for the concerns expressed in the articulation of those old ideals, no less than for the structural concerns so characteristic of our later age.

Notes

1. For a strong affirmation of Herschelian and Lyellian commitments, written in late October 1838, see Darwin (1839), pp. 615–625. For this dating, *see* Darwin (1986), p. 432.
2. The parenthetical parts of the following passage from a letter of Darwin (May 8, 1860) to Henslow are relevant here (Barlow 1967, p. 204): '... he [Adam Sedgwick] talks much about my departing from the spirit of inductive philosophy. – I wish, if you ever talk on subject to him, you would ask him whether it was not allowable (and a great step) to invent the undulatory theory of light – i.e. hypothetical undulations, in a hypothetical substance, the ether. And if this be so, why may I not invent a hypothesis of natural selection (which from analogy of domestic productions, and from what we know of the struggle for existence and of the variability of organic beings, is, in some slight degree, in itself probable) and try whether this hypothesis of natural selection does not explain (as I think it does) a large number of facts in geographical distribution – geological succession – classification – Morphology – embryology etc. etc. – I should really much like to know why such an hypothesis as the undulations of the either may be invented, and why I may not invent (not that I did *invent* it, for I was led to it by studying domestic varieties) any hypothesis, such as natural selection.'
3. Thus Lloyd (1983, p. 112) quotes the last three sentences only of the following postscript in a letter (22 May 1863) from Darwin to George Bentham (Darwin 1888, vol. 3, pp. 24–5): 'P.S. – In fact the belief in Natural Selection must at present be grounded on general considerations. (1) On its being a *vera causa*, from the struggle for existence; and the certain geological fact that species do somehow change. (2) From the analogy of change under domestication by man's selection (3) And chiefly from this view connecting under an intelligible pont of view a host of facts. When we descend to details, we can prove that no one species has changed [i.e. we cannot prove of any particular species that it has changed]; nor can we prove that the supposed changes are beneficial, which is the groundwork of the theory. Nor can we explain why some species have changed and others have not.'
4. Interestingly, Ian Hacking (1983, p. 277) reports:
   In an at present unpublished discussion note that I have just seen, Bas van Fraassen claims that causalism has its roots in Newton's search for *vera causa* (true causes) [sic] combined with the famous assertion, *hypotheses non fingo* (I do not make, or depend upon, hypotheses).

References

Barlow, N. (ed.) (1967). *Darwin and Henslow. The Growth of an Idea* London: John Murray.

Butts, R. (1970). Whewell on Newton's rules of philosophizing. In R.E. Butts and Davis, J.W. (eds), *The Methodological Heritage of Newton* pp. 132–149. Toronto: University of Toronto Press.

Butts, R. (1973). Reply to David Wilson: Was Whewell interested in true causes? *Philosophy of Science* 40: 125–128.

Curtis, R. (1986). Are methodologies theories of scientific rationality? *British Journal for the Philosophy of Science* 37: 135–161.

Curtis, R. (1987). Darwin as an epistemologist. *Annals of Science* 44: 379–408.

Darwin, C. (1839). *Journal of Researches into the Geology and Natural History of the various countries visited by H.M.S. Beagle.* London: Henry Colburn.

Darwin, C. (1859). *On the Origin of Species.* London: John Murray. Reprinted by Harvard University Press, 1975.

Darwin, C. (1888). *The Life and Letters of Charles Darwin.* (Edited by F. Darwin, 3 vols.) London: John Murray.

Darwin, C. (1986). *The Correspondence of Charles Darwin.* Volume 2: 1837–1843. Edited by F. Burkhardt and S. Smith. Cambridge: Cambridge University Press.

Ghiselin, M. (1969). *The Triumph of the Darwinian Method.* Berkeley: The University of California Press.

Giere, R.N. (1983). Testing theoretical hypotheses. In J. Earman (ed.), *Minnesota studies in the Philosophy of Science* 10: 269–298.

Good, G. (1987). John Herschel's optical researches and the development of his ideas on method and causality. *Studies in History and Philosophy of Science* 18: 1–41.

Hacking, I. (1983). *Representing and Intervening.* Cambridge: Cambridge University Press.

Herschel, J. (1830). *Preliminary Discourse on the Study of Natural Philosophy.* London: Longman.

Hodge, M. (1977). The structure and strategy of Darwin's 'long argument'. *British Journal for the History of Science* 10: 237–245.

Hodge, M. (1982). Darwin and the laws of the animate part of the terrestrial system (1835–1837): on the Lyellian origins of his zoonomical explanatory program. *Studies in the History of Biology* 7: 1–106.

Hodge, M. (1987). Natural selection as a causal, empirical and probabilistic theory. In L. Krüger, G. Gigerenzer and M. Morgan (eds), *The Probabilistic Revolution.* (2 vols,) Vol. 2, pp. 233–270. Cambridge, Mass.: MIT Press.

Hodge, M. (forthcoming). History, science, the earth and life. In M. Fisch and S. Schaffer (eds), *Essays on William Whewell.*

Hull, D.L. (1973a). Charles Darwin and Nineteenth-Century philosophies of science. In R.N. Giere and R.S. Westfall (eds), *Foundations of Scientific Method: The Nineteenth Century*, pp. 115–132. Bloomington, Indiana: Indiana University Press.

Hull, D.L. (1973b). *Darwin and his Critics.* Cambridge, Mass.: Harvard University Press.

Kavaloski, V. (1974). *The vera causa principle: a historico-philosophical study of a metatheoretical concept from Newton through Darwin.* Ph.D. dissertation, University of Chicago.

Kitcher, P. (1981). Explanatory unification. *Philosophy of Science* 48: 507–531.

Kitcher, P. (1982). *Abusing Science.* Cambridge, Mass.: MIT Press.

Kitcher, P. (1983). *The Nature of Mathematical Knowledge.* New York: Oxford University Press.

Kitcher, P. (1985a). Darwin's achievement. In N. Rescher (ed.), *Reason and Rationality in Science*, pp. 127–189. Washington D.C.: University Press of America.

Kitcher, P. (1985b). *Vaulting Ambition. Sociobiology and the Quest for Human Nature.* Cambridge, Mass.: MIT Press.

182

Laudan, L. (1981). *Science and Hypothesis. Historical Essays on Scientific Methodology*. Dordrecht: Reidel Publishing.

Laudan R. (1982). The role of methodology in Lyell's geology. *Studies in History and Philosophy of Science* 13: 215–250.

Laudan, R. (1987). *From Mineralogy to Geology. The Foundations of a Science, 1650–1830*. Chicago: The University of Chicago Press.

Lloyd, E. (1983). The nature of Darwin's support for the theory of natural selection. *Philosophy of Science* 50: 112–129.

Lloyd, E. (1984). A semantic approach to population genetics. *Philosophy of Science* 51: 242–264.

Lyell, C. (1830–1833). *The Principles of Geology*. (3 vols.) London: John Murray.

Powell, B. (1855). *Essays on the Spirit of the Inductive Philosophy, the Unity of Worlds and the Philosophy of Creation*. London: Longman.

Recker, D. (1987). Causal efficacy: the structure of Darwin's argument strategy in the *Origin of Species. Philosophy of Science* 54: 147–175.

Ruse, M. (1971). Natural selection in the *Origin of Species. Studies in History and Philosophy of Science* 1: 311–351.

Ruse, M. (1975). Charles Darwin's theory of evolution: an analysis. *Journal of the History of Biology* 8: 219–241.

Ruse, M. (1976). Charles Lyell and the philosophers of science. *British Journal for the History of Science* 9: 121–131.

Ruse, M. (1979). *The Darwinian Revolution: science red in tooth and claw*. Chicago: University of Chicago Press.

Suppe, F. (ed.) (1974). *The Structure of Scientific Theories*. Urbana: University of Illinois Press.

Thagard, P. (1978). The best explanation: criteria for theory choice. *Journal of Philosophy* 75: 76–92.

van Fraassen, B.C. (1980). *The Scientific Image*. Oxford: Clarendon Press.

Waters, C.K. (1986). Taking analogical inference seriously: Darwin's argument from artificial selection. In A. Fine and P. Kitcher (eds) *PSA 1986* 1: 502–513. East Lansing, Mich.: The Philosophy of Science Association.

Whewell, W. (1837). *History of the Inductive Sciences*. (3 vols) London: Parker.

Whewell, W. (1840). *Philosophy of the Inductive Sciences* (2 vols) London: Parker.

X

DISCUSSION:

DARWIN'S ARGUMENT IN THE *ORIGIN**

There are now in the literature several sustained attempts to show that Darwin's argumentation in the *Origin of Species* ([1859] 1964) fits and so vindicates some general philosophical proposal concerning scientific theorizing. There is, however, no consensus about that argumentation nor about the philosophical proposals. Most notably, Ruse (1975) has written on Darwin and the "received view" of the logical empiricists, Thagard (1978) on Darwin and inference to the best explanation, Lloyd (1983) on Darwin and the "semantic view", and Kitcher (1985) on Darwin and explanatory unification.

Recker (1987) has examined these writings and concluded that they all fail in fitting Darwin exactly to their favored general proposal. Hodge (1989) reached the same conclusion. Even more recently, Sintonen (1990), disagreeing with Recker, has claimed that a certain version of the semantic view does fit and is vindicated by the *Origin*, while Wilson (1992) has presented Darwin's theory as having the logical structure and observational evidence of a paradigm in Kuhn's sense.

The purpose of the present discussion is to explain why Sintonen's claim cannot be accepted, and, much more broadly, to indicate some conditions that any such claim must meet if it is to be historically accurate and philosophically cogent.

*Received May 1991.

Ultimately, the reason why Sintonen's claim must be declined is that it is grounded in a mistaken analysis of Darwin's argumentation in the *Origin*. Recker had urged—also mistakenly as it turns out (Hodge 1989)—that throughout the *Origin* Darwin is always arguing for the causal efficacy of natural selection, but that Darwin deploys three distinct argumentational strategies on behalf of this single causal efficacy thesis. Sintonen (see, especially, 1990, 689), disagreeing, holds, instead, that Darwin has two arguments: a short one to establish the *existence* of natural selection, and a much longer one to establish its *explanatory sufficiency* for a wide range of phenomena.

Although Sintonen gets closer to a correct analysis of Darwin's argumentation than Recker, his two-argument analysis is still incorrect. His mistake has arisen, as did Recker's, mostly from not seeing how Darwin conformed the structure of his argumentation to the *vera causa* (true cause) ideal. Both Recker and Sintonen give extensive discussions of this ideal as it was discussed by such authors as Herschel, Whewell and Mill in the 1830s and 1840s, but neither has seen (nor has Wilson) that in conforming his argumentation to that ideal, Darwin knowingly constructed *three* distinct, evidential cases, three component arguments, on behalf of natural selection: first, a case for its *existence* as a causal process going on in the world; second, a case for its *adequacy*, its competence to produce, adapt and diversify species; and, third, a case for its *responsibility*, for, that is, its having produced the species now living and the extinct species found as fossils. So, in sum, natural selection exists, it can have that sort and size of effect, and it has indeed formed the species that have originated so far.

The drift of these three arguments—or, better, evidential cases within Darwin's "one long argument"—can be recalled briefly. The first and second arguments are contained in the first eight chapters of the *Origin*, in its first edition. As for *existence*, species in the wild are subject to changes in conditions of life, and domesticated species show that any animals and plants exposed to changed conditions vary heritably. There is hereditary variation in the wild, then. There is also a struggle for life, for there is superfecundity and this entails a struggle for food, space and other limited requirements for life. In this struggle for life, there is a differential survival and reproduction of hereditary variants, for some hereditary differences affect chances of survival and reproduction. There exists in nature, therefore, a process of selective breeding, a process analogous to the selective breeding practiced by farmers and gardeners. As for *adequacy*, artificial selective breeding is known to be sufficient, competent, or adequate to produce, within a species, distinct races—of dogs, for example—adapted to distinct human ends. These races do not count, according to customary criteria, as distinct species. But natural selection has vastly longer to work and is much more comprehensive and discriminating. So, it can then produce races that would count as species, for they would be more permanently and more perfectly adapted and divergent in their organization, and hence infertile with one another, true breeding in their characteristics and without intermediate varieties. (There is more to Darwin's adequacy case, obviously, but this is the primary line of reasoning.)

The third argument is contained in the next five chapters (the ninth, that is, through the thirteenth). As for *responsibility*, the theory that natural selection has been the main agency responsible for bringing into being the living and extinct species is more probable, and so is to be preferred over any rival theory because it is better than any other at explaining several kinds or classes of facts about those species: biogeographical facts, embryological facts and so on.

Since I have set out elsewhere (Hodge 1977, 1987, 1989, 1991a and 1991b) the textual grounds and other advantages of this three-case, or three-component argument, explication of the one long argument of the *Origin*, and since I have indicated there, too, what can be learned from the growing, specialist literature on such topics in the history of the philosophy of science as the *vera causa* ideal, let me allude here only to some leading reasons for seeing this explication as historically instructive and philosophically suggestive.

As the papers just cited attempt to show, Darwin's argumentation is in keeping with Herschel's and Lyell's endorsement of this ideal as an appropriate ideal for geological science; his argumentation is, therefore, not in keeping with Whewell's rejection of the appropriateness of this ideal for that or any other science, nor, therefore, with Whewell's proposed replacement for this ideal in natural science, generally—namely, the consilience of inductions. More remotely, then, Darwin's commitment to this ideal is in descent from Reid's elaboration of what he took to be the implications of Newton's dicta about true and sufficient causes.

So much here for history. Philosophically, it is surely worth asking whether current theorizing about evolution by natural selection still takes the same form that Darwin gave his theorizing in 1859. For, as one of those papers (Hodge 1987) just cited urges, despite mathematical, Mendelian and molecular developments in this century, much the same structure of enquiry persists. There are questions about the definition of natural selection, and, beyond those definitional questions, there are empirical questions about its occurrence and prevalence (i.e., existence), about its consequences and capacities (i.e., what it suffices to effect), and about its past achievements (i.e., for what it has been responsible). These three clusters of empirical questions descend directly from the old *vera causa* evidential and explanatory concerns.

It would seem, therefore, that if the analysis of Darwin's argumentation hinted at here is correct, then it imposes constraints on two genres of task. First, anyone seeking to throw light on Darwin's argumentation—whether by means of semanticist, Kuhnian, or any other type of general philosophical proposal—must show how light is thrown on the three-case, or three-component-argument, structuring of that argumentation (see Hull 1989, 319). Second, anyone seeking to capture the form taken by current theorizing about evolution should be able to show how a similar structuring of evidential and explanatory concerns continues to be involved in establishing the empirical status of the theory of evolution by natural selection when this is construed, as it surely has to be, as a probabilistic causal theory.

REFERENCES

Darwin, C. ([1859] 1964), *On the Origin of Species*. Cambridge, MA: Harvard University Press.

Hodge, M. (1977), "The Structure and Strategy of Darwin's 'Long Argument'", *British Journal for the History of Science 10*: 237–246.

———. (1987), "Natural Selection as a Causal, Empirical and Probabilistic Theory", in L. Krüger (ed.), *The Probabilistic Revolution*, vol. 2. Cambridge, MA: MIT Press, pp. 233–270.

———. (1989), "Darwin's Theory and Darwin's Argument", in M. Ruse (ed.), *What the Philosophy of Biology Is: Essays Dedicated to David Hull*. Dordrecht: Kluwer, pp. 163–182.

———. (1991a), "Discussion Note: Darwin, Whewell, and Natural Selection", *Biology and Philosophy 6*:457–460.

———. (1991b), "The History of the Earth, Life and Man: Whewell and Palaetiological Science", in M. Fisch and S. Schaffer (eds.), *William Whewell: A Composite Portrait*. Oxford: Oxford University Press, pp. 255–288.

Hull, D. (1989), "A Function for Actual Examples in Philosophy of Science", in M. Ruse (ed.), *What the Philosophy of Biology Is: Essays Dedicated to David Hull*. Dordrecht: Kluwer, pp. 309–321.

Kitcher, P. (1985), "Darwin's Achievement", in N. Rescher (ed.), *Reason and Rationality in Natural Science*. Lanham, MD: University Press of America, pp. 127–189.

Lloyd, E. (1983), "The Nature of Darwin's Support for the Theory of Natural Selection", *Philosophy of Science 50*: 112–129.

Recker, D. (1987), "Causal Efficacy: The Structure of Darwin's Argument Strategy in the *Origin of Species*", *Philosophy of Science 54*: 147–175.

Ruse, M. (1975), "Charles Darwin's Theory of Evolution: An Analysis", *Journal of the History of Biology 8*: 219–241.

Sintonen, M. (1990), "Discussion: Darwin's Long and Short Arguments", *Philosophy of Science 57*: 677–689.

Thagard, P. (1978), "The Best Explanation: Criteria for Theory Choice", *The Journal of Philosophy 75*: 76–92.

Wilson, F. (1992), *Empiricism and Darwin's Science*. Dordrecht: Kluwer. In press.

XI

Knowing about Evolution

Darwin and His Theory of Natural Selection

EVOLUTION AND COGNITION

On the topical overlap between evolution and cognition, there are various clusters of questions one might distinguish. Here are two such clusters. First, one might ask how knowledge of evolution can be achieved. Humans, after all, have been accumulating reliable records about observed changes in animal and plant species for only a few thousand years, while evolution has taken place over tens, hundreds, even thousands of millions of years. The human period, wherein alone direct experiential access to evolution has been possible, is a tiny moment suspended between a much vaster past and, it is hoped, an indefinitely prolonged future. Questions can, then, be raised about how anything can be learned about evolution given these severe experiential limitations. Second, one can ask what insights about knowing itself can be gained from considering our mental faculties as products of this evolutionary process. Perhaps less confidence, or perhaps more, should be put in those faculties if they are thought to be legacies from ancestral, animal adaptations, rather than supernatural gifts from an omniscient God who has made man in His own image.

Currently there are some customary divisions of labor regarding these two clusters of questions. Biologists are more likely to address the first, psychologists the second. Only philosophers are likely to feel responsible for taking them all on in an integrated way. Typically, those divisions of labor were less sharp in the nineteenth century than in our own time; and it is striking how the young Darwin, most explicitly in his notebooks from mid-1837 to mid-1839, ranges freely across both clusters. Not that this ranging is surprising, when one appreciates that he was, from mid-1838 on, keeping two distinct but allied series of notebooks. One series (B–E) may be called the zoonomical notebooks, as the first notebook carries the title *Zoonomia* (meaning the laws of life), while the other series (M–N) is devoted to "metaphysics" (meaning not, as it did originally, the theory of being, but rather what it had come to mean in the eighteenth century: namely, the theory of mind, including reason, will, consciousness, habits, the moral sense, the social instincts, and so on, in man and other animals).[1]

More precisely, Darwin's concern with both clusters of questions reached a peak in the summer and autumn months of 1838. But there is a difference. He had, as a geological disciple of Charles Lyell, been for some six years engaging questions about how we, although able to observe only the present, can have scientific theories that count as knowledge, not as mere conjecture, about the vast prehuman past recorded in the rocks. By contrast, his concern with the respective roles of innate instincts and of learned habits in the acquisition of knowledge, and so in the progress of science, was a more recent preoccupation. It had arisen as he had come to novel conclusions about how new habits could initiate adaptive changes in bodily structures.[2]

While acknowledging this difference, however, it is well worth focusing on those months as an exceptionally revealing phase in Darwin's intellectual development. For it is obviously instructive to examine how he had come to reach the views he held at that time; while it is no less instructive to reflect on how he went on from there, in the immediate and in the more distant future. It is instructive not least because it was very soon afterwards, in the winter of 1838–9, that he came to his theory of natural selection in its first full formulation, and also because the public, published Darwin of 1859 and thereafter turns out not to have changed his mind to any serious extent on either of those two clusters of questions about evolution and cognition, to use this anachronistic phrasing once more; nor, indeed, would he ever change his mind on these matters.

AGREEING WITH LYELL AND HERSCHEL

During the *Beagle* years, the most consequential commitment Darwin made was to become a zealous disciple, as he put it, of Lyell's views in geology. The way Lyell's views were structured in the three vol-

Knowing about Evolution

umes of his *Principles of Geology* (1830–33) ensured that Darwin could not make that commitment without assimilating Lyell's own epistemological and methodological self-consciousness. This assimilation was aided and abetted by Darwin's admiration for John Herschel, Lyell's friend and Britain's leading astronomer and physicist. Darwin had read Herschel's *Preliminary Discourse on the Study of Natural Philosophy* (1830) before leaving on the voyage. He may well have been impressed then by Herschel's endorsement of Lyell's views. In the closing months of the voyage he met Herschel, now in South Africa for astronomical purposes; and they discussed geology as two knowing converts to Lyell's campaign.[3]

At the core of the consensus between Lyell and Herschel was the claim that geology could and should be, like celestial mechanics, a science that explained phenomena by reference to *verae causae* – that is, true, real, known, existing rather than hypothetical, conjectured, supposed causes. The consensus covered, moreover, the most contentious thesis of the *Principles:* namely, that this explanatory ideal could be satisfied only if very strong presumptions were made about the causes at work on the earth's surface throughout the temporal domain of geology, that is, throughout the vast aeons from the time when the oldest known fossil-bearing rocks were laid down through the present time and on into the future. The same causes, it was to be presumed, had been at work and with the same intensities and in the same overall circumstances, so that they have had and will continue to have the same sorts and sizes of effects. Only on this presumption could geology be a science of true causes, Lyell held. For a true cause, as Lyell and Herschel agreed, was one whose existence is evidenced independently of the facts it is invoked to explain. Here, they followed such explicators of Newtonian ideals of evidence and explanation as Thomas Reid in the previous century. The Cartesian vortices were adequate causes for planetary orbits. If they existed, then they would suffice to cause and so explain the orbits. But the trouble with the vortices was that the orbits themselves were the only facts that could be cited as evidence for the existence of these vortices. By contrast, Newton's gravitational force was evidenced by terrestrial phenomena, falling and swinging bodies, providing independent evidence distinct from the orbits in the heavens that it was invoked to explain. On the presumption of a uniformity of causes and laws, between the fallings and swingings down here and orbitings up there, the gravitational force could then be invoked as a true cause for those orbits.

It should be likewise in geology, Lyell and Herschel agreed. In this science, there should be a presumption of uniformity, not across space, from low to high, but across time, from the present that humans can observe to the vast prehuman, unobserved past that they were not present to observe. Geologists should refer the ancient changes recorded in the rocks to causes still active today and still adequate to produce, albeit often only over the long future ages, the same kinds and scales of effects. As their existence in the present can be confirmed independently of their action in the past, they are true causes, causes whose existence can be evidenced independently of their putative responsibility for the effects they are invoked to explain.[4]

Darwin's geological theorizing, from the middle of the voyage on, shows him to have knowingly embraced this *vera causa* evidential and explanatory ideal and to be knowingly conforming to it his theorizing on a whole array of phenomena: the extinction of large mammals in South America, the formation of coral islands, the distribution of erratic boulders, to name but a few. Nor would he ever abandon this ideal. Indeed, it would be tempting, at this point, simply to pass to his argumentation in the *Origin of Species* (1859) and to show how its structuring and so the composition of the book itself were deliberately conformed to the ideal. For, there, successive independent evidential cases are made for natural selection existing at present; for natural selection being adequate to produce over long ages new species from old; and for natural selection having been responsible for the production of those species now living and for those species that lived formerly and that have since become extinct.[5]

However, to pass directly from the private geological theorizing of the *Beagle* (1831–6) and post-*Beagle* London (1836–42) years, to the published biology of 1859, would be to miss out on the many intriguing biographical, historiographical, bibliographical, exegetical, and interpretative complexities that we are forced to confront if we concentrate, even if briefly, on the zoonomical and metaphysical inquiries as Darwin pursued and reflected upon them in those months of exceptional self-consciousness in the summer and autumn of 1838,

Knowing about Evolution

just before the theory of natural selection was first conceived (not, as legend and Darwin's later memories have it, suddenly in September or October 1838, but over several weeks in very late November and December).

DISAGREEING WITH LYELL

One recurrent challenge Darwin faced, before and after the voyage, arose because he was knowingly disagreeing with Lyell about several successive theoretical issues while being unwilling to break with Lyell's most general views about the presumptions and ideals required for geology to hold its place as a high-ranking science matching celestial mechanics in its epistemological and methodological credentials. Most conspicuously, in 1835 he rejected Lyell's theory about the causes for species extinctions (competitive defeats occasioned by ecological disruptions) in favor of another theory (whereby, Darwin thought, species senesce and die of old age like individuals); this was a theory that Lyell had discussed but criticized as not having evidential support appropriate to a *vera causa;* and so Darwin deliberately sought such support, most particularly in accepted generalizations about the known senescence of plant graft successions.

In 1836 and even more so in 1837, however, he made two further breaks with Lyell that raised the challenge set by the *vera causa* ideal in a much more extensive way. In 1836 he was first tentatively inclined to suppose that new species were not as Lyell had held, independently created and too fixed in their characters to diverge into descendant species; for Darwin decided in favor of some species at least arising in the transmutation of others. In dismissing and rejecting such transmutations, Lyell had insisted that anyone who was considering it seriously should consider too the entire theoretical system of Jean Lamarck, transmutation's most notorious advocate. Darwin, in the spring of 1837, now completely convinced of transmutation following his return to England and decisive new expert judgments on his extant bird specimens and extinct fossil mammal remains, did precisely that. For he decided to take Lamarck's side against Lyell and to elaborate a comprehensive systemic structure of theorizing, incorporating the transmutation of species along with the spontaneous generation of infusorian monads, and including man in the progressive development from those earliest, simplest beginnings. It was a system deliberately designed to match the structure of Lamarck's system as represented in Lyell's exposition (rather than in Larmarck's own very different one). It is this new system that opens Darwin's Notebook B, under the title *Zoonomia* (borrowed from his own, transmutationist grandfather's most famous book title) in July 1837.

Darwin had then done, systemically, what Lyell had held any convinced transmutationist should do. However, Darwin clearly accepted that several of his system's components were far from *vera causa* kosher, so that the system as a whole was epistemologically and methodologically imperfect.

Darwin's awareness of its imperfection is registered in his very phrasing. As he sketches the system in the first two dozen pages of the notebook, he says sometimes that "we see" or that "we know" that something generally is the case; but he says of other generalizations that we "suppose" them to be so. Moreover, as he works to improve his system in the following days, he is plainly attempting as far as he can to replace supposing with seeing, conjecturing with knowing. For he works to find which suppositions (about monads, for instance) may have to be rejected because they have false consequences; which other suppositions have hitherto undiscerned explanatory advantages; and for which can be found evidential support that is independent of their explanatory advantages.

In and of themselves, these ways of improving any system of theory were commonplace and customary enough, even banal, one might have said. But that is the point. Darwin, while knowingly pursuing a subversive program of inquiry, was concerned to see how far he could make his evidential and explanatory argumentation conform to accepted standards.[6]

A PROSPECTIVE PROJECT AND "MY THEORY" (1837–8)

In subsequent weeks and months Darwin did not develop the system sketched in July 1837–did not, that is, develop it as such, as a whole system. Instead, he came to articulate two goals that emerged from improvements made to components of this system. One goal was the theory of species propagation, to use his own term. The successive, reiterated production of new species from old was likened to the growth by budding and branching of a tree. It was a tree with many twigs dying (representing species extinctions) while others are dividing (representing species multiplyings and divergings). The clusterings of twigs and the gaps left by dead ones represented the lesser differences among species of distinct genera, or the greater differences among species of distinct orders or classes. In this arboriform process, species formations were credited to geographical separations and so isolations, with adaptive divergence made possible by sexual reproduction with inbreeding in changed conditions. These, as Darwin was arguing in 1837, were the circumstances that allowed for one species to give rise to two or more distinct descendent species. His account of these circumstances, and his arguments for their existence and efficacy, constituted "my theory" of species formation.

As for the promissory project, it should be emphasized that this project was only ever projected. It was talked about as a prospect, but it was never pursued, let alone completed. Moreover, it would be quietly abandoned by Darwin at the very time, and not coincidentally, when the theory of natural selection became the new "my theory" in the winter of 1838–9.

The promissory project was to make use above all of certain biogeographical facts. For Darwin promised himself that he would assemble instances of geographical series of congeneric species, that is, several congeneric species spread out with no spatial gaps between them, but remaining distinct in their characters because not interbreeding and adapted to slightly different local conditions.

Thanks to the commitment to species transmutations, such geographical series could be taken to be the result of a temporal succession of changes in species. What the project would do was, therefore, to assemble such instances of species change, so that generalizations could be inferred; so that, in Darwin's words, the laws of change could be established. Finally, inquiry would proceed into the causes of change, the causes lawfully responsible for these lawful changes. Now the structure and strategy of this promissory project, as Darwin returns to it time and again over more than a year, manifestly combined Darwin's loyalty to Lyellian geology with his assimilation of a standard view of what a successful science, like Newtonian physics, most obviously comprises. Lyell, as always moving from the accessible present to the inaccessible past, had often used horizontal, geographical inquiries as a way to reach vertical, geological conclusions. His long chapters on the geographical distribution of animals and plant species had been explicitly designed on those lines. Most particularly, he had studied what limited species in their spatial extent at present, in order to decide what had limited their temporal durations over aeons of the past. Biogeography (including competitive, ecological relations) was to indicate the causes of species extinctions. Darwin's project was likewise designed to move from knowable geographical facts to conclusions, otherwise unknowable, about the laws and causes of change over time. It would, then, exemplify, extend, and indeed vindicate Lyell's reform of geology so as to satisfy the *vera causa* ideal. Equally, in constructing a three-layered pyramid with lawful causes at the top, laws of change in the middle, and individual facts about change at the bottom, it would match those successes in the physical sciences where, for instance, planetary orbital facts were referred to planetary orbital laws (such as Kepler's), which were in turn explained by reference to causal laws (such as Newton's laws of force).[7] However, this promissory project remained just that and so was developed no further by Darwin.

A THEORY AND ITS EVIDENTIAL AND EXPLANATORY CREDENTIALS

What Darwin does concentrate on in the spring of 1838 is developing his theory of adaptive species formations. He retains his invocations of the special powers of sexual (as opposed to asexual) modes of reproduction, of geographical isolation, and of inbreeding in changed conditions. But he adds now a new understanding of how reproductive isolation can emerge. He eventually decides that an analogy with races within domestic species indicates that conspecific varieties will in time acquire an instinctive aversion to interbreeding. The analogy depends not on a comparison so much as on a contrast between wild species and domestic races. For Darwin is convinced that domestication itself vitiates such instincts. He is prepared to argue, therefore, that, by contrast, when marked and pro-

Knowing about Evolution

longed varietal divergence of bodily structure takes place in nature, divergence as marked as the divergence shown by domestic dog races, it would be accompanied by an instinctive aversion to inter-breeding. This reproductive isolation would then allow further structural divergence to proceed until interbreeding was no longer possible even if the instincts were not preventing it. Associated with this whole line of argument were two others. First, reports on the results of crossing an old breed with a more recent one in dogs, say, suggested that characters become more strongly inherited and less likely to blend on crossing as time goes on: Yarrell's law, as Darwin dubs it after the name of his informant. Second, as with reproduction, so quite generally, Darwin thinks, changes in structure follow changes in instinct which in turn follow changes in habits: so, conversely, new habits, on becoming inherited and instinctive, lead through their new inherited effects on bodily structures to changed structures.[8]

These additions to his theory of species formation do not make it, in Darwin's judgment, completely secure evidentially. Especially, he notes that the most hypothetical part of the theory is its assumption that, in the wild, varieties of long standing will eventually cease to interbreed. He acknowledges that the best he can do on behalf of this assumption is to argue, analogically, from the evidence that, were their instincts not vitiated by domestication, then races of domesticated species would not readily breed together. Indeed, he thinks this line of proof for transmutation in the wild is one of his most original insights.[9]

During these summer months of 1838, Darwin often reflects on the epistemological and methodological strengths and weaknesses of his theorizing. The reflections invoke the standard staples of the day. It is, he notes, a virtue in a theory if it connects many otherwise disparate phenomena; it is good if a theory allows for successful predictions; and purely hypothetical conjectures should be replaced wherever possible. By themselves, these remarks would hardly constitute a comprehensive and consequential interpretation for Darwin to be giving his theoretical insights. But he does indeed integrate them into a coherent stance that he is explicitly resolving to act upon in composing a book he is already contemplating, a book on his theory of species propagations. The resolution is, moreover, twofold, in that his theory is seen as having two distinct virtues. First, there is evidence for the theory itself, independent of its explanatory virtue. Included in this evidence for the theory itself will be, for example, the support it has from those inquiries that compare and contrast species in the wild with races under domestication. Second, the explanatory virtue is to be displayed by showing how many facts of many kinds, from comparative anatomy, biogeography, palaeontology, and so on, can be given unifying, connecting explanations by referring them to the theory of species propagations as elaborated in the arbori-form representation of species multiplications, divergences, and extinctions.[10]

Here, then, in the summer of 1838, well before he has his theory of natural selection,Darwin is showing himself already committed to the twofold view that will eventually condition how the *Origin of Species* will be composed in 1859. For that book will be a rewriting of Darwin's unpublished "Essay" of 1844, itself an expansion of his unpublished "Sketch" of 1842; and the "Sketch" and "Essay" are explicitly divided, as the *Origin* is less explicitly, into two principal divisions: a first division setting out the evidence for natural selec-tion, that is, for its existence and its ability to produce and diversify species; and a second division arguing that many facts of many kinds, from biogeography, comparative anatomy, and so on, are best explained by that theory. These facts thus indicate that natural selec-tion was most probably responsible for producing the extant and the extinct species.[11]

The summer of 1838 is also a time when Darwin, who opened his metaphysics (M) notebook in July of that year, reflects more ex-plicitly than ever before or since on large issues of philosophy. He reaffirms his theism very conspicuously, but insists repeatedly that God acts in nature through regular laws and causes and not in excep-tional, miraculous interventions. Darwin declares himself a material-ist, insofar as that means that the workings of the mind are lawful, caused consequences of brain structures. What is more, he declares himself a determinist. There may be the appearance of material events happening by chance rather than in accord with universal laws, but the appearance is misleading, he holds; events ascribed to chance must be assumed to be due to hidden actions of regular causes. So too with free will; there is the illusion that necessitating

Knowing about Evolution

causes are absent, but they should be assumed to be present, albeit hidden from our scrutiny.

Darwin's theism, materialism, and determinism were brought together in the conclusion that our possession of the very idea of the Deity is an inevitable consequence of the brain's organization and more remotely, therefore, of the laws responsible for that organization, laws instituted ultimately by God himself.[12] Accordingly, Darwin responded to William Whewell's *History of the Inductive Sciences* (3 vols, 1837) in one way that Whewell himself would never have welcomed. Having looked at it only briefly, it seems, the year before, on first acquiring it, Darwin now studied Whewell's book carefully in the late summer, through the autumn and almost up to Christmas 1838. Following Kant, Whewell had insisted that some very general principles, including those fundamental to science such as the principle of causation itself – the principle that every event has a cause–are presuppositions brought to the interpretation of experience, rather than conclusions drawn from experience. They are necessary and a priori rather than contingent and a posteriori. Darwin subsumed this view within his conviction that what is acquired as a habit can become an innate instinct by becoming hereditary. So principles that are necessary and a priori for us now were first encountered as general facts known a posteriori by our animal or even plant ancestors. In developing this account of human cognition, however, Darwin never considered that his own particular theories about plants or animals or coral islands were necessary or *a priori*. They continued to be empirically evidenced accounts of the causes producing particular kinds of effects in the living or in the physical world.[13]

Darwin registers various agreements and disagreements with Whewell's three volumes. For example, he likes Whewell's distinction between formal laws (like Kepler's) that introduce no causal powers and physical laws (like Newton's) that do so and can explain formal laws. This distinction fitted well with Darwin's understanding of how his promissory project would move beyond laws of change to causes of change. Whewell, however, made no dent in Darwin's old loyalties when he, Whewell, attacked Lyell's reforming of geology to meet the traditional Reidian *vera causa* ideal. Whewell argued that Lyell was mistaken in thinking that causes acting at present were in relevant ways better known than the causes whose effects the rocks recorded. Even more generally, Whewell argued that no successful sciences had succeeded by referring phenomena to known causes. If *verae causae* were taken, as they traditionally had been, to be known causes, then Lyell's understanding of how geology should secure its credentials as a science was, Whewell insisted, essentially misguided.[14]

Obviously, everything Darwin wrote about geology in subsequent years shows that he was entirely unmoved by Whewell's arguments and that he continued to side with Lyell and Herschel. Strikingly, an addendum to Darwin's *Journal of Researches,* written in November 1838, ends with an explicit reaffirmation of his conviction that to deviate from the Lyellian principles of geology would be to violate the very rules of inductive philosophy.[15]

In later publications, most notably his two-volume *Philosophy of the Inductive Sciences* of 1840, Whewell proposed an alternative to the traditional *vera causa* ideal. He called his new proposal the consilience of inductions. According to this proposal, one gives up any distinction between the evidence for a theory that is independent of its explanatory efficacy and evidence from that efficacy. For, Whewell argued, there can be no independent evidence in the sense usually intended. However, some dependent evidence can be strongly verifying of a theory. For, Whewell proposed, a theory is very unlikely to be false if it explains many facts of many kinds, including facts of a kind or kinds not contemplated when the theory was first conceived. Some writers fail to read Whewell correctly. Whewell knew perfectly well that explaining successfully many facts of many kinds had long been thought a virtue in a theory, as Herschel, for one, had emphasized. What was new in Whewell was the point about kinds of fact not contemplated when the theory was conceived.[16]

Consider next, then, two questions about Darwin. Did he learn about Whewellian consilience of inductions in 1838? And did he ever reject the old *vera causa* ideal and embrace that doctrine instead? As for the first, the answer is straightforward: No, he did not, because (despite the impression some historians have given) that doctrine is not taught in the *History* (1837) and so could not be learned from that source. As for the second question, yes, Darwin may well have read Whewell on consilience in the 1840s following publication of the

Knowing about Evolution

Philosophy. But did Darwin ever embrace and conform his evidential practices to that doctrine? One may fairly doubt it, for two reasons. First, most obviously in the *Origin*, Darwin is still distinguishing evidence for his theory independent of its explanatory efficacy from evidence that is not independent. Second, nowhere in that book or in any other, does Darwin ever tell the reader which kinds of facts were contemplated before and which after the theory was first conceived. So it is not simply that the *Origin*'s structure owes nothing to Whewell's distinctive teachings. In two ways, it was written in conformity with ideals Whewell sought to discredit and replace. The notion that the structure and strategy of the *Origin*'s argumentation is deeply indebted to Whewell's philosophy of science will doubtless continue to appeal to people, even though it simply cannot be reconciled with the great deal that is now known about the persistence of Darwin's adherence to Lyell's and Herschel's views from 1838 through to the writing and publishing of that book.[17]

A NEW THEORY MEETS AN OLD IDEAL

When Darwin came to his theory of natural selection, over a period of several weeks from late November through to early 1839, much was retained that was in the earlier theory of species formation. A decisive shift occurred, however, in the comparing and contrasting of wild species and domesticated races. Contrary to previous contrasts, both were now interpreted as adaptations. For both were explained as adapted by selective breeding. Further, both man's and nature's selection were thought to be able to produce adaptations even when working with chance or accidental variations. The two selections were thought to differ in degree: nature's was far more precise, prolonged, and comprehensive (affecting many characters, that is, rather than a few). So Darwin could say that species formed by nature and races made by man were produced by the same means, albeit very different in degree. A proportionality could then be argued for: nature's selection was to man's as species are to races. The causes are proportioned as the effects are. The powers and effects of man's selection being known, those of natural selection are to be inferred in accord with this proportionality. As for the existence of selection in the wild, a process of selective breeding is entailed by heredity, variation – especially in the changing conditions geology testifies to – and the superfecundity dramatized by Malthus.[18]

Thanks to its new selection analogy the new "my theory" had, for both adaptation and species formation, better *vera causa* credentials than had the earlier theory, and Darwin's notes show him cherishing it on that account.[19]

Where he is less explicit, indeed silent, is on how the new theory relates to the old promissory laws and causes of change project. It may well be that the old project drops away as the new theory comes into prominence, because the new theory is seen to make the old project redundant, even misconstrued. The old project would have used present geography as a foundation for knowledge of the course and causes of past change. The new theory finds another way to move from the present to the past, for it moves, by analogy, by proportion, from what man does with selection in the short run to what nature does in a vastly longer run. The gap between the observationally known present and the unknown past is crossed in a new way. But there is a peculiar feature of natural selection as a cause. It is lawful in that it arises in a regular fashion from the natural powers of heredity, variation, and superfecundity. It has, however, no law of its own; for there is no single, universal law that is to natural selection as the Newtonian inverse square law is to gravitational force. In that sense, natural selection is a cause that has no one law of its action. So the old promissory project's aim to find laws of change is hardly satisfied; it is rather circumvented, insofar as the new theory as to the cause of species changes includes no law for species changes. This lack of a law was, Darwin's later critics would urge, a difficulty. With gravitational theory, the law allows one to deduce what effects the cause will produce in specified circumstances, so that one can then decide whether these predictions are born out in fact. Natural selection does not seem susceptible to confirmation in this way.[20]

Any account of how Darwin's epistemological and methodological commitments were acted upon in the 1860s, especially in his public defenses of his theorizing, would properly require a paper in itself. But it may be worth making some brief comments here. First, any careful exegesis of Darwin's scientific argumentation will show, in my view, that he only ever had one ideal in mind, the one he saw Lyell and Herschel jointly upholding. What differs, then, is how fully,

Knowing about Evolution

in any case, Darwin thought that ideal could be met. He certainly thought the *Origin*'s exposition met it fairly well. But he was emphatic that his evidential case for pangenesis met it far less successfully, most obviously because while heredity, variation, and superfecundity were known, observable powers of plants and animals, the existence of the pangenes was not at all susceptible to anything like direct confirmation. The presentation of pangenesis accordingly does not match the *Origin* very closely at all.

Second, faced with objections to the arguments of the *Origin*, Darwin often appears in print and in private correspondence ready to retreat and even sometimes to abandon the claim that he is meeting the old *vera causa* ideal. For he seems to set aside that ideal as inappropriately demanding and to argue that some very reputable theories in physics, such as the wave theory of light, do not confirm to that standard but are nevertheless widely agreed to be little, perhaps none, the worse for not doing so. Natural selection should be then judged, he appears to suggest, by those other standards that have been invoked in making the case for, say, that optical theory. In making these moves, Darwin puts more emphasis on the virtue of explanatory unification and less on the virtue of evidence independent of that efficacy; and this emphasis takes him nearer to Whewell, further away from Lyell and Herschel.

That Darwin makes some such moves is manifest from the relevant familiar texts. But it is worth reading those texts with two questions in mind. First, how wholeheartedly does Darwin repudiate the appropriateness of the old *vera causa* ideal for his theory of natural selection? Is it not true that he continues to think that a defender of natural selection should hope that people will eventually obtain the evidence that would show natural selection to be a *vera causa?* For Darwin never suggests that he and his allies should not try to make stronger the evidential cases for the existence and adequacy of natural selection, the cases set out in the early chapters of the *Origin,* the chapters preceding those later ones devoted to the theory's explanatory virtue. Now, if Darwin had really come round to Whewell's own rejection of the old *vera causa* ideal in favor of the consilience of inductions, if Darwin had really undergone such a fundamental shift in his understanding of what makes the best scientific theorizing epistemologically and methodologically superior to the rest, then there would be far more dramatic shifts than are to be seen in his beliefs about where the way forward lay in making his theory a permanent, accepted contribution to science.

Finally, when one reviews the full sweep of Darwin's theorizing, whether in geology, biology, or psychology, there is one general conclusion that is surely difficult to resist. In his theorizing he was often, although not always, not merely markedly innovative but radical and subversive, and knowingly and deliberately so. But in his own eyes this theorizing did not require him to be correspondingly innovative about the epistemology and methodology of science. On the contrary, he seems anxious that his new theories be developed and presented within the familiar, reputable constraints satisfied by what his contemporaries count as the best science of the day. A moldbuster yes; but, in his own eyes not a metamoldbuster. Obviously, one could argue that Darwin was mistaken and misguided on this issue of how his scientific theories related to the contemporary philosophical theories about scientific theories. Perhaps, in effect if not in intention, he was truly, or at least has turned out to have been truly, a buster of molds at both levels. That is an exciting claim, but our excitement should not lead us to read back, into Darwin's own life and work, too many of the radical and innovative philosophical lessons that have been drawn from evolutionary biology in the century since Darwin. There is no danger that anyone will ever show Darwin to be a completely dull fellow. The notebooks, if not the published texts, show what a lively, imaginative, even wild and weird mind the man had. But there is no need to represent him as upsetting everything at every level. Nor, conversely, is there any need to make Darwin a canonical authority imposing limitations on our own reflections about evolution and cognition. The uses of Darwin for the purposes of philosophical inspiration can surely coexist peacefully, even fruitfully, with the historian's Darwin.[21]

NOTES

1. These notebooks are now available in Barrett et al. (1987). References here will be by notebook and manuscript page number, as in that edition. For recent accounts of these years in Darwin's life and work see the two recent biographies, which include extensive reference to the specialist

Knowing about Evolution

literature: Desmond and Moore (1992) and Browne (1995). The state of the art in Darwinian scholarship fifteen years ago is accessible in Kohn (1985). A main location for Darwinian studies since then is *The Journal of the History of Biology.*

2. Hodge (1982), Hodge and Kohn (1985), Richards (1987), Curtis (1987).

3. Herschel (1830), Ruse (1975).

4. See Hodge (1977, 1982), Ruse (1975), and Laudan (1987) for fuller accounts of Lyell's views in relation to Herschel's.

5. Hodge (1977, 1982, 1989)

6. For this whole section, see Hodge (1982) and Darwin's Notebook B 1-24.

7. For this section, see Notebooks B and C. For the last general exposition of the promissory project, see Notebook E 51–55.

8. See Notebook C and Hodge and Kohn (1985).

9. See Notebooks C 30, C 176-177, and D 69.

10. Notebooks B 104, C 62, C 76-77, D 69, D 71, D 117.

11. Hodge (1977, 1989).

12. See Notebook M, Richards (1987), and Manier (1978). See also Notebook C 166.

13. The indexes to Barrett et al. (1987) allow one to see when Darwin was reading Whewell. Curtis (1987) has a much fuller discussion than can be given here of Darwin's response to Whewell's epistemology for science. See also Curtis (1986).

14. Darwin's annotations of Whewell's *History* are given in Di Gregorio (1990) 866–8. Whewell's rejection of Lyell's reforms for geology is in the final chapter of Whewell's third volume. On Whewell and Lyell, see also Ruse (1975), Laudan (1987), and Hodge (1991).

15. Darwin (1839) 615–625. In this addendum, Darwin writes in a highly self-conscious and self-serving way about the epistemological superiority of his explanation for erratic boulders (rafts of floating ice) over a rival explanation (diluvial debacles) proposed by Louis Agassiz. Darwin (625) insists that, in his explanation, "only *verae causae* are introduced," and, what is more, "reasons can be assigned, for the belief that these causes have been in action" in the districts in question. By contrast, to resort to a deluge hypothesis before it is "absolutely forced on us," is, he claims, "to violate, as it appears to me, every rule of inductive philosophy."

16. See Laudan (1982) and Butts (1973).

17. Indeed, the notion is defended in this very volume, in Michael Ruse's enlightening essay. I can only insist that readers of our two pieces, in making up their minds about Darwin's debts to Whewell, should not overlook our agreements on other issues.

18. See Notebooks D and E and Hodge and Kohn (1985).

19. See, for example, Notebook E 71, 118, and Darwin's notes on a theologi-

cal book by John Macculloch printed in Barrett et al. (1987) 632–641, especially MS pp. 53 and 57.

20. See Kitcher (1985).
21. These last three paragraphs take us into two issues that cannot be done justice here. The first issue is whether Darwin's later epistemological and methodological reflections, in print and in private correspondence after the *Origin* was published, show him to have moved away from the views of Lyell and Herschel and toward those of Whewell. The second issue concerns how far J. S. Mill, Whewell, and other philosophers of science saw Darwin's theory of natural selection as satisfying their notions of what good scientific theorizing and reasoning looks like. The second issue I will leave aside here. It is best explored by starting with David Hull's essay in this volume, and then going to Hull (1995) and Hull (1975).

The first issue involves the question of when Darwin read and reflected on Whewell's views, especially about the consilience of inductions. Here Thagard (1977) makes a natural point of departure. But a separate question has to be raised about the timing of any shift that Darwin is thought to have made. Consider, for example, the well-known passage, quoted in this volume by Michael Ruse, from the introduction to Darwin's treatise *Variation of Animals and Plants under Domestication,* first published in 1868. That passage may indeed seem to invoke Whewell's views. However, in 1860 Darwin had already written in almost the same words, and to almost exactly the same effect, in a letter to S. P. Woodward. And yet three years on, in 1863, writing to George Bentham, he takes a line which is much closer to the old consensus between Lyell and Herschel. (For the letter to Woodward, see Burkhardt et al. 1993, 123; for the letter to Bentham, see Darwin 1888, vol. 3, 24–5 or the quotations from it given in Hodge 1977 and Hodge 1989). When these and other similar texts (for one other, see Hodge 1989) are all taken into account, it seems that the apparent inconsistencies in them can be largely, if not fully, resolved by distinguishing, on the one hand, Darwin's own view of the credentials he thought his theory ought to have and could have, and, on the other, the view that he thought readers would do well to take of the theory's credentials.

Thus, Darwin really did think that he had shown, through the analogy with man's selection, that natural selection could cause intersterile species, even if man's selection had not ever produced intersterile races of a domestic species. But many critics insisted that this line of evidence failed to establish natural selection as an adequate cause for intersterile species formation; and so, to that extent, the hypothesis that extant and extinct species had been produced by selection was just that, a mere hypothesis. To anyone who took this view, Darwin would insist that, since no one had shown that selection could not produce new species

Knowing about Evolution

from old, it was at least only fair to see how much this hypothesis could explain of geographical, embryological facts, and especially to see whether it could offer a more satisfactory explanation than the alternative hypothesis of separate creations of fixed species. So, in short, Darwin himself never really thought that he had only succeeded in formulating and defending a hypothesis that had no better *vera causa* credentials than the wave theory of light, which he knew was widely held to have almost none. But Darwin was happy to invite other people to see it that way, provided they played fair and compared its explanatory virtues with the special creationist alternative. In making that invitation, he knew enough to know that he was inviting other people to look at the theory more as Whewell looked at the wave theory of light, and less as Reid, Lyell, and Herschel had viewed the Newtonian theory of gravitational attraction. However, he never accepted his own invitation to look at the theory that way. He remained much too firmly convinced that his theory of natural selection satisfied the standards he had first embraced before he had even formulated the theory. (For a recent discussion of the structure and strategy of the argument of the *Origin* with references to earlier accounts, see Hodge 1992). Finally, let me emphasize what is implicit in this chapter as a whole: namely that Darwin was much more radical and innovative regarding the second cluster of questions (as distinguished in my first section) than regarding the first cluster. Curtis (1986 and 1987) sees him as equally so regarding both – an exciting but unsustainable proposal, I submit.

REFERENCES

Barnett, P. H., P. J. Gautrey, S. Herbert, D. Kohn, and S. Smith, eds. 1987. *Charles Darwin's Notebooks, 1836–1844*. Cambridge: Cambridge University Press.

Browne, J. 1995. *Charles Darwin*, volume 1. New York: Knopf.

Burkhardt, F., D. M. Porter, J. Browne, and M. Richmond, eds. 1993. *The Correspondence of Charles Darwin, volume 8* Cambridge: Cambridge University Press.

Butts, R. 1973. "Reply to David Wilson: Was Whewell Interested in True Causes?" *Philosophy of Science* 40: 125–128.

Curtis, R. C. 1986. "Are Methodologies Theories of Rationality?" *The British Journal for the Philosophy of Science* 37: 135–161.

1987. "Darwin as an Epistemologist." *Annals of Science* 44: 379–408.

Darwin, C. 1839. *Journal of Researches into the Geology and Natural History of the Various Countries Visited by H.M.S. Beagle*. London: H. Colburn. Reprinted in facsimile (1952), New York: Hafner.

1859. *On the Origin of Species* London: John Murray. Reprinted in facsimile (1975), Cambridge, Mass: Harvard University Press.

1888. *The Life and Letters of Charles Darwin*. 3 vols. Edited by F. Darwin. London: John Murray.

Desmond, A. and J. Moore. 1992. *Darwin*. London: Michael Joseph.

Di Gregorio, M. A. 1990. *Charles Darwin's Marginalia*, volume 1. New York and London: Garland Publishing.

Herschel, J. F. W. 1830. *A Preliminary Discourse on the Study of Natural Philosophy*. London: Longman. Reprinted in facsimile (1987), Chicago: University of Chicago Press.

Hodge, M. J. S. 1977. "The Structure and Strategy of Darwin's 'Long Argument.'" *The British Journal for the History of Science* 10: 237–246.

1982. "Darwin and the Laws of the Animate Part of the Terrestrial System (1835–1837): On the Lyellian Origins of His Explanatory Program. *Studies in the History of Biology* 6: 1–106.

1989. "Darwin's Theory and Darwin's Argument." In M. Ruse, ed., *What the Philosophy of Biology Is: Essays Dedicated to David Hull*, 163–182. Dordrecht: Kluwer Academic Publishers.

1991. "The History of the Earth, Life and Man: Whewell and Palaetiological Science." In M. Fisch and S. Schaffer, eds., *William Whewell: A Composite Portrait*, 255–289. Oxford: Clarendon Press.

1992. "Discussion: Darwin's Argument in the *Origin*." *Philosophy of Science* 59: 461–464.

Hodge, M. J. S. and Kohn, D. 1985. "The Immediate Origins of Natural Selection." In D. Kohn, ed., *The Darwinian Heritage*, 185–206. Princeton: Princeton University Press.

Hull, D. 1973. *Darwin and His Critics*. Cambridge, Mass.: Harvard University Press.

1995. "Die Rezeption von Darwins Evolutionstheorie bei britischen Wissenschaftsphilosophen des 19. Jahrhunderts." In E- M. Engels, ed., *Die Rezeption von Evolutionstheorien im 19. Jahrhundert*, 67–105. Frankfurt: Suhrkamp.

Kitcher, P. 1985. "Darwin's Achievement." In N. Rescher, ed., *Reason and Rationality in Science*, 127–189. Washington D.C.: University Press of America.

Kohn, D. 1985. *The Darwinian Heritage*. Princeton: Princeton University Press.

Laudan, L. 1981. *Science and Hypothesis: Historical Essays on Scientific Methodology*. Dordrecht: D. Reidel.

Laudan, R. 1987. *From Mineralogy to Geology: The Foundations of Science, 1650–1830*. Chicago: University of Chicago Press.

Lyell, C. 1830–33. *The Principles of Geology*. 3 vols. London: John Murray. Reprinted in facsimile (1991), Chicago: University of Chicago Press.

Manier, E. 1978. *The Young Darwin and His Cultural Circle*. Dordrecht: D. Reidel.

Richards, R. J. *Darwin and the Emergence of Evolutionary Theories of Mind and Behavior*. Chicago: University of Chicago Press.

Knowing about Evolution

Ruse, M. 1975. "Darwin's Debt to Philosophy: An Examination of the Influence of the Philosophical Ideas of John F. W. Herschel and William
Whewell on the Development of Charles Darwin's Theory of Evolution." *Studies in the History and Philosophy of Science* 6: 159–181.

Thagard, P. 1977. "Discussion: Darwin and Whewell." *Studies in History and
Philosophy of Science* 8: 353–356.

Whewell, W. 1837. *History of Inductive Sciences.* 3 vols. London: J. W. Parker.

XII

Generation and the Origin of Species (1837–1937): A Historiographical Suggestion

I

Bernard Norton's friends in the history of science have had many reasons for commemorating, with admiration and affection, not only his research and teaching but no less his conversation and his company. One of his most estimable traits was his refusal to beat about the bush in raising the questions he thought worthwhile pursuing. I still remember discoursing at Pittsburgh on Darwin's route to his theory of natural selection, and being asked at the end by Bernard what were Darwin's views on heredity. I answered with the conventional waffle to the effect that the theory concerned the populational fate rather than the individual production and transmission of heritable variation, so that whatever views Darwin had on heredity had only a subsidiary place in his theorizing. Bernard was not fooled. 'I would have thought', he said, 'that in order to understand anyone's theorising about evolution it would be necessary to look at his views on heredity'.

As a general maxim, the comment obviously embodies an unavoidable truth. Where difficulties arise is in acting on the maxim when we concern ourselves not merely with one particular theorist, but with a long run of changes in evolutionary theorizing over several decades. One run of a century proves especially challenging. In 1837 Darwin opened his *Notebook B* under the heading 'Zoonomia', the laws of life. A year and a half later, in *Notebook E*, a successor to *B*, he reached his first formulations of natural selection. In 1937, Theodosius Dobzhansky published *Genetics and the Origin of Species*, the single most decisive text in the inauguration of a new, synthetic theory of evolution. All historians of evolutionary biology as it developed in the century after Darwin would insist that any historiography adequate to their task must clarify how views of evolution and views of heredity were brought into diverse, shifting interactions in this period. The present paper aims to contribute to the appreciation of the nature of this challenge.[1]

There will be no disagreement today that one traditional response to this challenge is quite unacceptably Whiggish. That response, to caricature it crudely, says that Darwin's own Darwinism was deficient on two counts when it came to heredity. For Darwin still accepted 'soft' heredity, which tends to make natural selection redundant in accounting for adaptive change, and he still accepted 'blending' heredity, which renders natural

1 In writing this paper I have been much helped by discussions with Peter Bowler, Frederick Churchill, Lindley Darden, Jim Grisemer, Jon Harwood, Ernst Mayr, Robert Olby, Onno Meijer, Jan Sapp and Phillip Sloan.

selection ineffective. So, what the historian has to explain is how new commitments since Darwin, commitments to 'hard' and 'particulate' heredity, came, in the new synthesis, to vindicate natural selection as both necessary and sufficient for bringing about evolution.

What is less apparent is that finding a more satisfactory historiography than that allowed by this Whiggish response may require us to think about some very general features of biological science as it has developed over the last two centuries. In case this declaration seems needlessly hyperbolic, consider the framework explicitly presupposed in Ernst Mayr's book, *The Growth of Biological Thought* (1982). Mayr's treatment of evolution in relation to heredity in the century from Darwin to Dobzhansky is conditioned directly by the most fundamental distinctions he draws within biological thought itself. There are, Mayr holds, 'two biologies', *evolutionary* biology, the subject of his text, and *physiological* biology, a subject appropriate to a separate volume. Moreover, within *evolutionary* biology, Mayr distinguishes *evolution* as such from *variation and its inheritance*. Accordingly, Mayr takes us from Darwin to Dobzhansky under the heading *evolution*, postponing such heredity theorists as Francis Galton, August Weismann, Hugo De Vries and H. J. Muller for the section on *variation and its inheritance*, and keeping physiology for another occasion altogether.[2]

Now, Mayr's 'two biologies' thesis conforms well with the old notion that, as a modern science, biology had two sources in the last century. On the one hand, according to this notion, there was natural history, including geology and geography no less than entomology, with its locations in museums, voyages and mining, and its concern with species, their histories and affinities; while on the other hand there was physiology, associated with medicine, microscopes and dissection, and directed not at species, but at the structure and functioning of individual organisms.

However, there are now compelling reasons for avoiding any such dualistic historiography of biological science in our efforts to understand the developments that link the young Darwin with Dobzhansky a century later. For whatever else may be counted as a paradigm instance of evolutionary biology, it is surely incontestable that what Darwin was doing in his 'Zoonomia' theorizing is entitled to be so designated. But what Darwin was doing turns out to involve *physiology* no less centrally than *natural history*. Moreover, at a minimum, we may require of any historiography of the century after 1837 that it bring out how certain figures, such as the Weismann of the 1890s and the Muller of the 1920s, were successors to the Darwin of the notebooks. Especially, we would need to understand at a minimum how all three were linked by a preoccupation with the *theory of life and death* and the *theory of sex*, a preoccupation that belonged at least as much to physiology as to natural history.

This preoccupation was intrinsic to Darwin's concern with the *theory of generation*, and to that extent, then, Weismann and Muller were Darwin's successors in generation theory. To recognize the continuity in this intellectual succession requires us, however, to break with another historiographical tradition. According to this tradition, the eighteenth century had the *theory of generation*, the nineteenth the *idea of evolution* and

2 E. Mayr, *The Growth of Biological Thought: Diversity, Evolution and Inheritance*, Cambridge, Mass., 1982. A very recent paper greatly enhances our understanding of such issues: J.A. Carron, ' "Biology' in the Life Sciences, *History of Science*', (1988), **26**, pp. 223–268.

the twentieth the *science of genetics*. Peter Bowler's recent text, *Evolution: The History of an Idea* (1984), for instance, treats fully of generation theories in the Enlightenment and of genetical doctrines in our own century. But its one hundred and fifty pages in the Darwin (1837) to Dobzhansky (1937) narrative require no 'cell, theory of' or 'sex' in his index.[3] So, once again, we have here a framework for the writing of history that would no longer be appropriate, were we to start thinking about heredity and evolution theorizing after Darwin as continuing Darwin's own preoccupation with generation and— through generation—life, death and sex. For, whatever else a historiography of the century after Darwin has to include, it must have a place in its narrative for the efforts of the Weismann of the 1890s and the Muller of the 1920s to ground evolutionary biology in a cell-theoretic understanding of life, death and sex.

The proposals to be made in the present paper have, then, two sources that have been yielding strikingly consilient results of late. On the one hand, there are the newer views of Darwin's zoonomical explanatory programme, its scope and aims; and on the other hand there are newer views of what was going on in the last century and a half among investigators into heredity, hybridization, fertilization and so on, such topics as are conventionally and anachronistically included within the history of genetics.[4] There may be a certain timeliness as well as irony, therefore, in the suggestion that the Darwin industry, often not unjustly notorious for its narrowmindedness, can play some part in prompting broader views than have been customary as to what developments in late nineteenth- and early twentieth-century genetics were integral to evolutionary theory. For example, take the Symposium on 'Relations between Theories of Heredity and Evolution (1880–1920)' at the 1977 International Congress of the History of Science. Three theses explicitly guided that symposium: first, that theories of heredity were integrated—even sometimes coextensive—with theories of evolution half a century before the 'new synthesis' of the 1930s; secondly, that the 1880–1920 period has needed attention in its own right, free from any distracting demands of the Darwin and Dobzhansky industries; thirdly, that the principal issue whereon heredity and evolution theorizing converged in that period

---

3 P.J. Bowler, *Evolution: The History of an Idea*, Berkeley and Los Angeles, 1984. But see also two subsequent books by Bowler (notes 4 and 13).

4 For a sampling of these newer views of Darwin's programme, see P.R. Sloan, 'Darwin's Invertebrate Program, 1826–1836: Preconditions for Transformism,' in D. Kohn (ed), *The Darwinian Heritage*, Princeton, 1985, pp. 71–120, and 'Darwin, Vital Matter, and the Transformism of Species,' *Journal of the History of Biology*, (1986), **19**, pp. 369–445; M.J.S. Hodge, 'Darwin and the Laws of the Animate Part of the Terrestrial System (1835–1837): On the Lyellian Origins of his Zoonomical Explanatory Program,' *Studies in the History of Biology*, (1983), **6**, pp. 1–106; 'The Development of Darwin's general biological theorizing,' in D.S. Bendall (ed), *Evolution from Molecules to Men*, Cambridge, 1983, pp. 43–62, and 'Darwin as a Lifelong Generation Theorist,' in D. Kohn (ed), *The Darwinian Heritage*, pp. 207–244. For a sampling of recent work on the history of theories of heredity and related topics, see P.J. Bowler, *Mendelism: The Emergence of Hereditarian Concepts in Modern Science and Society*, forthcoming; F. Churchill, 'From Heredity Theory to Vererbung: The Transmission Problem, 1850–1915,' *Isis*, (1987), **78**, pp. 337–364; R.C. Olby, *Origins of Mendelism*, 2nd edn., Chicago, 1985; 'The Emergence of Genetics,' in R.C. Olby, G.N. Cantor, J.R. Christie and M.J.S. Hodge (eds), *Companion to the History of Science*, forthcoming, and 'Historiographical Problems in the History of Genetics,' *Rivista di Storia della Scienza*, (1984), **1**, pp. 25–38; J. Farley, *Gametes and Spores: Ideas about Sexual Reproduction, 1750–1914*, Baltimore, 1982. Extensive bibliographic guidance is also available in S.G. Brush, *The History of Modern Science. A Guide to the Second Scientific Revolution, 1800–1950*, Ames, Iowa, 1988.

was the inheritance of variation.[5] However, if the suggestion made in the present paper is correct, then the Darwin industry relates to the Dobzhansky industry through links that require us to go beyond that inheritance of variation issue, as a principle of historiographic integration for heredity theory and evolution theory, beyond it to the issues that life, death and sex had long been raising for the theory of generation.

II

There should be no surprise that these longstanding issues were always central to Darwin's theorizing. For reasons tracing to the new mechanical philosophy and the new heliocentric astronomy of the seventeenth century, theorizing about the origins, diversity and adaptation of species had long been located between the theory of generation, on the one hand, and the theory of the earth, on the other. Think of Buffon, whose *Époques de la Nature* of 1778 tackled these problems about species in extending and integrating his theory of generation and his theory of the earth of 1749.

With Darwin, such a location for his zoonomical theorizing about species origins derived from those authorities and institutions that were decisive for his intellectual development over a whole decade up to 1837. On generation, there were Edinburgh medicine and Robert Grant and his own grandfather Erasmus Darwin; and, no less, there were Cambridge botany and Henslow whose teaching and writing on plants—contrary to conventional stereotypes of English parsons and naturalists—was deliberately emulating French trends in being not solely *systematic* but equally *physiological*. On the theory of the earth, there were Cambridge and Sedgwick and then Lyell. A leading feature of Darwin's life as a scientific theorist, from the voyage years (1831–1836) on, was his habit of working to integrate these two legacies, the Grantian and the Lyellian, by concentrating on their areas of intersection. This habit is manifested during those voyage years, both in his reflections on species' extinction—reflections of consequence for all his later thinking about species, including their origins—and in his conjectures concerning coral islands, where his Grantian concern with colonial invertebrate animal propagation, and his Lyellian concern with the elevation and subsidence of the earth's crust, are brought into intimate conjunction.[6]

Recent studies have now documented how Darwin's apprenticeship to his various physiological mentors conditioned not only his zoonomical programme of 1837, but the entire course of his career as a biological theorist through his publications of the 1860s and beyond. Just as work on the early Newton has discredited all attempts to distinguish between a Newton of the *Principia* and Newton of the *Opticks*, including the *Queries*, so it is now evident that the Darwin of the *Origin of Species* (1859) and the Darwin of *Variation of Animals and Plants under Domestication* (1868), including the *pangenesis* chapter, are one issue of one source, the zoonomical theorist of the 1830s.

Charles Darwin's very use of Erasmus Darwin's title—'Zoonomia'—to designate his enquiry should remind us that the younger Darwin's was the first generation to take the

5 E.G. Forbes (ed), *Human Implications of Scientific Advance: Proceedings of the XV International Congress of the History of Science, Edinburgh, 10–15 August 1977*, Edinburgh, 1978. See Symposium No. 8.

6 The evidence of what is said in this and the next paragraph is set forth in the papers on Darwin op. cit. (4).

cultivation of a science of the laws of organic life, or a science of organized bodies, as a commonplace. This new science of life was indeed what others, notably Lamarck, had called *biology*. What is more, precisely because this was the 1830s, no distinction was being made as yet between biology, as the science of life, and the science of physiology, in the sense of general as contrasted with human physiology. In later decades, notably the 1850s and 60s, various scientists would have their own reasons for construing physiology in various intellectual and institutional ways. One construal, associated especially with Claude Bernard, would deliberately break with older precedents to contrast physiology with biology so as to secure autonomy for physiology as a distinct, experimental science. Complementing that construal would come those contrasts between physiological and morphological science familiar in the last decades of the nineteenth century.[7]

When it is said, here, that Darwin's zoonomical theorizing was as much physiological as natural-historical, there is, therefore, a twofold thesis being urged: first, that Darwin's theorizing was physiological in the dominant 1830s sense, when physiology was not distinguished from biology as then characterized; and, secondly, that consequently the work of subsequent theorists, which was physiological in the later senses, could be, in fundamental ways, nonetheless continuous with Darwin's own zoonomical enterprise.

The content of this two-fold thesis does not depend, then, on any essentialist insistence on some spurious invariance as to the character or domain of physiological science. Rather, it requires only a recognition of the continuities manifested by certain persistent issues, as those issues arose at various intersections between theorizing about generation and theorizing about the origin of species, beginning, for our purposes here, with Charles Darwin's own earliest work.

Before engaging problems concerning the extinction and origin of species, Darwin had reflected and observed (literally microscopically) much about 'associated life' and generation by 'gemmules' in colonial invertebrates such as Bryozoans (as now classified). The life of such colonies is co-ordinated, he concluded, by a 'central living mass' of continuous growth; while the 'gemmules' often arise from prior 'granular' matter in the parent. He had also learned to think of the entire asexual issue of a plant bud, with that bud considered as an individual organic being, as limited in its total duration by the finite amount of life in that bud.[8]

Drawing on such conclusions, Darwin developed, from 1835 to 1837, a generational theory of species extinction. The theory was explicitly founded on the premise that all

7 The history of changing ideas and institutionalizations of physiology can be studied in J. Schiller, *Physiology and Classification: Historical Relations*, Paris, 1980; T. Lenoir, *The Strategy of Life: Teleology and Mechanics in Nineteenth-Century German Biology*, Dordrecht, 1982; J. Lesch, *Science and Medicine in France: The Emergence of Experimental Physiology, 1790–1855*, Cambridge, Mass., 1984, and J.V. Pickstone, 'Science in France,' *History of Science*, (1988), **26**, pp. 201–211. See also Carron op. cit. (2). For correctives to common stereotypes of what natural history comprised in the 1830s, see P.F. Rehbock, *The Philosophic Naturalists. Themes in Early Nineteenth-Century British Biology*, Madison, 1983. The complexities in comparing physiological and morphological strategies of explanation are elucidated for one particular case in F. Churchill, 'Hertwig, Weismann, and the Meaning of the Reduction Division Circa 1900,' *Isis*, (1970), **61**, pp. 429–458.

8 The evidence for what is said in this and the next five paragraphs is set out in the papers on Darwin op. cit. (4).

generation, sexual and otherwise, involved division; and on the argument that any number of individuals tracing to the division of a single individual would have a limited duration of life tracing to that individual; and that a sexual succession of individuals making up a species could, therefore, have a limited life as does a plant bud graft succession or polyp animal colony.

Consequently, Darwin began his species transmutation theorizing, in 1837, as an account of the generational causes whereby a species could avoid becoming extinct by senescence without producing any offspring species. On his explicitly teleological reasoning, the distinctive properties of sexual as contrasted with asexual generation were what made possible such a transmutation under changing conditions in the long run; just as sexual generation could rescue an asexual graft bud succession from senescence without issue in the short run. It was the maturation in the product distinctive of sexual reproduction that allowed heritable organizational innovations in changing conditions, Darwin argued. The other distinctive feature—the crossing of two parents—was not innovative; but it was retentive, Darwin soon came to think, in ways needed for the permanent constitutional embedding of those adaptive variations in maturation that were induced by gradual changes in conditions. Darwin now saw, in the ontogenetic replacement of hermaphroditism by separate sexes, vestiges of the phylogenetic origins of the means necessary for the progress that had led to mammals and man.

This retentive function for crossing—frustrated by selfing—did not require the earlier appeal to sex as circumventing extinction by senescence; and such circumvention was no longer needed anyway when Darwin came to trace extinction not to a diachronic limitation on the duration of life, but to a synchronic limitation on the quantity of life and so, following Lyell again, to competitive ecological defeats rather than endogenous generational exhaustions. Life, as the vivification of organization, may be, Darwin now suspected, available unlimitedly and exogenously, and so setting no restriction on generational production; the limitations coming rather, as Malthus had taught, from the finite food and other resources needed for nutrition and growth. When Darwin developed his theory of natural selection it incorporated this Malthusian view of the limitations on the quantity of life.

Within two or three years of reaching that theory, any remaining contrasts between sexual and asexual generation apparently broke down in Darwin's eyes. He constructed his theory of pangenesis, in 1840–1841, it seems, in the wake of this new conclusion, constructing it as a theory, therefore, of the common powers and matter responsible for all generation, sexual and asexual, from growth and healing to multiplication by sperm and egg. The theory, as a pan-micro-ovulational theory, was apparently integrated early on with the cell theories of the late 1830s as expounded in particular in an English translation of Johannes Müller's celebrated *Physiologie*, where an identity in the cellular constitution and generational powers of unfertilized sexual ova and asexual buds was upheld.

Natural selection and pangenesis gave Darwin not a teleological but a dysfunctional view of generational variability. The greater variability with sexual rather than asexual generation, like the greater variability under domestication rather than in the wild, was tentatively traced to a greater disturbance of the nutrition intrinsic to growth. For growth itself, and so all generation, he presumed, would be completely conservative and so accompanied by no variability in the absence of such nutritional changes as disturbed its proper, normal replicative function. So, even though such disturbances from changed conditions tend, like the occasional outbreeding of an inbred stock, to enhance vigour, they are frustrations, not executions, of the conservative, reproductive function common to all generation, sexual no less than asexual.

III

Even this brief glance at these life, death and sex issues, as they were integral to Darwinian biology at its very inception, can prompt considerations that are indispensable for any historiography for the century from 1837 to 1937.

To bring out what such considerations might be, it is appropriate to glance at Weismann and his American admirer, Muller. Now, these two would feature prominently in any Whiggish account of what happened after Darwin and before Dobzhansky. So, it will be as well to emphasize here that they are not being featured on the present occasion in order to perpetuate, even in revised form, the traditional Whiggish routine, whereby Darwinism triumphs through the rise of hard and particulate heredity. Rather, Weismann and Muller are appropriate choices precisely because, by seeing them as successors to Darwin, the zoonomical generation theorist, one can see why it is illegitimate to continue interpreting them, in the Whig way, principally as vindicators of Darwin the selectionist.

Before turning to Weismann there are two general comments called for in clarifying his relationship to Darwin's zoonomical enterprise. The first concerns myths of synthesis, particularly of a synthesis between cytology and evolution. Scientists have always loved myths of synthesis, and historians should, therefore, always be wary of uncritically assimilating those myths. One such myth is developed very instructively in the introduction to the second edition (1900) of E. B. Wilson's treatise on the *Cell in Development and Inheritance.*[9] There Wilson has evolution coming out of natural history and only being integrated in the 1870s and 80s with cytology as coming out of medical and physiological traditions; so, as Wilson sees it, Weismann's thesis of the continuity of the germ-plasm is offering the best prospect of a successful synthesis between evolution and the theory of cells, the two leading generalizations about life to come out of the century then just closing.

The obvious trouble with this Wilsonian historiography is that it reads out of the record the widespread quest to integrate evolutionary theory with the old cell theory of the 1830s, a quest which was alive and well in the 1860s. Darwin's evolutionary biology, as publicly pursued in that decade, had arisen as such a synthesis. There have been many *new syntheses* since but, we have to insist, they have not been new in being syntheses.

As a second general comment one must very much agree with Frederick Churchill in insisting that Darwin's public theorizing of the 1860s was far from exceptional, being, rather, typical of an evolutionary–generation–theoretic genre, then exemplified also by such synthesizers as Haeckel and Spencer.[10] Despite divergences—Haeckel's was more a

9 E.B. Wilson, *The Cell in Development and Inheritance*, New York, 1900, pp. 1–13.

10 F. Churchill, 'Sex and the Single Organism: Biological Theories of Sexuality in Mid-Nineteenth Century,' *Studies in History of Biology*, (1979), 3, pp. 139–178.

morphological, Spencer's more a physiological programme of explanation—the central, common feature of such syntheses was that the theorist was seeking an integrated structure of argumentation that was to move all the way from the presence of life—as entailing *cellular growth* and *nutrition,* that was presumed *inherently conservative*—to the adaptive diversification of species and higher taxa. Evolutionary biology, here, was therefore cell-theoretic, growth-theoretic and generation-theoretic. Conversely, cell theory, and so life theory and generation theory were, for the Darwins, Haeckels and Spencers, evolutionary. Consequently, these mutually integrative implications were presupposed by any subsequent biologists who were pursuing what such synthesizers were pursuing in the 1860s.

That August Weismann was such a biologist has been made manifest in many recent studies, especially those by Churchill, which have elucidated Weismann's instructive agreements and disagreements with all three of these 1860s synthetic theorists. Such elucidations, complemented by the studies of Baxter, Farley, Mayr and Robinson, have brought out the decisive connections among the themes dominant in Weismann's *Essays upon Heredity and Kindred Biological Problems* (as collected in the authorized English edition) of 1892 and his *Das Keimplasma: Eine Theorie der Vererbung* of the same year (English translation, 1893).[11] It is the connections among these themes that allow us to see how Weismann's evolutionary biology was continuous in its preoccupations with Darwin's original zoonomical program, even when Weismann was breaking with Darwin's premises, let alone his conclusions.

As he himself emphasized explicitly, Weismann's two most prominent contrasts— between *germ-plasm* and *soma* and between the *reducing* division and *ordinary* cell division—had implications for evolutionary theory because they were grounded by him in his explicit views of life, death and sex in *unicellular* as contrasted with *multicellular* organisms. His rejection of the rejuvenescence theory of sex was dependent on his view that unicellular life was potentially immortal, and only ended in death by accident; whereas senescence and death in multicellular organisms were there by adaptive design, thanks to natural selection for reproductive optimizations. Even in multicellular

11 A. Weismann, *Essays upon Heredity and Kindred Biological Problems,* eds E.B. Poulton, S. Schönland and A.E. Shipley, 2 vols, Oxford, 1891–1892; *The Germ-plasm: A Theory of Heredity,* trans. W. N. Parker and H. Rönnfeldt, London, 1893. In addition to the papers of Churchill op. cit. (4) and (7), see his 'August Weismann and a Break from Tradition,' *Journal of the History of Biology,* (1968), **1**, pp. 91–112; 'Weismann's Continuity of the Germ-Plasm in Historical Perspective,' *Freiburger Universitätsblätter,* (1985), **24**, 107–124. I am grateful to Jon Harwood for making available to me a copy of the double issue of this journal devoted to Weismann under the title *August Weismann (1834–1914) und die theoretische Biologie des 19. Jahrhunderts* and edited by Klaus Sander. See, further, Churchill, 'Weismann, Hydromedusae and the Biogenetic Imperative: A Reconsideration,' in T.J. Horder, J.A. Witkowsky and C.C. Wylie, (eds), *A History of Embryology,* Cambridge, 1986, pp. 7–33; A. Baxter and J. Farley, 'Mendel and Meiosis,' *Journal of the History of Biology,* (1979), **12**, pp. 137–173; Farley, *Gametes and Spores* (op. cit. 4); E. Mayr, 'Weismann and Evolution,' *Journal of the History of Biology,* (1985), **18**, pp. 295–329, also (in German) in the volume edited by K. Sander and, as 'On Weismann's Growth as an Evolutionist', in E. Mayr, *Toward a New Philosophy of Biology. Observations of an Evolutionist,* Cambridge, Mass, 1988, pp. 491–524; G. Robinson, *A Prelude to Genetics. Theories of a Material Substance of Heredity: Darwin to Weismann,* Lawrence, Kansas, 1979. It is the indispensable achievement of Churchill and of Farley, in his invaluable book, to have given us the insights into nineteenth-century ideas about *reproduction* that historians of 'evolutionary biology' and of 'genetics' have long needed.

organisms, the life of the germ-plasm was a continuation of the immortal life of their unicellular ancestors, and so distinct from the mortal life of the soma. Sex is, then, independent of multicellular life and death and arose, originally, by natural selection in virtue of the advantages of amphimixis: the production of variation through the combination of distinct ancestral germ-plasms.

Weismann's position on sex was, then, unlike either the early or the late Darwin's. The greater variation accompanying sex is not explained as a function of the impressionability of maturing organization in the offspring, nor as a dysfunction of nutrition in the parent, but as the function of the halving and combining of germ-plasms before and after fertilization. This conclusion left Weismann still looking for the ultimate sources of the differences in the combining germ-plasms. Here he did join Darwin and many others in appealing to the dependence of all growth on nutrition, and to a consequent susceptibility of growth to replicative redirection consequent upon nutritional redirection. For Weismann invoked such suppositions both in his proposal that germ-plasm differences traced to environmental influences on germ-plasm growth; and in his later proposal of germinal selection, which has different elements within the germ-plasm differing in their struggle for the nutritional requirements for growth and so multiplication.[12] In such ways Weismann continued to pursue what he saw as one primary aim in any theory of heredity, namely, to relate heredity to growth.

Even this momentary recollection of these central Weismannian themes enables us to see the need for a number of the distinctions that are required, if we are to make progress in integrating the historiography of evolution with the historiography of heredity theory in this late nineteenth- and early twentieth-century period.

One distinction that, in various forms, has been gaining favour of late is between theories of heredity that confine themselves to problems concerning the *transmission* of traits, and theories that do not have that limitation but offer an account of *development* as well as transmission.[13] This distinction has, surely, to be invoked with care. It may be that the late (Mendelian) Morgan saw, even valued, the early theory of the gene as an exemplary instance of the first division of this distinction, and that his opponents preferred to seek theories in the second, more comprehensive division.[14] However, in other cases than the Morgan one, this distinction—between narrow transmission and broader

12 On germinal selection see not only Weismann's essay, *On Germinal Selection as a Source of Definite Variation*, trs T.J. McCormack, 2nd edn, Chicago, 1902, but also his final, grand evolutionary synthesis: *The Evolution Theory*, trs J.A. and M.R. Thomson, 2 vols, London, 1904, a synthesis Weismann continued to refine, not least in relation to Mendelism, up to the third German edition of 1910.

13 This distinction is put to use, especially, in Bowler *Mendelism* (op. cit. 4) and *The Non-Darwinian Revolution: Reinterpreting a Historical Myth,* Baltimore, 1988. Meanwhile it is prominent in I. Sandler, and L. Sandler, 'A Conceptual Ambiguity that Contributed to the Neglect of Mendel's Paper,' *History and Philosophy of the Life Sciences*, (1985), 7, pp. 3–70, and J. Harwood, 'Genetics and the Evolutionary Synthesis in Interwar Germany,' *Annals of Science*, (1985), 42, pp. 279–301, and 'National Styles in Science: Genetics in Germany and the United States between the World Wars,' *Isis*, (1987), 78, pp. 390–414.

14 G. Allen, 'T.H. Morgan and the split between embryology and genetics, 1910–1935,' in Horder *et. al.* (eds) *A History of Embryology*, pp. 113–146. The relations between genetics and embryology in the decades around 1900 have been clarified importantly in several papers by J. Maienschein. See, especially, her 'Preformation or new formation—or neither or both' in the volume edited by Horder *et. al.* pp. 73–108, and 'Heredity/Development in the United States circa 1900,' *History and Philosophy of the Life Sciences*, (1987), 9, pp. 79–93.

276

transmission and development theorizing—is not likely to prove so illuminating, because there will be other distinctions cutting across it.

One such further distinction is that between growth and development. On the new cell theory so widely accepted in the 1880s, growth involved cell multiplication by cell division, while development involved, in addition to cell multiplication and so growth, cell differentiation. Growth could be, then, completely conservative, a mere making of more of the same; while development involved always the making of differences. Now, Weismann's germ-plasm theory not only offered to explain heredity as conservative because of the conservative perpetuation of the intact germ-plasm in germ track growth, and as variational because of the combinatorily innovative halving and conjunction of the germ-plasm complements in the maturing and fertilization of sex cells; it also offered to explain the differentiation in development as due to somatic growth involving, by contrast, an unconservative distribution of the germ-plasm determinants inherited from the parents.[15]

This Weismannian contrast—between developmental growth in soma and the growth without development that perpetuates, with halvings and doublings, the germ-plasms—can allow us to see the historiographic utility of yet another distinction. This time the distinction is in interpretations of variation; for one may distinguish developmentalist interpretations from all others. On a developmentalist interpretation, variations are construed as lawful additions or losses to a development that conforms to some developmental law. Thus, in the exemplary case of Robert Chambers' *Vestiges of Creation* (1844), the variations required for phylogenetic progress arise typically as ontogenetic advances that conform, as all ontogenetic progressions do, to a developmental law, in this case a law Chambers formulates by blending Meckel's law of parallelism with von Baer's replacement for it, the law of progressive specification, as represented in Carpenter. Here, then, phylogeny as ontogeny writ large is developmental because both ontogeny itself and ontogenetic variation conform to the same law.[16]

Now, developmentalism, as paradigmatically exemplified by Chambers, is the main theme in Peter Bowler's important new book, *The Non-Darwinian Revolution* (1988).[17] For, in this book, Bowler argues that there was indeed a shift, in the nineteenth century, from static creationism to evolutionism. But despite Darwin's own non-developmentalist proposals as to the course and causes of evolution, it was predominantly a shift to developmentalist evolutionism. Bowler holds that only with the later Mendelian revolution did a new non-developmentalist view of evolution, akin to Darwin's, arise in the

15 The precise teaching of Weismann and his successors on the place in growth of the germ tracks and the soma is greatly illuminated in J.R. Grisemer and W.C. Wimsatt, 'Picturing Weismannism: A Case Study of Conceptual Evolution,' forthcoming in a volume edited by M. Ruse and provisionally entitled *What Philosophy of Biology Is*.

16 On this subject generally (including Weismann on ontogeny and phylogeny), see S.J. Gould, *Ontogeny and Phylogeny*, Cambridge, Mass., 1977, and, for Chambers and Carpenter and their contemporaries, see D. Ospovat, 'The Influence of Karl Ernst von Baer's Embryology, 1828–1859; A Reappraisal in Light of Richard Owen's and William B. Carpenter's Paleontological Application of "Von Baer's Law",' *Journal of the History of Biology*, (1976), 9, 1–28.

17 See the books of Bowler cited in n.13. I am preparing a review of Bowler's book for this journal.

1930s, with a new grounding in novel conceptions of heredity and variation. Bowler's elaboration and defence of his thesis, a refreshing and mostly correct thesis in my view, often fits very well with the suggestion being made in the present paper. However, there may be an appearance of disagreement that should be resolved here. Bowler tends to talk of *growth* and of *development* as if they are inseparable subjects if not indistinguishable concepts. The result is that, for Bowler, a biologist who is separating questions of heredity and variation, and so evolution, from developmentalist presumptions of lawful purposiveness is also separating them from the interpretation of growth. Now, this result shows, I submit, that we do best not to follow Bowler in this conflation. By avoiding it, we can accept fully, as Bowler himself is inclined to, that, as in the case of Darwin, an interpretation of heredity and variation, and so evolution, can be grounded in the theory of growth—in the sense of nutrition and generation, with or without the multiplication of individuals—and yet that interpretation need not be a developmentalist one. Thus, in Darwin's thinking, as Bowler emphasizes, variations arise whenever there are disturbances and disruptions of the normally replicative processes of growth involved in development; but there is no requirement that these departures from the replicative tendency in growth should conform to any laws for developmental progressions.

Strikingly, as Bowler brings out, Weismann presents a less clear case of non-developmentalist presuppositions. It is true that, on Weismann's account, heredity is located in growth, and that this is not the growth involved in the developmental differentiation of the soma. So, it is tempting to conclude that, although disagreeing fundamentally with Darwin's pangenesis, Weismann is even less of a developmentalist, concerning heredity, variation and so evolution, than Darwin. For Weismann, like Galton but unlike Darwin, no longer sees inheritance as a power or principle that is intrinsically opposed to variation as a contrary power or principle; for Weismann, again like Galton, interprets heredity as the passing on by parents to offspring of ancestral legacies of determinants that are being conjoined and combined by sexual reproduction in any one generation. The resemblance of offspring to parent is, then, not due to the product being given its character by the replicative powers of inheritance residing in the entire parent organization, but owing to the offspring possessing many of the same determinants that the parent received from ancestors shared by parent and offspring.[18] However, as Bowler rightly insists, Weismann remained loyal to the Haeckelian law of recapitulation, whereby ontogenies recapitulate phylogenies; so, a complete integration of heredity, variation and evolution theory would, for Weismann, have to explain why phylogeny is restricted to variational innovations that can arise as additions and condensations in ontogenetic changes. Bowler is, then, quite right in judging that Weismann's overall position may be closer to Chambers' and further from Muller's or Dobzhansky's than one may be tempted to assume. Nevertheless, if we distinguish, as we should, between growth and development, Bowler's salutary judgement about Weismann still leaves open the question of whether the theory of generation issues—of life, death and sex as bearing on growth—do not persist so as to be transformed but not eliminated by any Mendelian revolution. And we

18 The significance of Galton's thinking in this regard has been stressed by Olby, *Origins of Mendelism*, pp. 55–64, following the analyses of Ruth Cowan. See, especially, her 'Francis Galton's Contributions to Genetics,' *Journal of the History of Biology*, (1972), 5, pp. 389–412.

shall see that they did indeed persist to thrive long after 'gemmules' and 'pangenes' had given way to 'genes'; long after, that is, *generation* had given way to *genetics*.

IV

Rapid though it has been, our consideration of the distinctions needed in relating Weismann to the early and late Darwin can alert us, therefore, to certain continuities and discontinuities between Weismann and Muller that must be respected by any historiography for a Darwin to Dobzhansky narrative.

Familiar historiographies for genetics would insist that in moving from Weismann in the 1890s to Muller in the 1920s, one is moving across a 'watershed', a great divide dated 1900, the year of a 'rediscovery of Mendel' that supposedly led to the 'founding' of a 'new science of genetics'. However, the most recent work by specialists in this period around 1900 is full of reasons for scepticism about these longstanding conventions. Our theme—of generation-theoretic approaches to the origin of species—does not directly engage all of these reasons for scepticism. But it engages enough of them to suggest which of the revisionist insights already available may be most indispensable to the historiographical challenge occupying us here.

One such insight is that the best exegesis of Mendel's own paper shows that it presents a cell-theoretic proposal as to the production of *variable* versus *constant hybrids*, and that in doing so it proposes nothing like a general theory of heredity in sexual as contrasted with asexual reproduction; nor anything like a theory of the gene.[19] We have, therefore, to appreciate the magnitude of the developments in the period from 1900 to 1920, in seeing how far from Mendel's own position theorists of Mendelism, such as Muller, had moved in that time. For they had moved to identifying Mendelian heredity as coextensive with sexual reproduction and to identifying genes as coextensive with chromosomal structures.

One way to see what such moves entailed for generational theorizing about species origins is to consider where Mendel himself appears to be located. For he has now been placed in that long line of people since Linnaeus who were concerned with the possibility that, thanks to sexual generation, new species can arise from old, in the hybridization of earlier species. Such new species would, then, arise as constant hybrids.[20] So, Mendel's conclusions from his findings with variable Pea (Pisum) hybrids could not constitute for him a theory of how sexual generation makes species production possible when it does. For he was limited to concluding that variable hybrids are formed when reversible cellular mixings of the material dispositions for different characters are effected in crossing, while constant hybrid production requires irreversible mixings.

One thing that was done in the decade and a half after Mendel's paper became widely studied was, then, the virtual elimination of the constant hybrid side of this contrast and the pursuit of the other, to the point where Mendelism, as a theory of reversibly mixing,

---

19 See, especially, Olby, *Origins of Mendelism*, pp. 196–258, and L.A. Callender, 'Gregor Mendel: an Opponent of Descent with Modification,' *History of Science*, (1988), **26**, pp. 41–75.

20 Olby, *Origins of Mendelism*.

segregating and recombining factors, became nothing less than a theory of sexual reproduction, having along the way, and not coincidentally, been made to yield a theory of sex
determination.[21] For those who favoured chromosomal interpretations of heredity,
this development resulted in a new cytological account of the growth—including the
meiotic growth—distinctive of sexual generation and a new cytological theory of sex
determination.

It was in pursuing the implications of such developments that Muller, the most
ambitious cytological and Mendelian theorist in the decade from 1910 to 1920, was led
to the theory of life itself as entailing growth and to the theory of sex as not entailing
rejuvenescence. These are the themes made explicit, especially, in three conference
papers: 'Variation due to change in the individual gene' (1921), 'The gene as the basis of
life' (1926) and 'Some genetic aspects of sex' (1931).[22]

To integrate his gene theory with his theory of how life entailed growth, Muller drew
on L. T. Troland's conjectures concerning autocatalytic and heterocatalytic powers in
crystalloid and colloid substances.[23] As Arnold Ravin has emphasized, Muller's argument here moves from present powers to conclusions about the earliest steps in evolution.[24] Living, involving growing, involves autocatalysis; but the autocatalyst, being
mutable, can change structure so as to have new effects in heterocatalysis while continuing to propagate itself. Now, such a system could itself only result from biological
evolution involving reproduction, variation and natural selection; an evolution that
must, Muller argues, itself have started with a simple, autocatalytic agent, a gene.
Through this reasoning, therefore, evolutionary change is reconciled with the conservative actions intrinsic to life as involving growth.

As to sex, Muller concludes that it is not necessary but is highly beneficial for evolution. That it is not necessary he deduces from Weismannian premises: that there is no
ultimate protoplasmic need for rejuvenescence of the germ-plasm through sexual union,
and that mutational variation in hereditary determinants is not consequent on panmixia,
the relaxation, that is, of natural selection with free interbreeding. Hence, fertilization
and so sex is not needed to rejuvenate flagging protoplasm, nor is crossing and so sex
needed to induce mutations.

The full benefit of sex is deduced from anti-additivity, indeed anti-multiplicativity,
premises. Advantageous gene variations can spread in a population far quicker with sex

21 For a contemporary version of this point, see H.H. Newman, *Evolution, Genetics and Eugenics*, revised
edn, Chicago, 1925, p. 321: 'though Mendel's laws appear to be merely laws of hybridization, they have a
much wider application: *they are really the laws of sexual reproduction*.' For Mendelism and sex determination, see Farley, *Gametes and Spores* and S. Gilbert, 'The Embryological Origins of the Gene Theory,' *Journal
of the History of Biology*, (1978), **11**, pp. 307–351. For the state of discussion before Mendelism see P. Geddes
and J.A. Thomson, *The Evolution of Sex*, revised edn, London, 1901.

22 The three papers appeared, respectively, in *The American Naturalist*, (1922), **56**, pp. 32–50; *Proceedings of the International Congress of Plant Science*, (1929), **1**, pp. 897–921 and *The American Naturalist*,
(1931), **64**, pp. 118–138. The first and parts of the other two are reprinted in H.J. Muller, *Studies in Genetics*,
Bloomington, 1962. For Muller's thinking at this period the indispensable biography is E.O. Carlson, *Genes,
Radiation and Society: The Life and Work of H.J. Muller*, Ithaca, 1981.

23 On Troland and Muller's debts to him, see R.C. Olby, *The Path to the Double Helix*, London, 1974,
pp. 112–113, 146–7 and 435–6.

24 A.W. Ravin, 'The Gene as Catalyst; the Gene as Organism,' *Studies in the History of Biology*, (1977),
1, pp. 1–47.

because they can thereby enter into combinations whose advantage may exceed the sum, indeed the product, of the advantages that those mutational differences each confer. In this benefit, sex, whose very 'essence' is 'Mendelian recombination', is, therefore, Muller urges, accessory to the primary process of gene mutation. So, the understanding of sex in evolution, as enhancing rates of mutation spread, is dependent on the new views, owed to Muller himself most prominently, of gene mutation as a process generating and introducing gene variation at a given rate in any population.[25]

Sewall Wright had his own very different reasons for assimilating Troland's conjectures as to life in relation to autocatalytic and heterocatalytic processes. Always active in pursuit of physiological genetics, and drawing on his conclusions from that pursuit when formulating his evolutionary theories, Wright sought a coherent structure of theory that moved all the way from genes, as involved in such catalytic processes, to the advantages of sex, and beyond that the advantages of genetic drift, in speeding the production and spread of optimal genic permutations.[26]

Dobzhansky's *Genetics and the Origin of Species* brought together many traditions of research: some in natural history, some in experimental genetics, some of these traditions being Western, some Russian. Among its many debts, none was more decisive than those to the new views of genic mutation principally articulated by Muller, and those to the theoretical population genetics of selection in interaction with drift, as expounded by Wright who had often been agreeing and disagreeing with Fisher in working out where he stood.[27] So, although there is in Dobzhansky's text no overt treatment of the generation-theoretic issues of life, death and sex, he is linked, through such direct and indirect sources as Muller, Wright and Weismann, to a historical succession descending from the published Darwin and, beyond him, the Darwin of the original zoonomical explanatory programme.

V

A discussion as impressionistic as this, and so broad in the brush strokes it deploys, may appear at best suggestive rather than conclusive. It will, then, be as well to suggest, before closing, what specific recommendations for historical practice would seem to be implied by this discussion. Five, in particular, may be especially appropriate on the present occasion.

25 Muller, 'Some genetic aspects of sex,' section 1, which is reprinted in Muller, *Studies*, pp. 469–473. On these new views of mutation, see R.C. Olby, 'La Théorie génétique de la selection naturelle vue par un historien', in *Colloque R.A. Fisher et l'Histoire de la Génétique des Populations* in *Revue de Synthèse*, (1981), 103–4, pp. 251–290.

26 Ravin, 'The Gene,' discusses Wright's debts to Troland. More generally, see W.B. Provine, *Sewall Wright and Evolutionary Biology*, Chicago, 1986. Wright went beyond the theory of life to a panpsychical theory of mind. His philosophical views, compared and contrasted with R.A. Fisher's, will be treated of in a paper now in preparation by John Turner and myself.

27 For the many sources of Dobzhansky's synthesis, see, E. Mayr and W.B. Provine, (eds), *The Evolutionary Synthesis: Perspectives on the Unification of Biology*, Cambridge, Mass., 1980. His relations with Wright are covered in Provine, *Sewall Wright*. The first edition of *Genetics and the Origin of Species*, New York, 1937, has been reprinted recently (1982) with an introduction by S.J. Gould, the publishers as before being Columbia University Press.

First, those of us who have been raised to think of ourselves as historians of evolutionary biology should not think that we are getting off the subject when the trail leads into physiological terrain, no matter how physiology is understood at the period and in the context concerned.

Secondly, it may sometimes be more enlightening to think of many nineteenth-century scientists as working on ground already occupied by eighteenth-century theorists of generation, rather than thinking of them as making contributions to the twentieth-century sciences of cytology, embryology, genetics and so on.[28]

Thirdly, there may be nothing ultimately indefensible about the usual battery of distinctions that has dominated the historiography of this subject over the last thirty or so years: particulate *versus* blending inheritance; soft *versus* hard heredity; discontinuous *versus* continuous change; Mendelians *versus* Biometricians; experimentalists *versus* naturalists; genotype *versus* phenotype; transmission *versus* development, especially. However, the time may have come to concentrate no less on issues, such as the life, death and sex issues, that cut across the comparisons and contrasts that this battery has so valuably prompted in the past.[29]

Fourthly, although the suggestions of the present paper have been made in an intellectualist mode, their possible social–historical implications may be no less useful in opening up fresh interpretations of familiar subjects. The ideological and institutional history of evolutionary biology has tended hitherto to concentrate on such developments as the eugenics movements, where the role of hereditarian and selectionist thinking is manifest. Through the issues of life, death and sex, evolutionary biology will surely prove to have cultural sources beyond those encountered when the selective breeding theme is made central.

Fifthly, we shall have to do as I have not done here, and that is pay proper attention to the way the issues we are following are sometimes expressly applied to empirical inquiries (as they are in Weismann); while at other times (as in Muller) they are made explicit only when synoptic texts are constructed for occasions (such as conferences) when theoretical rather than empirical stocktaking is in order. However, when all such complications are allowed for, it will surely prove fruitful to take seriously the topical and thematic continuities presented by the issues of life, death and sex over the century after Darwin opened his *Notebook B*.

---

28 It is telling that late nineteenth-century biologists often wrote on eighteenth-century generation theory, especially in connection with the revived issue of preformation and epigenesis.

29 For example, it is instructive to see how Churchill's treatment of the Weismann-Spencer controversy over 'soft' heredity leads to the themes stressed here: 'The Weismann-Spencer Controversy over the Inheritance of Aquired Characters,' in E.G. Forbes (ed), *Human Implications of Scientific Advance. Proceedings of the XVth International Congress of the History of Science, Edinburgh 10–15 August 1977*, Edinburgh, 1978, pp. 451–468. For a very full account of Lamarckian theorizing and other alternatives to the Darwinian selectionist tradition in this period, see P.J. Bowler, *The Eclipse of Darwinism: Anti-Darwinian Evolution Theories in the Decades around 1900*, Baltimore, 1983. Jan Sapp has been showing that a further contrast—*nuclear* versus *cytoplasmic* inheritance—can provide new insights into the intellectual and institutional history of genetics from 1900 on. See his paper 'The Struggle for Authority in the Field of Heredity, 1900–1932: New Perspectives on the Rise of Genetics,' *Journal of the History of Biology*, (1983), 16, pp. 311–342 and his book, *Beyond the Gene. Cytoplasmic Inheritance and the Struggle for Authority in Genetics*, New York and Oxford, 1987.

XIII

BIOLOGY AND PHILOSOPHY (INCLUDING IDEOLOGY): A STUDY OF FISHER AND WRIGHT

1. INTRODUCTION: HISTORIOGRAPHY, BIOLOGY AND PHILOSOPHY

The task undertaken in this paper will make better sense if I relate it to two familiar rationales for doing any work in the history of science.[1] They are rationales, indeed, that are matched in doing the history of many things besides science. The first is that if there is no collective quest for critical history, then myths and legends flourish, so that we have to settle for Rusk's version of Vietnam or Thatcher's invocations of the Victorian era. The second is that the future is unavailable and the present transient, so that the past is the only long run accessible to us. If we wish to understand how something — science, the economy or whatever — goes over the long haul, it is to the past that we must turn.[2]

Now, there are currently under active discussion a number of proposals about the long run of evolutionary theorising in biology, Three, in particular, will concern us here.

First, several writers, including Garland Allen and John Greene, have urged that the original Darwinism of Darwin's supporters should be read as mechanistic and materialist. The question has been raised, therefore, as to how subsequent evolutionary theorists stand in regard to that mechanistic materialism. Allen suggests that some have progressed beyond it to holistic materialism, others to dialectical materialism. Not surprisingly, however, Allen's and Greene's shared assumption has been challenged. Most recently, it has been challenged by Robert Richards in his comprehensive history of evolutionary theories of mind and behavior. Anyone discussing such figures as R. A. Fisher (1890—1962) and Sewall Wright (1889—1988) has to decide, then, whether they conform to the Allen historiography or, rather, to Richards's alternative.[3]

Second, Peter Bowler has argued that there was an evolution revolution in the nineteenth century, but that it was a non-Darwinian revolution, in that there was no widespread acceptance of Darwin's own views about the ramifying, directionless course and opportunistic, selectional causation of evolution. Later, Bowler holds, there was, in the twentieth century, a Mendelian revolution that moved evolutionary theory away from nineteenth-century, non-Darwinian views; especially away from non-Darwinian developmentalism, the interpretation, in other words, of evolution as a process analogous to individual development. So, on Bowler's account, it was this Mendelian revolution that prompted a delayed Darwinian revolution, by allowing something like Darwin's own non-developmentalist position to become widely adopted for the first time. We are invited, accordingly, to ask how far any mid-twentieth-century evolutionary theorists conform to Bowler's thesis.[4]

Third, there are now several volumes devoted to what has been called the Probabilistic Revolution, that shift in thought whereby chance and chances, matters stochastic and statistical, rose from marginal insignificance in the early seventeenth century to pervade everything from science and politics to sport and commerce in our own time. Evolutionary theorising has often involved probabilistic thinking, so that any study of particular evolutionary theorists offers an occasion to confirm or correct various views now current as to how probabilistic thinking in the sciences has developed over the long run.[5]

Fisher and Wright turn out to be highly instructive figures when related to such general historiographical issues as are raised by these three proposals. Most especially, we can be enlightened if we approach them, as biographers have: that is, with no predeliction for dividing their work from their lives, or for dividing their scientific work from their philosophical and even their ideological sympathies and commitments.[6] We can decline such predelictions and take a more comprehensive and integrated view, moreover, without making any presumption of explanatory priority in favor of one element or another in the biographical picture; without, say, presupposing that scientific positions are always dictated by prior religious beliefs, or that methodological reflections are always *post hoc* rationalisations of prior scientific convictions.

The desirability of not making any such presumption in the present instance is apparent once we consider what understanding of the term philosophy is appropriate here. In one sense, a scientist's philosophy is what is manifested in his or her practice of construing in a distinctive way very general questions about some scientific subject. Fisher is well known, for example, for comparing and contrasting his Fundamental Theorem of Natural Selection with the Second Law of Thermodynamics. To examine the presuppositions Fisher was making in such comparisons and contrasts is to examine his philosophy. In another sense, a scientist's philosophy is manifested in any position he or she takes on such issues as mind in relation to body or determinism and free will. To examine Fisher and Wright's views on these traditional issues is to examine their philosophies. Finally, there is a third sense of philosophy, one that includes ideology. Anyone's philosophy, whether scientist or otherwise, is being examined, in this sense, when we look to see how he or she stands in relation to liberalism or to socialism, the exemplary ideologies or social philosophies. It is a measure of how pervasive are philosophical themes, in the thinking of Fisher and Wright, that for our purposes here we need to consider all three senses of the term philosophy.

Professional scientists in the twentieth century do not usually develop and publish explicit philosophical views. To that extent Fisher and Wright are not typical, however, we should not let that point mislead us into thinking that their philosophical reflections were seen by them as peripheral excursions confined to moments when they were off duty. Indeed, we have to avoid begging any questions as to how their philosophy was related to their science. The relationship was, indeed, very different in each case; and that difference is something to be paid close attention, rather than given some characterisation grounded, *a priori*, in general preconceptions as to how professional science and amateur philosophy are likely to be associated with one another.

Let it suffice here to emphasise that Fisher and Wright were both serious amateurs when it came to philosophy. Wright prepared for his biographer Provine a table of interests and achievements for himself and others in the Wright family. As with mathematics and theoretical biology, Wright gave himself four stars underlined for philosophy. His autobiographical remarks in his philosophical writings make it clear that the reading decisive for his views on metaphysical issues was mostly done in the period 1912—14, and so before he had completed his doctoral studies at Harvard. Only much later, however, did he give those views a public airing. To the surprise of his audience, he devoted his Presidential Address to the American Society of Naturalists, in 1952, to a discourse that ran all the way from the hierarchy of organisms — including cells and species as organisms — to a dual aspect panpsychist analysis of mind, will and freedom.[7] Likewise, Fisher's biography, his correspondence and his reviews, his articles and his

234

lectures, show that he was actively reading and reflecting on philosophical subjects from his student days on. By the 1930's he was making a full debut as a philosophical writer: in 1932 in a Herbert Spencer Lecture on the *Social Selection of Human Fertility* and in 1934 in an article, on 'Indeterminism and Natural Selection', for the first volume of *Philosophy of Science*. These publications had a sequel in 1950 in his Eddington Memorial Lecture: *Creative Aspects of Natural Law*.[8] Neither Fisher nor Wright felt any reticence or embarrassment regarding philosophy. For neither of them, then, did the designation of some view as philosophical carry any derogatory connotations. Quite the reverse. Reviewing Wright's first, full-length exposition of his evolutionary theory — the 1931 paper, 'Evolution in Mendelian Populations' — Fisher explained that, apart from "the scientific conclusions" established independently by workers in several countries, Wright "makes some philosophical observations on the nature of the evolutionary process, which are of great interest, although necessarily more personal and subjective". Fisher then gave a sympathetic but critical account of Wright's conclusions about the respective roles in evolution of factors making for heterogeneity and factors making for homogeneity: the conclusions later known as Wright's shifting balance theory of evolution.[9] Fisher's categorising of these views as philosophical implied no castigation. The two men were still on very friendly terms at this time. When engaged in discussing, even without agreement, the conditions most favorable for adaptive evolution, both saw themselves as supplementing and complementing secure scientific results with more uncertain conclusions that were properly called philosophical but none the worse for that.

2. A DECADE OF COLLABORATION AND AMICABLE DISAGREEMENTS: 1924—34

A study of Fisher's and Wright's biology and philosophy does best to begin with Fisher. Once we have a sense of Fisher's views, we can ask how and why Wright agreed and disagreed with him as he did. To proceed thus is historically and biographically appropriate, because Fisher was one step ahead of Wright at a crucial moment; so that Wright was making up his mind about Fisher's thinking rather than the other way round.

The crucial moment came in 1922. For it was then that Fisher published his paper 'On the Dominance Ratio'. Despite its misleadingly narrow title, this paper gave, in fact, a general treatment of population genetics. It was, moreover, the first paper to analyse the genetics of populations in the way that was quickly to become standard. For Fisher here considers such causes of change as selection and random sampling error insofar as they influence the statistical distribution of gene frequencies. So, a population is treated as a collection of genes, with each gene having a certain frequency because present in a certain proportion of individuals; and it is enquired what the statistical distribution of those gene frequencies is. Thus, if that distribution is a normal distribution, so called, as represented by the familiar bell curve, then many genes will be present in about half the individuals, while only a few will be present in either a great majority or a small minority. Such an analysis allows, therefore, for a quite general and abstract representation of the variability of a population.

Evolution, on such a representation, can be analysed as change in the distribution of gene frequencies. For, under Mendelian assumptions, the distribution is stable in a large population with random mating and no mutation, no selection and no migration. Accordingly, in this paper, Fisher did what no one had done before. He asked how such factors as dominance relations, mutation, selection and random extinction of genes in finite populations would affect the distribution; and he derived expressions for the effects of various mutation rates or selection intensities and so on. He also hinted at a conviction that he would never give up: namely, that adaptive evolution is most effectively produced in a large, randomly breeding population subject to sustained natural selection of very small heritable differences.

The strength of this conviction was already apparent in Fisher's review, the year before, of A. L. and A. C. Hagedoorn's book *The Relative Value of the Processes Causing Evolution*. However rare mutations may be, they can still be an "important factor in maintaining the variability of species", Fisher insisted, because "their frequency of occurrence will increase proportionately to the number of individuals in the species". A "very small rate of mutation" will suffice in "a large population" both to "continually supply new forms for the action of natural selection" and to "counterbalance" very easily the loss of variability through random extinctions of genes.[11]

The importance of large population size, for Fisher's understanding of the respective roles of random and selectional influences on the fate of mutant genes, is apparent also in a short, popular piece written at this same time. In 'Darwinian Evolution of Mutations', Fisher explained how he saw the new Mendelism allowing for a vindication of Darwin's selectionism. For this vindication, two novel findings are decisive, Fisher urged. First, the ordinary differences between parents and offspring are due not to new influences producing new characters, but to the segregation of existing Mendelian genes yielding new gene combinations. Second, new genes do arise but only very infrequently; however, heredity being particulate not blending, the maintenance of heritable variability is ensured by segregation, and so does not require a high rate of mutation. Any new gene arising by mutation — "in a single individual of population consisting of some thousands of millions" — will have its populational fate determined partly by chance, partly by selection. In an early phase, even if the gene is subject to favorable or unfavorable selection, chance will play the major role, because the initial frequency is so low. However, if the mutant gene survives this phase, and if it is advantageous on average, then selection will ensure that its frequency continues to rise, even though the advantage is very small. Hence, then, the benefit of sexual reproduction: mutation is "a leap in the dark", much more likely to fail adaptationally than succeed, especially if it has a large effect. The benefit of sex comes from maintaining the variability of a species with the minimum of mutations; with, that is, the "greatest stability of the reproductive processes". For consider any population "differing in a great many Mendelian factors, as all sexual populations are found in nature to do;" here "a single mutation may enable thousands of genetic combinations to be tested, and if any of these should happen to be very advantageous, it will by selection become the predominant type." Thus is "manifest" the benefit of "Mendelian inheritance of sexually reproductive organisms", most especially when "complex adaptations have to be made to a slowly changing environment," as Fisher argued.[12]

These three publications of 1921—1922, taken together, constitute an opening initiative by Fisher in his incipient campaign to establish a Darwinian view of evolution on the new foundations provided by his own prior synthesis of biometry and Mendelism. Readers of his earlier publications could have discerned his Darwinian loyalties, but only in 1922 could they have recognised the main strategies in the new campaign.

Two years after the dominance ratio paper appeared, Fisher and

Wright met and talked in the United States. Wright had not seen the paper, but Fisher sent him an offprint on returning to England. Wright then set about making his own analysis of the statistical distribution of gene frequencies. His mathematical techniques were very different from Fisher's. Integral and differential calculus had been Fisher's main resources. Wright used his own method of path coefficients that he had pioneered in previous papers on inbreeding, outbreeding and selection. However, despite these differences in the mode of mathematical attack, the two men converged on explicit agreements, often developed in correspondence, about the quantitative treatment of mutation rates, selection and so on as determining the statistical distribution of gene frequencies. By 1931 they were satisfied that the last mathematical discrepancies had been resolved.[13]

By this time Fisher had published his book of 1930, *The Genetical Theory of Natural Selection*, and Wright his long 1931 paper on 'Evolution in Mendelian Populations'. Now, Wright was composing a very full review of Fisher's book, while Fisher was writing a short review of Wright's paper. What the book, the paper and the two reviews, together with the correspondence, make plain is that despite that mathematical consensus on those quantitative topics, there was no agreement at all as to the conditions most favorable for adaptation and progress in evolution. For while Fisher was arguing, as he had ten years before, for a large randomly mating population subject to selection, Wright concluded that the optimal condition was a large species broken up into small partially-isolated local populations with a considerable degree of inbreeding within these populations, and low rates of inter-breeding among them. In his review of Fisher, Wright summarised Fisher's account of the transformation and multiplication of species in evolution, and declared Fisher's "conception of evolution" to be "pure Darwinian selection". Distancing himself from the "indefinitely large population" of Fisher's "scheme", Wright urged that, on his own scheme, there would be "a rapid differentiation of local strains, in itself non-adaptive, but permitting selective increase or decrease of the numbers in different strains and thus leading to relatively rapid adaptive advance of the species as a whole". Thus between "the primary gene mutations, gradually carrying each locus through an endless succession of allelomorphs, and the control of the major trends of evolution by natural selection", he would "interpolate a process of largely random differentiation of local strains".[14]

It is well known that this disagreement between Fisher and Wright was to prove enormously influential and fruitful over the next half century and more. What needs to be emphasised here is its independence from the mathematical consensus between the two men. The disagreement is not traceable to differences in their mathematical techniques. On the contrary, it existed despite the mathematical consensus they had reached by different technical routes. Nor can personality conflicts be invoked at this juncture. Later, starting in 1934, Fisher and Wright fell out personally and irreversibly, so as to be incapable of discussing any differences of opinion amicably. But, before that falling out, there was a whole decade of collaboration and amicable disagreement. Fisher's biographers Box and Bennett and, even more definitively, Wright's biographer Provine have established the character and content of the interaction over this decisive decade; and the account given here draws throughout on Provine's writings. What our comparative and contrastive biographical task next requires of us is therefore plain enough. We need to look at the biology and philosophy of both men in the years before that decade. By doing that we can hope to understand better the remarkable mixture of agreement and disagreement the emerged in those years.[15]

3. FISHER: SCIENTIFIC AND PHILOSOPHICAL THEMES OVER HALF A CENTURY

It is now well established that the undergraduate Fisher (1909—12) had already developed four clusters of interests — apart from those in his degree subject, mathematics — that were to last him the rest of his life: Mendelism, biometry, Darwinism and eugenics. It is also clear that those commitments were being actively integrated even at this early age. Nor are the main lines of the integration in doubt. Mendelism and biometry were not to be opposed to one another, as they were still by many in England, including the leading Mendelian, Bateson and the leading biometrician, Pearson. Rather, as Fisher saw it, the Mendelian account of hereditary factors, to be known in a few more years as Mendelian genes, was to be reconciled with the biometric estimates of the correlations among the quantitative characteristics of relatives, whether parents and offspring, siblings or cousins. Moreover, the principal premise in this reconciliation was already the one that would be made in 1916, when Fisher wrote the paper of 1918, 'On the

Correlation of Relatives on the Supposition of Mendelian Inheritance':
the premise, that is, that the continuous quantitative variation in a
strongly heritable character, such as height appeared to be in some
human populations, is due to many Mendelian factors each of small
effect.

Given this reconciliation of Mendelism and biometry, the integration
of Mendelism and Darwinian selectionism could take, also, the form it
would take later: that is, with Mendelian factor differences providing
for the heritable variation that would be the material for selective
breeding, as practised by man on the farm or entailed by the struggle
for existence in nature.

For the young Fisher, this Mendelian selectionism offered, in turn, to
integrate evolution with eugenics. Taking the human species to be
originally a gradual, natural product of Mendelian heredity worked on
by Darwinian selection, he could ask how far selection among different
human societies may make for social progress in the future.[16]

Now, although this much about the undergraduate Fisher is plain, at
least in outline, we have to be wary of finding all his beliefs and attitudes, even of the early 1920's anticipated ten years before. To give just
three examples, a discriminating history of his thinking over the decade
after Cambridge would have to emphasise how his eugenic views
changed markedly as individual selection within societies joined selection among societies as a principal concern; it would have, too, to
emphasise how his Mendelism expanded from being merely a theory of
factors subject to linkage and segregation and so on, to being a theory
of genes that included, following Muller especially, new conclusions
about the nature and rates of gene mutations; and how his selectionism
was developed in contrasting gene frequency changes — and so gene
frequency distribution changes — due to selection, and those due
to such chance accidents of differential survival and reproduction
("Hagedoorn effect" Fisher would call it first) as would eventually, in
later years, be dubbed random genetic drift.

We also have to be wary of tracing Fisher's interests to a single
motivational source. Some years ago, I agreed with some others in
denying that Fisher was, in any strong sense, an evolutionary biologist;
recommending that we think of him, instead, as a Mendelian biometrician whose sole ultimate motivation was to establish a new Mendelian
eugenics. I am now convinced that this recommendation is inconsistent
with the full biographical picture, and that it is wrong in suggesting that

the two characterisations — evolutionary biologist and Mendelian eugenist — might be exclusive of one another. Fisher was both of these things, and he was not either of them to a lesser degree because he was the other. On the contrary, his commitment to the understanding of the causes of evolution and his commitment to the improvement of society — or at least to the prevention of deterioration in English society — were complementary and mutually reinforcing, although also sufficiently distinct if not separable that their sources require distinct, albeit integrated, biographical treatments.[17]

To see something of what such a treatment might involve, it will be appropriate to concentrate briefly on Fisher's selectionism. The depth and intensity of Fisher's preoccupation with selection is familiar to all who have read in, or even about, the man. However, it will be urged here that this depth and intensity were even more extreme than has yet been recognised. For a start, recall Fisher at high school choosing a set of Darwin's volumes as a prize award. This moment alerts us to Fisher's lifelong tendency to identify evolution as a topic with Darwin as an author, in that there will be for him no making up one's mind about anything concerning evolution without making up one's mind about Darwin's view of evolution. Fisher's passionate, perennial concern with evolution and Darwin constitute good grounds for insisting that he was an evolutionary biologist. The identification of evolution and Darwin entailed, for Fisher, another identification, because he always read Darwin as making one supreme contribution to evolution as a topic: namely, natural selection. So evolution, Darwin and natural selection turn out to be, for certain purposes, one topic. Now, Fisher did open the very Preface of his book *The Genetical Theory of Natural Selection* of 1930 by declaring evolution and natural selection to be distinct, and by proposing that natural selection be studied in its own right and independently from its explanatory deployment as a cause of evolution. However, in making natural selection an independent topic, Fisher did not make evolution one, too.

4. ADAPTATION IN A TWO-TENDENCY UNIVERSE

As dozens of Fisher's publications show, he was always of the view that evolution was primarily an adaptive process — a continuing adjustment of complex structures and actions to environmental change — so that the primary constraint on a theory of evolution was that it explain adaptation as a process and so as a product. Natural selection ("Darwinism") and the inheritance of acquired characters ("Lamarckism") were the only two theories to really offer to meet this constraint, Fisher liked to argue; but since Darwin's day Lamarckism had been shown to be inconsistent with what is known of variation and heredity. There can, then, be no acceptable thinking about evolution that is not primarily thinking about natural selection; a causal process, as Fisher recalled, that had been valued, and rightly, by the very first Darwinians as a "known cause", one that is evidenced independently of the facts it was invoked to explain.[18] Fisher's writing conforms consistently to this limitation, in that whenever evolution is the subject, the theory of natural selection — as the only acceptable theory of adaptation, and so of evolution — is what has to be discussed. Fisher accordingly once opened a discussion on the theory of natural selection by going so far as to say:

Theories of Evolution are of two kinds, those that, in Professor Watson's words, "are explanations primarily of adaptation and only secondarily of the origin of species", and those which fail to account for adaptation. To the first class belongs the theory of the inheritance of acquired adaptations called Lamarckism, and the theory of the natural selection of innate adaptations. For these two theories evolution *is* progressive adaptation and consists of nothing else. The production of differences recognisable by systematists is a secondary by-product, produced incidentally in the process of becoming better adapted.[19]

Now, we can not discern just when and how Fisher first reached this extreme adaptationist-selectionist position. The evidence suggests, however, that it was always there, in that it did not replace any earlier, different position; it was, rather, that as his thinking about evolution developed over the Harrow and Cambridge years, this outlook came to be more and more an entrenched foundation for any subsequent reflections. What we can be more sure of is that, from 1912—22, two presences in his life were to work upon Fisher so as to reinforce, even more strongly, this incipient foundation. Of the two, the first in time was statistical mechanics as represented by an eminent master of that branch of physics, James Jeans, Fisher's principal teacher at Cambridge for a postgraduation year (1912—13) devoted to statistical mechanics and quantum theory. The second was the Darwinian heritage and legacy itself, as represented by Leonard Darwin, a childless mentor who came, in effect, to adopt Fisher as the grandson of Charles Darwin that he never had himself, but whose place Fisher quickly came to take within a year or two of moving from Cambridge to London.

This conjunction — of statistical mechanics and Darwinian patronage — proves to be such a telling one for any effort to understand Fisher's entire biological and philosophical outlook, throughout his whole life, that it is as well to anticipate, here, something of the subsequent analysis, in order to convey in outline its larger significance. One strategy in doing this, in dramatic terms, is to say that Fisher's universe is going to be, ultimately, a two-tendency universe. There will be the entropic tendency which, if not countered, takes everything toward more probable, and less organised, states. This is the way down. Unique in its countering of such degenerative, deteriorational trends is the counter-entropic tendency entailed by selection as it takes anything subject to its workings toward less probable and more organised states. This is the way up.

The dramatisation can be made more personal, because the way down and the way up have each a principal interpreter; so that Fisher's natural science has ultimately a two-hero history: Ludwig Boltzmann (1844—1906), as leader of a triumvirate also including James Clerk Maxwell and Willard Gibbs, and Charles Darwin (1809—1882) senior member of a partnership with Wallace. For these are the nineteenth-century authors of the two great probabilistic insights that have, as Fisher saw it, set the decisive precedents for all subsequent thinking about the inanimate and animate creation, including man himself.

To talk of creation and of man as a creature, who is himself creative, is entirely in order here. Fisher never wavered in the Anglican, Christian faith he was raised in, and he was eventually to preach occasional sermons in his college's chapel at Cambridge during his years there as Professor of Genetics. His science always sought causes, and causation he would view, quite explicitly, as probabilistic and creative, indeed creative because not deterministic but indeterministic. His indeterminism was to be developed to integrate his Christianity, with its theory of creation and creativity, and his science, his Boltzmannian and Darwinian science. Likewise, in integrating his Christianity with his eugenics: the indeterminism allows for man to emulate God's own work by freely intervening in social life, in the institution of those eugenic measures that alone can ensure the permanency of English civilisation; for these measures — suitable family allowances, say — can offset the social decline otherwise arising from the diminished fertility that accompanies the greater contribution to society made by those gifted individuals who have risen to take their rightful place in the middle classes. Here, Fisher joins his scientific and political forces with Major Leonard Darwin, whose zeal and energy on behalf of the eugenics movement were often directed to the membership of the Eugenics Society whose President he was from 1911 to 1928. Gratefully and enthusiastically did Fisher, for his part, discharge his filial and grand-filial roles as eugenist and evolutionary biologist in the years culminating in the publication in 1930 of *The Genetical Theory of Natural Selection* with its dedication to the former President. What is more, for three more decades until his death in 1962, Fisher was always building upon, and never repudiating, the outlook on the universe and on science represented in his career by this early conjunction of James Jeans and Leonard Darwin.

5. A SUGGESTED READING OF FISHER'S THEORY OF NATURAL SELECTION

To confirm this two-tendency and two-hero reading of Fisher's world and Fisher's thought, it proves useful to do three things, before leaving Fisher for Wright. The first is to offer some suggestions as to what may be implicit, although not explicit, in Fisher's understanding of evolution by natural selection.[20] Here, necessarily one cannot document directly the evidence for the interpretations offered, but one can attempt to show how those interpretations for various topics cohere into a plausible overall view of Fisher's position. The second is to look at moments when Fisher brings into explicit conjunction his thoughts about statistical mechanics — and related subjects such as thermodynamics and the kinetic theory of gases — and his scientific analyses of evolution by natural selection. Finally, the third is to follow Fisher himself in his overtly philosophical reflections, where, again, that conjunction is often brought to bear on his Christianity and his eugenics.

Before moving to the first of these tasks, there is a point to be made about all three. Throughout, the discussion will be abstract and general, in that rather little contact will be made with the precise details of what Fisher had to say about, say, mutation rates in *Drosophila* flies or mimetic resemblances in *Papilio* butterflies or the measurement of the average excess and average effect for any gene substitution. This lack of contact is in no way meant to suggest that such detailed matters are irrelevant. On the contrary, I would stress that one virtue in an abstract and general discussion is that it may allow for light to be thrown on such specific matters. Fortunately, Fisher's treatments of several such relevant matters have received extensive attention and clarification in a number of recent publications.[21] The present discussion is, therefore, designed to complement those publications, by being adapted throughout to what they have already done.

Turning now to the first task, it requires us to appreciate what may have been implicit in Fisher's privileging of natural selection as the sole counter-entropic agency at work in the natural world. One implication would have been that the optimal conditions for natural selection — and therefore, for Fisher, for evolution — would be the conditions wherein other, entropic factors are least frustrating of the work of natural selection. This consideration most concerns mutation and random genetic drift. For these are both disordering processes. Mutation is a disordering failure of exact replication in the genes of an individual. Random drift is a disordering failure in a population to exactly replicate its genetic structure.

Mutation is, obviously, a necessary condition, in the very long run, for the continued work of natural selection, because it is only by gene mutations that new genes, rather than new permutations of old genes, can arise. However, thanks to inheritance being particulate, not blending, very low rates of mutation can suffice to provide adequate material for selection, and the larger the population the lower the rate that suffices. Moreover, just as low rates are least disordering so are small effects; so, by the early 1920's, Fisher is emphasising that empirical findings about mutations make it reasonable to assume that mutations are low in their rates, small in their effects, very rarely advantageous and mostly initially masked in their effects because initially recessive. For their being all these things allows natural selection to be vindicated as the only cause that can determine the direction of evolution.

Random drift is an inevitable concomitant of selection, in that any population that is subject to selection is subject to drift. But the larger it is the less subject will it be to drift without being less subject to selection. Drift may be important in determining what mutations survive beyond initial rarity. But, beyond that, drift has no contribution to make to the adaptive adjustment of the whole of a large population to gradual environmental change.

So far what has been said concerning Fisher's thinking about genes and gene mutations may seem overly simple. But that is precisely what is to be grasped: namely, that Fisher prided himself as having seen that in its essentials this topic was a simple one, for the purposes of the theory of natural selection. For, after all, consider how few are the assumptions he feels it necessary to establish in Chapter One of the *Genetical Theory* — the chapter on 'The Nature of Inheritance' — before proceeding to the second chapter on 'The Fundamental Theorem of Natural Selection'. His strategy in that first chapter is, primarily, to introduce only the most basic assumptions about inheritance that were not available to Darwin and that can, now that they have been secured by recent genetical research, vindicate Darwin's theory, natural selection. The strategy, thereafter, is, first, to reach the most comprehensive statement possible about natural selection — the Fundamental Theorem — without involving mutation at all, as a special topic; and, then, second, to introduce mutation in three chapters, one on dominance as something that some mutations eventually acquire thanks to selection, and two on variation as determined by mutation and selection together. A further chapter on sexual reproduction and sexual selection can complete the quite general treatment of natural selection, before it is applied to mimicry and finally to man and society.[22]

The workings of natural selection can, then, be exhibited, throughout, as necessary and sufficient for all adaptive progress, right up to the moment, that is, when the conditions for a permanent human civilisation are confronted. For while natural selection can be shown, in Fisher's view, to ensure the progress from barbarous social life to civilised society, its continued action, if not actively countered by eugenic measures, will lead to an inevitable deterioration in any civilised society, thanks to the heritability of fertility and the social promotion of the less fertile.

Now, it is here that the earlier adaptationist stance has to be transcended. For, in order to understand the causes, consequences and conditions for countering this social promotion of the less fertile, Fisher has to engage the economics of social promotion and so of property, wealth and class in a society such as his own; and he has no way to represent these economic features of a civilised society as constituting an environment for the society to which it can be seen as adapting itself. Nevertheless, these features serve as a constraint on the eugenist, because Fisher takes the conservative liberal view that the best way forward is not to move to a different economic system, but to reverse the correlation between social promotion and infertility by introducing economic inducements, such as tax measures, that would provide sufficient incentives, in the existing economic system, for those individuals, whose heritable characteristics have enabled their social promotion, to raise more children and so keep up the populational frequency of the genes responsible for those characteristics. So, in this context, certain distinctive elements in Fisher's selectionism are plainly present: in particular, selection is worked through some influence exercised throughout a population, in a mass selection, therefore; however, selection is not acting counterentropically to produce and maintain highly improbable, complex adaptations to changing environmental circumstances.

Is Fisher's Mendelian selectionist eugenics for serving the permanence of civilisation in society a difficulty, then, for the reading given here of his adaptationist-selectionism? Surely not; for Fisher himself dwells explicitly on the reasons for a distinctive treatment of selection in civilised rather than barbarous societies. For one thing, differential fertility takes over from differential survival as decisive for the direction of change, once the shift is made out of barbarity into civilisation. This contrast entails, in turn, the contrast concerning adaptation as an adjustment to environmental circumstances; for whereas adaptation is always for survival and reproduction in, and so to, some particular circumstances, fertility as such is not interpreted as an accommodation to anything particular and circumstantial.

Such discontinuities, between the eugenics for civilised society and the genetical theory of natural selection for plant, animal and pre-civilised human life, suggest, therefore, two points about the interpretation of Fisher's biology and his eugenics. The first is that his evolutionary biology can not be reduced to any straightforward projection onto all nature of his eugenics. The second is that his eugenics had its own motivational sources in what is called here his conservative liberalism, with its meritocratic, possessive, individualistic assertion of middle class values within the existing economic organisation of society. An interpretative quest for the underlying unities in Fisher's thought must, therefore, look beyond his hereditarian, Mendelian selectionism as an element obviously common to his eugenics and his evolutionary biology. One is in a better position to discern what else has to be brought into such an interpretative quest, once a survey has been made of the comparisons and contrasts Fisher makes between his physics and his biology.

6. STATISTICAL MECHANICS, THERMODYNAMICS AND THE THEORY OF GASES, AND GENETICS AND THE THEORY OF SELECTION

The comparisons and contrasts that must concern us here are more diverse and complex than one might think at first. For a start, we have to notice when Fisher is invoking the precedent of statistical mechanics and when, rather, thermodynamics or, again, the kinetic theory of gases or, more generally, what he calls the kinetic theory of matter. Furthermore, he has various reasons for bringing these different precedents into his biological expositions; so that there is no way to reduce his invocation of them to a single formulation comprehending all cases.

One very early comparison appears in his undergraduate lecture, given to the Cambridge University Eugenics Society, on "Heredity". Biometrical work can yield beautifully certain results, he says. Dealing only in observations, biometricians avoid "all the difficulties of abstract theories"; moreover, although their observations are probably full of small errors, they seem able "to squeeze the truth out of the most inferior data". In every case the probable error can be calculated and, while possible error is unlimited, the probability of large errors is demonstrably very small. Fisher was "recently impressed" with this powerful feature of the theory of probabilities. Put a kettle on the fire and it will probably boil; not certainly, for it may freeze. The odds against this last outcome are very large, but it "remains a possibility, or so my 'theory of gases' tells me".[23]

Here, then, we have the young Fisher impressed with a commonplace analogy between probability in the theory of observational errors and in gas theory. A few years later, less commonplace arguments are appearing. In 1915, he and his co-author C. S. Stock, writing for the *Eugenics Review*, are castigating those who have opposed Darwinism to Mendelism, overlooking that whereas Darwinism is concerned with evolution, Mendelism concerns heredity rather than evolution as such. The mistake has led to serious misunderstandings regarding eugenics, and even ill-founded American legislation.

Raising as so often the issue of generality, Fisher insists that Mendelism is unlikely ever to "cover even the field of heredity"; while, by contrast, Darwin's theory does cover the field, in that it "explains and co-ordinates new facts". Mistakes, what is more, are more likely to arise from new specialised work such as the Mendelians pursue, than from a well-established theory such as Darwin's. Eugenists should hold to general principles; with Mendelism they are vulnerable to objections; on Darwin's ground they are not. The importance to eugenists of the broad principles of the *Origin* is paramount. Darwin's teaching, even if it were all that were available, says Fisher, "would not only allow but compel us to formulate eugenic concepts and proposals". The nature of this compulsion is clarified by two invocations of the kinetic theory of gases.[24]

The first is introduced to considering selection itself. Selection is the main cause of changes in the constitution of a mixed population. The existing and possible agencies of selection now, as always, provide the most fruitful field of eugenic research. "These agencies acting at large amidst a multitude of random causes", any one being the predominant influence on some particular individual, "nevertheless determine the progress or decadence as a whole". Likewise, then, in the "kinetic theory of gases", where the "several molecules are conceived to move freely in all directions with greatly varying velocities" but with a "statistical result that is a perfectly definite measurable pressure". The common feature here, then, shared by selection theory and gas theory is the reliability and predictability of the outcome when the individuals are numerous and the causes acting upon them independent. Allied to this lesson is the second. "Controversy may rage round the nature and properties of the atom yet our knowledge of general principles enables us to calculate gas pressures with accuracy". Here, we are "independent of particular knowledge about separate atoms, as in eugenics we are independent of particular knowledge about individuals". Of course such knowledge is important, interesting and useful, but it is, Fisher declares, "unnecessary alike for a general theory of gases and for a general theory of eugenics".[25]

This drive for generality in conforming to the precedents set by the theory of gases, is most explicit in Fisher's explanation, in his 1922 dominance ratio paper, of what he saw as his achievement in his 1918 paper on the correlation of relatives. That earlier paper attempted, he explained, "an examination of the statistical effects in a mixed population" of a "large number" of Mendelian factors. Now, apart from a continuing belief that the biometricians' results discredited Mendelian inheritance, there was another widespread misunderstanding regarding any such attempt. For it was generally believed that "the variety of the assumptions to be made about the individual factors" — assumptions about the dominance of particular factors, about the size of their effects, about their proportion in the population, about dimorphism and polymorphism, and about linkage, and also assumptions about "more general possibilities" such as homogamy, selection and environmental effects — made "it possible to reproduce any statistical resultant by a suitable specification of the population". It was important, therefore, Fisher recalled, "to prove that when the factors are sufficiently numerous, the most general assumptions as to their individual peculiarities lead to the same statistical results".[26] The stage is thereby set for the analogy with the theory of gases:

> Although innumerable constants enter into the analysis, the constants necessary to specify the statistical aggregate are relatively few. The total variance of the population in any feature is made up of the elements of variance contributed by the individual factors, increased in calculable proportion by the effects of homogamy in associating together allelomorphs of like effect. The degree of this association, together with the quantity which we termed the Dominance Ratio, enter into the calculation of the correlation coefficients between husband and wife and. between blood relations. Special causes, such as epistacy, may produce departures, which may in general be expected to be very small from the general simplicity of the results; the whole investigation may be compared to the analytical treatment of the Theory of Gases, in which it is possible to make the most varied assumptions as to the accidental circumstances, and even the essential nature of the individual molecules, and yet to develop the general laws as to the behavior of gases, leaving but a few fundamental constants to be determined by experiment.[27]

From what we know of James Jeans's work on the theory of gases, he would have made sure that Fisher, as his pupil, acquired a keen sense of the latitude permissible, regarding assumptions about the circumstances and the properties of gas molecules, when one is carrying out these kinds of derivations. Fisher's understanding of this analogy with his 1918 derivation of the correlations of relatives may well trace, therefore, rather directly to what he had learned in that postgraduation year at Cambridge.

The 1922 paper contains in its own summary a further analogy: "the distribution of the frequency ratio" for different hereditary factors is — in the absence of selection and random survival effects and so on — a stable one like that of "velocities in the Theory of Gases". Years later, Fisher was more explicit about the change of viewpoint that this analogy made possible. The frequencies of different genotypes define the gene ratios for the population, so "it is often convenient to consider a natural population not so much as an aggregate of living individuals as an aggregate of gene ratios; such a change of viewpoint" being like that familiar in gas theory where specifying the "population of velocities" is often more useful than specifying "a population of particles".[28]

7. THE ANALOGIES AND DISANALOGIES IN THE GENETICAL THEORY

The analogy between the two stable distributions was duly developed further by Fisher, so that by 1930 in the *Genetical Theory* he was using it in contrasting the different implications of the blending theory of inheritance and of the particulate theory. The two theories give quite different interpretations of the fact that brothers and sisters, whose parentage and so entire ancestry is identical, may differ greatly in their hereditary constitutions. A blending theorist sees here evidence of new and frequent mutations associated — witness the greater resemblance of identical twins — with temporary conditions at conception and gestation. "On the particulate theory it is a necessary consequence of the fact that for every factor a considerable fraction, not often much less than one half of the population, will be heterozygotes, any two offspring of which will be equally likely to receive unlike as like genes from their parents". Given "the close analogy between the statistical concept of variance and the physical concept of energy", the heterozygote may be thought of "as possessing variance in a potential or latent form", so that rather than being lost on the mating of homozygous genotypes "it is merely stored" in a form allowing it to reappear later.[29]

A population mated at random immediately establishes the condition of statistical equilibrium between the latent and the apparent form of variance. The particulate theory of inheritance resembles the kinetic theory of gases with its perfectly elastic collisions, whereas the blending theory resembles a theory of gases with inelastic collisions, and in which some outside agency is required to be continually at work to keep the particles astir.[30]

We are now, finally, prepared to appreciate the significance for Fisher of the analogy he is going to draw between his Fundamental Theorem of Natural Selection and the Second Law of Thermodynamics. For the Fundamental Theorem relates the rate of increase in fitness, that selection produces in a population, to the genetic variance in fitness that selection consumes in producing that effect, genetic variance that would otherwise be conserved; just as the Second Law concerns the progressive loss of energy available for work.

Fisher's own main statement of his theorem in 1930 reads thus: "*The rate of increase in fitness of an organism at any time is equal to its genetic variance in fitness at that time*", with the word "species" used instead of "organism" in a recapitulation at the end of the chapter.[31] Now, much has been done since Fisher to clarify how he or anyone else may have derived this law from his or other assumptions. What concerns us, here, however, are not the possible derivations, but Fisher's contrasts and comparisons between his theorem and the second law of thermodynamics. For this purpose, two glosses are valuable. The first has been offered recently by Price. Given how Fisher's own derivation proceeded, says Price: "What Fisher should have written is something like this: 'In any species at any time, the rate of change of fitness ascribable to natural selection is equal to the additive genetic variance in fitness at that time'". For, according to Price, what Fisher's theorem says is that "natural selection (in his restricted sense meaning only additive effects) at all times acts to increase the fitness of the species to live under the conditions that existed an instant earlier".[32]

A second gloss was offered at the time by Wright who suggested that the simplicity of Fisher's formulation must not mislead us. For, he holds, Fisher's special sense of "genetic variance" means that dominance relations, epistatic effects and the effects of mutation and migration are not comprehended by the term and nor therefore by the theorem.[33]

Fisher himself stressed that the theorem is "exact only for idealized populations" with no fortuitous fluctuations of genetic composition. And he was keen to give an estimate of the size of the effect of these fluctuations, that is "a standard error" for the expected "rate of increase in fitness". He satisfies himself that this error will be small because such fluctuations will be small compared to the average rate of progress. Once again, he has his gas analogy. The "regularity" in the average rate of progress is guaranteed for the same reason that a bubble of gas obeys the gas laws. A visible bubble may contain billions of molecules, more than most animal and plant populations contain individuals, but "the principle ensuring regularity in the same", in that even if there are large fluctuations from one generation of a population to the next, over many generations the deviations will be small. By stating the theorem in the form he has, and specifying the relation between progress in fitness

— as measured by m, Fisher's Malthusian parameter — and its standard error, the objection can be met that natural selection "depends on a succession of favorable chances". Natural selection only depends on this in the same sense that the very reliable income of a casino owner does. There is every difference between "a succession of favorable deviations from the laws of chance" and the "continuous and cumulative action of these laws"; and it is on such a continuous and cumulative action that "the principle of Natural Selection relies". Beyond discrediting that old objection to natural selection, putting the theorem in the form Fisher gives it also allows him to make his multiple comparisons and contrasts with the thermodynamics.[34] First come the comparisons:

It will be noticed that the fundamental theorem proved above bears some remarkable resemblances to the second law of thermodynamics. Both are properties of populations, or aggregates, true irrespective of the nature of the units which compose them; both are statistical laws; each requires the constant increase of a measurable quantity, in one case the entropy of a physical system and in the other the fitness, measured by m, of a biological population. As in the physical world we can conceive of theoretical systems in which the dissipative forces are wholly absent, and in which the entropy consequently remains constant, so we can conceive, though we need not expect to find, biological populations in which the genetic variance is absolutely zero, and in which fitness does not increase. Professor Eddington has recently remarked that "The law that entropy increases — the second law of thermodynamics — holds, I think, the supreme position among the laws of nature." It is not a little instructive that so similar a law should hold the supreme position among the biological sciences.[35]

Then — after a hint that the way down and the way up might one day be brought within some even more general synthesis — come the contrasts, including, last of all, the contrast of the very greatest significance:

While it is possible that both may ultimately be absorbed by some more general principle, for the present we should note that the laws as they stand present profound differences — (1) The systems considered in thermodynamics are permanent; species on the contrary are liable to extinction, although biological improvement must be expected to occur up to the end of their existence. (2) Fitness, although measured by a uniform method, is qualitatively different for every different organism, whereas entropy, like temperature, is taken to have the same meaning for all physical systems. (3) Fitness may be increased or decreased by changes in the environment, without reacting quantitatively upon that environment. (4) Entropy changes are exceptional in the physical world in being irreversible, while irreversible evolutionary changes form no exception among biological phenomena. Finally, (5) entropy changes lead to a progressive disorganisation of the physical world, at least from the human standpoint of the utilization of energy, while evolutionary changes are generally recognised as producing progressively higher organization in the organic world.[36]

There, then, is a climactic text that takes on its full significance when we associate the broad sweep of its claims with Fisher's entire life and thought. Within two years after this he had been explicit in the characterisations of Boltzmann and of Darwin that accompanied this entire conception of science. Of Boltzmann, Fisher said:

Perhaps the most dramatic development [in the theory of gases] was when Boltzmann restated the second law of thermodynamics, the central physical principle with which so many of the laws of physics are interlocked, in the form that physical changes take place only from the less probable to the more probable conditions, a form of statement which seemed to transmute probability from a subjective concept derivable from human ignorance to one of the central concepts of physical reality. More concretely, perhaps, we may say that the reliability of physical material was found to flow, not necessarily from the reliability of its ultimate components, but simply from the fact that these components are very numerous and largely independent.[37]

Of Darwin he wrote that it was "his chief contribution, not only to Biology but to the whole of natural science" to have identified a process whereby "contingencies *a priori* improbable, are given, in the process of time, an increasing probability, until it is their non-occurrence which becomes highly improbable".[38]

8. THE BIOLOGY AND PHILOSOPHY OF A SCIENTISTIC ROMANTIC

One virtue of the two-tendency and two-hero reading of Fisher's universe and of Fisher's science is that it saves us from certain simplifying fallacies that we might otherwise commit. It would be tempting, for instance, to see Fisher's work in evolutionary biology as essentially the work of an alien invader, as the work, that is, of a man trained in mathematics and physics who would impose the methods, concepts and presuppositions of hard, exact science on a subject not naturally amenable to such a treatment. To be saved from this fallacious view of Fisher as evolutionary biologist, one needs only to see how characteristic of his entire outlook is his writing on such traditional Darwinian natural history topics as bird courtship or insect mimicry. For the obvious generalisation from Fisher's preoccupation with that kind of natural history topic is correct. He was not only in descent from Darwin, he rightly saw himself in descent from a Darwin who had had among his decisive ancestors the naturalists of truly olden days, indeed the parson naturalists and natural theologians of those olden days.[39]

Another virtue of this reading of Fisher is that we are saved from reading him as someone who sought to advance biology principally by making it like physics in being mathematical. For, obviously, there would be two things wrong with such a reading. First, Fisher is often concerned as much with contrasts between evolutionary biology and mathematical physics as he is with comparisons. Second, it is not merely the mathematical form of equations, measurements and derivations in mathematical physics that Fisher sees as enlightening for evolutionary theorising. It is, more often and more fundamentally, the conceptual and, indeed, philosophical implications of the content of the physics itself, rather than its mathematical form.

It is in fact interesting to note that Fisher's mathematics, that is, principally, his mathematical statistics — including what he saw as its lessons for the understanding of statistical inference and, beyond that, inductive logic — has little discernible relevance to his thinking about genetics and evolution. This may be surprising. It would be reasonable to conjecture, for instance, that his early work on the statistics of small samples might somehow have conditioned his approach to the theory of natural selection. However, it is hard to find any evidence for this. In fact, when Fisher wrote on the work of his hero Gosset ('Student'), he discussed Gosset's pioneering insights into small sample statistics and later noted that Gosset was also right-minded, by Fisher's lights, on evolution, being a good selectionist; but Fisher made no link between these two elements in Gosset's thought.[40] It would be rewarding to find links, in Fisher, between his mathematics and his biology; links, that is, other than those — such as the obvious one provided by statistics itself — that are mediated by topics in physics. However, so far they have remained elusive, and have therefore to be left aside here, as belonging among future possibilities rather than among biographical results already achieved.

It will be apparent already that any account of Fisher's eventual philosophical synthesis of physics, biology and eugenics could equally well take as its guiding theme his Christianity, his Darwinism, his theory of civilisation or his statistical view of nature and man. For our purposes here, however, one choice recommends itself as especially appropriate. That is Fisher's indeterminism. In particular, this choice is appropriate because Fisher himself was most explicit, in his philosophical writings, in presenting his indeterminism as informing his outlook on everything else. We do well, therefore, to examine how and why he could see it functioning in such a comprehensive way in his philosophical life.

Before embarking on that examination, it will be as well to have some sense of Fisher's style and sympathies as a philosophical thinker. One distinguishing feature emerges, for example, from any glance at his intellectual biography. He combined within him both what we may call scientism and what we have to call romanticism. By scientism here is meant a commitment to give science an ultimate authority on all subjects: in a word, to set no limits to the domain nor to the authority of science. On the other hand Fisher has been called a romantic, and rightly, because of this taste, both in thought and deed, most obviously in his twenties, for pursuing flights of fancy and fantasy so infused with sentimentality, nostalgia and idiosyncrasy that anyone thinking of himself or herself as a realist and a rationalist child of the Enlightenment would have to view Fisher's life and thought as distinctly alien.

Two youthful compositions can take us from an exemplar of Fisher's scientism to a no less exemplary instance of his romanticism. Already showing that drive for generality in any scientific principles he embraces, Fisher opened his 1912 paper on "Evolution and Society" by declaring that Darwin's theory requires very little to be assumed about the "nature of species". The "same process" of selection goes on in "languages, religions, habits, and customs, rocks, beliefs, chemical elements, nations, everything else" that may be "stable" or "unstable". For natural selection all that is needed is that the "suitable to survive" shall do so, while the "unsuitable, unstable" do not. So, the selectional history of the habit of smoking or bridge playing or family prayers — Fisher's own examples — can be dealt with in the "same very general manner" as the evolution of the worms parasitic in some parrots. Fisher does not go on in fact to deliver on this promise as such, but devotes the rest of the paper to intersocietal selection in man; concluding that this selection tends to favor social specialisation and regimentation, but that we may "still hope that magnificent qualities and capabilities of the best type of man will render specialisation unnecessary" and that "the small, spirited nations were right" to believe "liberty was better than regimentation".[41]

There may be no sure echo here of Nietzschean romanticism. But Fisher was already a devoted member of a circle of friends who gave each other names taken from Nietzsche's *Thus Spake Zarathustra*; and within three years he had published a piece on 'The Evolution of Sexual Preference', in the *Eugenics Review*, that climaxed in a final three sentences (crossed out later by Fisher in his copy) where "we pass, like Nietzsche, beyond Good and Evil", with morality ceasing to be "arbitrary and dogmatic" but taking "its place as a particular formulation of the requirements of the Highest Man — of our ultimate judgments of human value". This finale is reached by starting with a defence of Darwin's theory of sexual selection, and then arguing that all aesthetic and ethical generalisations have natural evolutionary origins in sexual selection, so that aesthetics and ethics themselves are grounded in the judgments made by individuals in their choice of mates. Eventually, there arises a struggle between ethical and aesthetic valuations, with ordinary people rating beauty superficial and moral worth fundamental. But "in the deepest minds" the idea of beauty links itself to "nothing less than the mystical appreciation of human personality", the "highest plane", and "the source" for all other, lower valuations. It is, here, then, that Good and Evil are surpassed.[42]

So, whatever else Fisher's philosophy has to do for him, it has to give him a universe where liberty and beauty can hold a place in the life of Christian and a eugenist. By 1932, certainly, and very likely several years before that, Fisher had seen his way to securing such a philosophy, by starting from what he took to be a finding of science itself: namely, that the Boltzmannian way down and the Darwinian way up can, indeed should, be taken as freeing us from determinism and all its implications.

9. THE BIOLOGY AND PHILOSOPHY OF AN INDETERMINIST

It is essential, then, to appreciate how scientistic is Fisher's very approach to philosophy. His 1934 paper on 'Indeterminism and Natural Selection' is quite explicit: many older philosophical positions have been discredited by the progress of science. The Greeks constructed plentiful philosophical theories, but without sufficient knowledge of empirical science these philosophical exercises of the imagination were doomed to later implausibility. In modern times philosophy has taken empirical science into account; but there is often a lag, in that doctrines live on in philosophy that are no longer sanctioned by the best empirical sciences. So it is with determinism, Fisher argues. He goes to some length to argue that seventeenth-century science may have made determinism reasonable, but that developments in nineteenth-century science have made it no longer inevitable or even credible. Once again it is the Boltzmannian and Darwinian developments that are jointly decisive. "The historical origin and experimental basis" of "physical determinism" show that "this basis was removed" with the "kinetic theory of matter", while its "difficulties" grew with the evolutionists' admission that "human nature, in its entirety, is a product of natural causation".[43]

Fisher is explicit that no twentieth-century views on "indeterminacy" as discussed in quantum physics, add any new extra argument, only a welcome supplement to the nineteenth-century foundations for indeterminism. The reason is that the indeterminism Fisher defends holds only that reliability and predictability in aggregates, such as gases, require no determinism — only numerousness and independence — in the motions of the component elements. Nor is there any evidence, from other grounds, of component determinacy in nineteenth-century science. What twentieth-century physics has done then is merely, once again, to find no such evidence since. Appealing to generality, as ever, Fisher can, therefore, argue that indeterminism is the more general assumption, in that it avoids the additional special presuppositions about component elements that determinism makes.[44]

So enamored is Fisher of his indeterminism that he is prepared to present it as serving a vast range of philosophical needs, without entailing any serious drawbacks at all. Thus, in the eighteenth-century man, as freely and rationally choosing and acting, seemed beyond all possible scientific explanation because scientific explanation was thought to require deterministic causation. Conversely, then, when evolutionary biology brought man within science the plausibility of the old deterministic exclusion of man was challenged. With an indeterministic view that does not oppose the laws of nature to the laws of chance, natural and human science, including social science, can be unified. Marriages remain free, individual human choices and actions, but the statistical lawfulness of variations in aggregate marriage rates shows that in this, as in everything else, man and society come within a single scientific enterprise seamlessly joined with natural science.

Fisher can accordingly reduce to one issue the implications for social policy of the great shift from Laplacean deterministic science to Boltzmannian-Darwinian indeterminism. The possibility of science requires only numerousness and independence in the component elements in the aggregate being studied. Human societies are all obviously numerous enough in the components; but the issue of independence remains to be understood. Social organisation and public opinion are the two main features of society limiting independence of individual actions. However, Fisher's eugenics is not frustrated by these features. Indeed it can be made effective by being accommodated to them. For a start, heredity — the transmission from parents to children of the relevant traits — proceeds apart from any limits on independent action set by social organisation or public opinion. Now, in addition to heredity, the eugenist needs only to consider one other item: social promotion or the assignment of individuals to occupational grades and any correlation it may have with heritable fertility or infertility. The social promotion of infertility, this correlation, appears from historical sociological research to be an inevitable feature of every civilised society where social class is present, wealth is a factor in social promotion and children are an economic burden. Any measure that countered this last feature would, then, be effective in reversing the aggregate correlation by providing incentives for numerous social promotions to be won by more rather than less fertile individuals.[45]

The great nineteenth-century developments in kinetic theory show, therefore, that indeterminism is entirely consistent with the orderliness in the natural and social worlds and with success in the quest for knowledge, including causal knowledge, of those worlds. As for causation, far from diminishing this concept, indeterminism enhances it. With determinism everything is determined all the way forward from some indefinitely remote moment in the past. So, there is no room to think of subsequent events as causing anything that was not already determined. By contrast, with indeterminism, events can be understood as originating new lines of causation, and of making a difference by making things happen that would not otherwise have done so. The asymmetries in human and animal remembering, striving and aiming make no sense with determinism. If the future is as fixed as the past, memory should not refer uniquely to the latter; nor striving and aiming to the former.[46]

By enriching the concept of causation as it does, indeterminism clarifies the concept of creativity and so the relations between science, on the one hand, and art, ethics and religion on the other. Within science causation is creative in a strictly scientific sense. In a game of chance, we can predict all the possible forms of the result, stating in advance their probabilities of occurrence, without foreseeing just which

XIII

will occur. Likewise for any event, it can be creative if it can be causal, in that it can be the one that makes the difference in bringing about a new event what could not be predicted from earlier events. Beyond causality, however, the concept of creativity may imply an emotional stance. Being merely new in time, like a new penny, is not enough to exemplify creativity in this fuller sense; there must also be novelty in the "nature and potentialities" of what is produced. "This is intended when we apply the word to the work of a scientist, or an artist; that his work matters in itself, and to the future of his art or science". To be creative in this sense work "must have value, intellectual or aesthetic, moral or social value; consequences which excite wonder, or admiration".[47]

Fisher's view is, therefore, that any truly causal process is creative in the scientific sense; it may but need not be creative in the full sense. One cannot complain that the causal process invoked by some scientific theory is not creative in the full sense. That depends on our feelings about the consequences.

10. THE DISPELLING OF MECHANISM, MATERIALISM AND PESSIMISM

On this ground, Fisher welcomes the insistence of the philosophers Henry Bergson and Jan Smuts on talking of "creative evolution"; and of Alfred North Whitehead's (one of his teachers in mathematics) emphasis on nature as a creative process. However, where Fisher disagrees with Bergson and Smuts is over their view that Darwinian evolution by natural selection is not creative in the appropriate senses. Bergson has denounced Laplacean mechanism as inconsistent with our very consciousness of temporal duration. And he has argued that Darwinian or Lamarckian evolutionary biology is no less mechanistic, and so no less inconsistent with what is most indisputable in our experience. But, Fisher holds, Bergson has mistaken what it is that we need to be liberated from in Laplacean science. It is not mechanism, whatever that might be, but determinism. Make the shift to Boltzmann and Darwin properly understood, and one has all the philosophical dividends that Bergson is rightly seeking, but seeking in the wrong place. So Bergson's version of evolution, which invokes an *élan vital*, a vital impetus peculiar to living things, is philosophically unnecessary as well as unacceptably mysterious and empirically unsound in its claims about the role of mutations in evolution.[48]

Here, as always, Fisher rejects a choice between mechanism and vitalism. The distinction between deterministic and indeterministic causation can be defined with rigor, Fisher insists, but no one has satisfactorily distinguished a "mechanistic from a vitalistic organism". The only issues Bergson can ultimately raise are, therefore, the ones that are all resolved by an indeterministic view of science including Darwinian evolutionary theory, with its new, genetical theory of natural selection.[49]

Likewise with Smuts's call for holism. Smuts demands a holistic, creative process in the animal or plant germ, somehow eliciting the mutations needed for evolution. But Fisher, having praised the "wisdom and width of his more essential views" and the "religious feeling" he brings to his interpretation of evolution, urges that Smuts has no need to find holistic, creative processes inside zygotes. Properly understood, says Fisher, natural selection puts the creative causation at the outside boundary of every organism. For it is in the interactions of organisms with their environments, and so in the successes and failures of all the functions of all their organs, that creative causation is found. For conscious beings, this boundary is where consciousness itself resides, on the boundary between our inner subjective life and the outward objective sources of sense experience. More generally, the theory of selection seems "holistic" to Fisher, in its emphasis on the "mutual reaction of each organism with the whole ecological situation in which it lives — the creative action of one species on another".[50]

Faced with the charge that Darwinian evolutionary biology is materialistic, pessimistic and somehow "soulless" in its vision of nature, Fisher replies that he can not agree; because he sees nature as full of creative activity and of signs that attempts to lessen evil and promote progress can succeed. While Lamarckism reminds him of the doctrine of salvation by faith, in that strivings are decisive, Darwinism joins works to faith in making actual achievements, doings and dyings, decisive. It is then "hard" to see "anything unedifying or disquieting" in a theory of evolution that rests on the actual "performances" of organisms. As for evil, it is, on Fisher's view, "relative" in that "it changes its nature with evolutionary progress and with the changing structure of human society".[51] The torture of prisoners and the wreaking of man-made ecological disasters rightly concern us more now than the sins proscribed in the Old Testament. As Fisher might have argued, today, the philosophy of an indeterminist Boltzmannian and Darwinian scientist coheres well with the ideals of Amnesty International and Green-peace.

But even if indeterminism can discredit all charges of mechanism, materialism and pessimism, is there not still a defense needed of our own free wills as individuals? Sometimes Fisher contents himself with merely asserting that we know our wills to be free in the only senses required for a defense of active policies for progressive social change. However, he had struggled with free will in letters to Leonard Darwin and returned to it in correspondence with the neurophysiologist C. S. Sherrington later. In 1934, he had argued that a mere illusion of free will could have no survival value and so was inconsistent with crediting all our human nature to natural selection. But that thesis had still left a difficulty: namely, "to allow the evolutionary process, which depends upon the permanent and therefore deterministic properties of genes, to take any part in the development of such a capricious quality as the possession of powers of individual choice". Here is Fisher's "attempt to set out a possible relationship" between the evolutionary process and individual choice:

The development of a given genotype (even in given environmental conditions) is indeterminate in that undirected chance happenings intervene at all stages, each such event having perhaps permanent or increasing consequences, as development proceeds, on the integration of the nervous system and the formation of character.

Individual action, e.g., choice, is always in part predetermined by the genotype, in part by the subsequent effects of physically fortuitous developmental happenings in the past, and in part undetermined and ascribable to fortuitous contemporary happenings.

Both the course of development, and the instantaneous state of the nervous system, are such as to amplify the effects of initially minute (quantum) events, so as to have molar consequences.

This general principle of amplification has been of importance to survival, in some way at present obscure, perhaps connected with the organization of the whole bodily mass into individual unity, perhaps in orienting its reactions towards the future (as purpose or intention), and has evolved to its present high degree by reason of its survival value. It, though not the particular modifications which it favors, is determined by the genotype.

It is open to a man, religiously inclined, to assert that the primary elements of indeterminacy in development and choice are fortuitous only in the physical sense, being in reality divinely guided, much as the apparatus of games of chance were regarded as guided by the Goddess Fortuna.[52]

This is Fisher in 1947. In 1922 he had been one step ahead of Wright in population genetics; but in the philosophy of the freely choosing mind of the individual, Fisher was unwittingly retracing a path that Wright had already taken before they had ever met. This makes an appropriate moment, therefore, to switch to the young Wright.

11. WRIGHT: ADJUDICATIONS FOR SCIENCE AND WITHIN SCIENCE

It would be tempting, but also misleading, to introduce the young Wright by looking at those biographical landmarks that match up best with corresponding moments in Fisher's early life. It would be misleading because their early lives were too different in decisive ways. To be sure, Wright, like Fisher, read Darwin at school and apparently became from then on in favor of Darwin's theory of gradual evolution by natural selection. But this parallel with Fisher is a limited and superficial one, in that Wright was eventually to owe his views on the place of selection theory in evolutionary biology to educational and research contexts that had no equivalent in Fisher's life, a life it has to be remembered that included no formal instruction in biological science after high school.

Wright was many years in the making as a qualified zoologist. The years (1906—1911) at Lombard College, ending with a bachelor's degree, took in a wide sampling of biological subjects and various texts on evolution. A year (1911—12) at the University of Illinois was preceded and followed by summer visits to Cold Spring Harbor. So, when Wright went to Harvard's Bussey Institution in 1912, to work with Castle toward a PhD in genetics (awarded in 1915), he was extensively prepared in the relevant disciplines, but had had no occasion to work out explicitly and precisely where he stood on the issues dividing evolutionary theorists at the time.

However, it seems that there was never for Wright a choice to be made between Mendelism and Darwinism, or between Mendelism and biometry. He grew up, so to speak, seeing the three as consilient; this consilience being more easily assumed in the U.S., where the English opposition between the Mendelian Bateson and biometrician Pearson had no close equivalent. At the Bussey Institution, such a consilience was especially strongly represented in the views of Castle and East.

The question arises, then, as to whether eugenics formed for Wright, as it did for Fisher, a fourth commitment to go with a youthful embracing of Mendelism, Darwinism and biometry. Wright's attitude to eugenics at this time is not easily discerned from direct documentation.

However, it seems most likely, from the stance he would later adopt, that it was never an enthusiasm, nor something, conversely, that he was bent on opposing. More generally, Wright does not seem to have actively aligned himself with any political doctrines or groups. The overall impression one has is that he felt most comfortable with what would then have been, for an academic son of an academic family, broadly the middle ground. For a time, when Wright was an undergraduate, his father evidently moved out of the middle ground toward socialism, as upheld by Eugene Debs; but the son was less enthusiastic.[54]

Likewise with religion; it seems, although here we have even less direct documentation, that there were in Wright no strong feelings, religious or hostile to religion, for him to bring to his science in general, or to his understanding of evolution in particular. And yet, before he had finished his first year at Harvard, Wright was reading Bergson's *Creative Evolution* — presumably in the English translation published the year before, in 1911 — and being prompted to embark on a philosophical quest that eventually led him to construct a comprehensive metaphysics of mind and matter of his own, a dual-aspect panpsychism, wherein the ultimate, inner reality of everything from ions to men, is held to be mental, while science studies only those external appearances and observable actions that constitute the material aspects of things.[55]

What can be said, therefore, about the motivation of this philosophical quest, if it has no obvious political or religious rationale? Can we anticipate, here, what a closer analysis of Wright's intellectual biography might reveal, so as to identify those themes that must run through any discussion of the relations between his science and his philosophy? Such a task is harder in Wright's case than Fisher's. However, there is one theme that it may be useful to announce in advance: the theme of adjudication. For Wright's philosophical quest started from a quite general issue: the conflict, especially as Bergson had identified it, between the Laplacean determinism presupposed by much scientific thought, and the phenomenon of consciousness itself. Now, Wright saw this conflict as a quite general one. Although biologists were peculiarly involved with the relation between mind and matter — unlike physicists who could usually ignore minds and psychologists who could sometimes ignore matter, as Wright reflected — the only resolution of the conflict that was satisfactory for a biologist would be one that comprehended all natural science. What is more, Wright's eventual resolution of it had as an explicit corollary that natural science takes no account of the inner, ultimate mental reality of things, and can, therefore, proceed deterministically — at least in the sense of determinable probabilities — without conflicting at all with any convictions that may be held about consciousness and minds. Unlike Fisher's indeterminism, therefore, Wright's panpsychism is not designed to show that the content of the best science is such as to satisfy any legitimate metaphysical demands that we may make of it. Quite the contrary, for Wright's panpsychism is designed to show that science, limited as it is to the material aspects, can not be expected to meet any metaphysical demands. Only the philosophy of panpsychism itself can meet those demands.

There is, then, in Wright no equivalent to Fisher's vindicative attitude toward Darwinism, the commitment, that is, to vindicate Darwin's theory both scientifically and philosophically. Within evolutionary biology, Wright sees himself as an adjudicator. Where Fisher seeks to rethink — and vindicate — Darwin in the light of the new genetics, for Wright it is not Darwin, or even natural selection, but rather evolution itself that has to be rethought in the light of the new genetics. That rethinking is what the statistical theory of population genetics has to contribute. In such a rethinking, the claims of rival theories — De Vries's mutationism, Wagner's isolation theory, Eimer's orthogenesis and Darwin's selectionism — have to be adjudicated, by establishing under what conditions this or that trend, agency or tendency predominates in conditioning the course of evolution. For this reason, Wright did not write under any such title as Fisher gave his book, but addressed himself to the more comprehensive topic: 'Evolution in Mendelian Populations' and 'The Roles of Mutation, Inbreeding, Crossbreeding and Selection in Evolution'.

The way Wright carried out his adjudicational business regarding evolution was conditioned, above all, by a feature of his career that, once again, has no equivalent in Fisher's. Fisher had gone straight to his vindicative business, in the sense that he had always brought his developing knowledge of the new genetics — knowledge of genes and mutations and so on — directly to bear on the theory of evolution by natural selection. Wright, on the other had, was to come at evolution through the intermediary subject of livestock breeding, its theory and practice. Before rethinking evolution in the light of the new genetics of the first twenty years of the century, he had worked at rethinking livestock breeding in that light; only subsequently did he elaborate his theory of evolution as a generalisation and accommodation of his livestock breeding conclusions to the resolution of what he saw as the main challenges in understanding evolution itself. Many years later, Wright would indicate that he saw cultural evolution in man as analogous to organic evolution, so that his shifting balance theory of organic evolution was to have its analog in understanding cultural evolution. With Fisher, it was enlightening to ask what he saw as common to all processes of selection. With Wright, one needs to ask what it is that is seen as common to livestock breeding, organic evolution and cultural evolution. The answer is explicit and reiterated by Wright himself: the decisive feature is cumulative change, that is change involving persistence as well as variability. We have, therefore, to think of Wright as always theorising by adjudicating among diverse proposals concerned with making intelligible — to use his phrase — processes of cumulative change. Such phrases remain mere phrases, needless to add, until associated with particular domains. We should look without delay, then, at Wright's livestock breeding work.

12. FROM BREEDING TO EVOLUTION

There is now a standard way to reconstruct the origins of Wright's shifting balance theory of evolution, and although there is no reason to question it one does need to ask precisely what it can and can not do for us biographically. The standard reconstruction traces directly to Wright's own retrospective insights. These insights, unlike many scientists' autobiographical reflections, are unlikely to be seriously mistaken for two reasons. First of all, Wright's habit, throughout his life, was to refer back constantly to his earliest publications and his earliest researches. He was always keeping track, far more than most scientists do, of where he had been and how he had reached whatever was his present position. Second, where Wright's memories have been checked against independent evidence they have shown themselves remarkably accurate. We can therefore have reasonable confidence concerning his account of what was decisive in forming his views on evolution.

On one point, however, we have to stay critically alert. Wright's own most detailed version of that account was given originally as a talk, and as a talk before a meeting of the American Society of Animal Science,

266

formerly the American Society of Animal Production. It was then addressed to an audience with a special interest, in every sense, in livestock breeding. And sure enough, Wright called the talk: 'The Relation of Livestock Breeding to Theories of Evolution', although it is his own theory alone that is discussed in any detail. Now, in that talk Wright distinguished four lines of his early work, prior to 1925, that were decisive for his theory. Given the title of his talk, one might think that Wright was recalling all four as somehow coming out of his livestock breeding studies during his years (1915—1925) with the U.S. Department of Agriculture. But, as Wright himself makes plain, two of the four trace to his years at the Bussey and so before he had any special, professional concern with breeding. A reconstruction of the whole story would have, therefore, to ask how the Bussey years' work conditioned the U.S.D.A. years' thinking and, then, how the work of both periods was drawn on in the typescript exposition of the evolution theory composed in 1925, but not published until 1931 in a modified form.[56]

Consider the two Bussey conclusions, as set out in Crow's lucid retelling of Wright's recollections. First, Wright was convinced by Castle's selection experiments that selection was effective in producing a divergence between hooded rat strains that were almost completely white or almost completely dark; but he was also impressed by the need to stop the experiment eventually on account of the infertility produced. The conclusion seemed to be that mass selection can be effective in changing one trait, but that intense selection also brings other, unwanted traits. Second, Wright was convinced, by his own doctoral study of hair color and pattern in guinea pigs, that genes in combination often had unpredictable effects; and that selecting for individual traits may not produce the desired combination. Obviously, then, Wright entered the U.S.D.A. and took up his breeding researches already convinced that mass selection of individual traits was often efficacious but also not reliably optimal.[57]

Consider next, then, the two U.S.D.A conclusions, again as set out in Crow's version. First, the study of inbreeding and crossbreeding in guinea pigs confirmed that inbreeding leads to a decline, but it also increased random differentiation among different lines and fixed distinct combinations — not only of superficial traits such as colors, but of internal physiological traits. Among those distinct combinations were occasional favorable ones. Second, the history of livestock breeding, of

Shorthorn cattle especially, showed the advantage of combining selection with inbreeding within herds followed by the exporting of sires from superior herds to others, with a consequent benefit to the whole breed. Now, one can accept that — as Crow puts it — by replacing "guinea pigs" and "cattle" in these four conclusions with "natural populations," one has the essential core of the shifting balance theory.[58] However, what still remains to be asked is why Wright saw the problems posed for breeders as sufficiently like the problems addressed in evolutionary theory, that a solution could be transposed in this way. It is here, as already indicated, that we have to concentrate on Wright's preoccupation with the issues he introduces through such telling terms as *cumulativity, plasticity, balance, levels* and the *interpolation* of a process of mostly random differentiation of local strains between the primary gene mutations and the control of major evolutionary trends by natural selection.

To look ahead, these various issues come down to two main clusters that can, perhaps, be seen as ultimately one. A first cluster is introduced right at the opening of the third and final section of the 1931 paper, the section — 'The Evolution of Mendelian Systems' — that attempts to draw conclusions about evolution from the findings of the previous two sections — those on the 'Variation of Gene Frequency ' and 'The Distribution of Gene Frequencies and Its immediate Consequences'. In attempting to draw such conclusions one may assume, says Wright, that causes making for variation are favorable, those reducing variation unfavorable. However, evolution being cumulative change, rather than mere change, "fixation in some respects is as important as variation in others". Livestock breeders, says Wright, liken their work to modelling in clay, speaking of moulding the type toward the chosen ideal. The analogy is a good one, suggesting as it does that in both cases "a certain intermediate degree of plasticity" is required.[59]

A second cluster was often insisted upon when Wright contrasted his views with Fisher. Wright's early studies of factor interaction concluded that genotypes relate to phenotypes "by a very complex network of biochemical and developmental reactions", with each character usually affected by many gene substitutions, each substitution having many pleiotropic effects and the intervening processes involving nonadditive interactions. This "viewpoint" thus contrasts with the "common treatment of organisms as mosaics of unit characters (Fisher's norm)" when considering their evolution. So, on Wright's viewpoint, "evolution becomes a much more intelligible process if based on natural selection among interaction systems rather than among alleles at each locus separately". Without strong linkage, the allelic selection is all that is possible with natural selection among individuals in a panmictic population where Fisher's fundamental theorem would apply, because, here, recombination quickly lead to the dissolution of any particular gene combinations. The more intelligible process is, however, possible with inbreeding and with selection at two levels. With the Shorthorn cattle there had been selection by particular breeders in building up their herds, and by breeders in general among herds as sources of sires. The first favors "the allele at each pertinent locus that gives the most favorable effect on the average of all combinations with such alleles at other loci". This, says Wright, is "*genic* selection"; and the resulting patterns are "fixed, more or less, by close inbreeding". Selection among such herds as sources of sires is "selection among the diverse interaction systems that happened to have been arrived at". This, he says "is *organismic* or (genotypic) selection". It was recognising "that the two-level process was much more efficient than mere individual selection" that led him to see "whether an analogous two-level process might not occur in nature".[60] Wright does not make his point explicitly in this way, but we can say, in summary, that the interpolation, in the conception of the organism, of complex gene interactions, mediating between gene substitutions and phenotype character differences, required in the conception of population structure, the interpolation of inbreeding, random drift and selection in local races, subgroups of small population size, between the reproduction of individuals and the changes taking place in the species as a whole.

13. BALANCE, LEVELS AND THE INTELLIGIBILITY OF EVOLUTION AS CUMULATIVE CHANGE

As an exegetical policy it proves most instructive to concentrate on the 1931 paper. It is not that Wright's shorter, less demanding and more familiar 1932 paper is misleading in its representation of his views.[61] It is, rather, that the 1931 paper is more explicitly related to those general issues that provide the common grounding of his thinking about livestock breeding and evolution. The analogies and metaphors elaborated through the 1932 device of the adaptive topography allowed the exposition of the shifting balance theory to be sundered from those issues, and expressed briefly in pictorial epitome without reference to the framework assumptions that are so prominent in 1931.

Of the more than five dozen pages of the 1931 piece, only a dozen and a half — all but two of them at the end — are explicitly devoted to evolution. It is easy enough, then, as well as indispensable for our purposes, to see what are Wright's principal proposals. A fundamental lesson that he wants to instil is this: "The problem is to determine how an adaptive evolutionary process may be derived from such unfavorable raw material as the infrequent, fortuitous and usually injurious gene mutations". This is the problem because the "basic cumulative factor in evolution is the extraordinary persistence of gene specificity", while the "basic change factor is gene mutation, the occasional failure of precise duplication"; and because the conclusion from the "present status of genetics" is that "any theory of evolution" must be based on the properties of Mendelian genes together with the "statistical situation in the species". That these two resources are decisive follows from "the fact that the evolutionary process is concerned, not with individuals, but with the species, an intricate network of living matter, physically continuous in spacetime," and with its responses to external conditions, which relate to "the genetics of individuals only as statistical consequences of the latter".[62] For this restriction to statistical consequences follows, Wright argues, from the discrediting of Lamarckian and other theories that would relate changes in the environment to genetic change in the species through their direct physiological consequences.

A constrast with Fisher is already apparent, then. Where Fisher stresses that heritable variation has turned out to be, in the new genetics, exactly what a selection theorist needs to free him from all the difficulties he had with blending inheritance, Wright takes the wider view: as an evolutionary theorist, a theorist of cumulative change, including adaptation, Mendelian mutations are, in and of themselves, distinctly unpromising. The study of the statistical situation in the species, to use his phrase, has, therefore, a big challenge to overcome.

Wright proceeds to clarify and meet this challenge with argumentation that has no equivalent in Fisher. More particularly he begins his argumentation by giving himself resources that have no equivalent. Their significance is well worth emphasising, therefore, since they can provide light on Wright's entire caste of mind as an evolutionary theorist.

The resources relate, once again, to the theme that cumulative

270

change requires factors that make for persistence as well as factors making for innovation. No understanding of Wright's shifting balance theory can ignore this theme, precisely because the balance proclaimed in the very name of the theory is a balance between persistence and innovation.

Wright lists nine pairs of such factors in two columns: 'Factors of Genetic Homogeneity' and 'Factors of Genetic Heterogeneity'. Some pairings are straightforward enough, such as the first — gene duplication and gene mutation — and the fifth — linkage and crossing over. With others there are complications. Environmental pressure (s, for selection, in the equations) is a homogenetic factor opposed to individual adaptability, because individual adaptability allows similar genotypes to develop into dissimilar phenotypes and so to reduce the diminution of variance due to environmental pressures exerted as selection. But then individual adaptability also shows up as a homogenetic factor opposed to local environments of subgroups (s_1 in the equations); for it also allows for dissimilar genotypes to develop into similar phenotypes even in differing local environments. Clearly, therefore, for some of these factors, whether they are homogenetic or heterogenetic will depend on the circumstances and on what else is going on.[63]

Obviously, too, we have here a conceptual framework that differs totally from Fisher's with its division of all causation into two categories: the reliably, if not strictly invariably, counterentropic (natural selection alone) as opposed to all the rest, which are entropic. This contrast with Fisher is confirmed when we turn to the second of Wright's resources. For Wright goes on to explain the role of these factors in evolution by considering, in certain cases, the difference that was made to all subsequent evolutionary processes by their original introduction in the history of life. Thus he considers the evolutionary process in, first, viruses where the gene is the organism and, second, bacteria and blue green algae, which have, as he holds, gene aggregates but no mitosis or meiosis, and where, therefore, "the conditions are not favorable for an extensive cumulative process". Not surprisingly, Wright, citing East, urges that the "most important factor in transcending the evolutionary difficulties inherent in the characteristics of gene mutation" is "biparental reproduction". But while this provides a rich field of variation, "by itself it provides rather too much plasticity"; for along with the adaptability it gives to the species comes the consequence that "a successful combination of characteristics is attained only to be broken up in the next generation" by meiosis itself.[64]

Now, "the principle that a balance between factors of homogeneity and of heterogeneity" is more favorable for evolution than either by itself is well illustrated by the effects of an alternation of a run of asexual generations with an occasional sexual generation. This works very well both in nature, in many species, and as a technique deployed by plant-breeders, because beneficial combinations generated sexually can be preserved and multiplied asexually. However, asexual reproduction is often absent, especially in higher animals. The consequence of its absence is definitive of the very "purpose" of his paper, Wright explains, which is "to investigate the statistical situation in a population under exclusive sexual reproduction" so as to clarify "the conditions for a degree of plasticity in a species" that makes "the evolutionary process an intelligible one".[65]

This characterisation of the very purpose of Wright's paper is accordingly followed by an examination of the various homogenetic and heterogenetic factors as they influence any population. Examining their consequences in large populations, small populations and medium ones, Wright argues that intermediate population size is optimal, large populations being subject mainly to slow and often to reversible changes with selection, while small ones are too liable to decline and extinction from random fixation of disadvantageous genes. By contrast, with a population of intermediate size, a "continuous and essentially irreversible" change seems ensured "even under completely uniform conditions", the direction being "largely random over short periods but adaptive in the long run".[66]

This conclusion sets the stage, in turn, for the analysis of what happens in a large population divided into partly isolated subgroups of small size. Here, even though the subgroups are small the results are much as they are for populations of intermediate size. There is a "partly nonadaptive, partly adaptive radiation among the subgroups". Those coming up with "the most successful types" would "presumably flourish and tend to overflow their boundaries while others decline, leading to changes in the mean gene frequency of the species as a whole". And, in this case, "the rate of evolution should be much greater" than is the case of a single undivided population. This last comment shows, then, that Wright's notion of what is optimal includes not only the constraints of cumulativity and adaptation but also of quickness in achieving permanent, adapted productions. Once again, it is plain the these constraints on any theory of evolution are carried over by Wright from his livestock breeding analyses.[67]

It is plain, too, that the smallness of the partially isolated subpopulations is decisive for Wright for reasons that have to do both with random drift, in the sense of fortuitous or random or indiscriminate differential reproduction, and with inbreeding, the breeding together of relatives. The mathematical treatment of random drift and inbreeding tends to assimilate the two together, but their significance for Wright's overall theory can be obscured by that assimilation. Inbreeding is a homogenetic, fixational factor in any one population, allowing for complex interactive gene systems to be held together. Random drift, on the other hand, as is emphasised in the 1932 paper, counteracts the tendency of the intrapopulationally homogenetic factor, selection, to hold the population at the nearest adaptive peak; for drift allows the population to cross an adaptive saddle and so come within the selectional influence of a higher peak.

These complex reasons for Wright's emphasis on the advantages of smallness in subgroups are worth noting here, because in the ensuing debates between Wright and Fisher and, more broadly, among their respective followers and champions, there was a tendency to concentrate on only one element in Wright's overall proposal: namely, that drift will be tending to cause nonadaptive divergence among subgroups and so, too, perhaps, eventually among congeneric species resulting from any eventual speciations. A fuller understanding of what Wright was about requires concentration, as always, on his preoccupation with persistence and variation as conditions for cumulative change. The precise wording of familiar sentences from the closing summary paragraph of the 1931 paper can put us on to a better reading of the paper as a whole. As "a process of cumulative change" evolution "depends on a proper balance of the conditions, which, at each level of organisation — gene, chromosome, cell, individual, local race — make for genetic homogeneity or genetic heterogeneity of the species". In a large, subdivided population, there is "a continually shifting differentiation" among the local races "intensified by local differences in selection but occurring under uniform and static conditions" and inevitably producing "indefinitely continuing, irreversible, adaptive and much more rapid evolution of the species".[68]

14. THE BIOLOGY AND PHILOSOPHY OF A MONISTIC DUAL-ASPECT PANPSYCHIST

Wright's 1931 paper included a short subsection, of a single paragraph, on " 'creative' and 'emergent' evolution", where he explained how he saw his scientific account of evolution relating to his philosophical views. The discussion is so abbreviated, however, that to go to it directly is to risk missing the items that were decisive for Wright's intellectual development. Fortunately, in three other papers, written in 1953, 1964 and 1975 respectively, he dwelled autobiographically on the origins of the position encapsulated in 1931: 'Gene and Organism', 'Biology and Philosophy of Science' and 'Panpsychism and Science'.[69] What these three papers make plain is that there are two ways to go astray in understanding the relations between philosophy and biology in Wright's life and work. The first would be to read his views within biology as somehow deriving as a consequence from more comprehensive philosophical views. This would be a mistake because, as Wright himself insists, his philosophy was designed to leave his biology following ideals of method, evidence and explanation that were already commonly if not universally accepted among working scientists. However, the second mistake is to infer from this that Wright's philosophical concerns, the questions he asked, the answers he considered, the reading and reflecting he did, were without influence in conditioning his life as a biological theorist. For in decisive ways it is pretty certain that they were influential. In making such a claim for this kind of influence, it will be as well to admit to yet another underdetermination thesis. Just as scientific theories are held by many to be underdetermined by any factual evidence for them, so it seems are scientific and philosophical views by each other. Whatever influences run back and forth still leave those views less than strictly determined. But to acknowledge that limitation is still to find room for a biographically defensible case for conditioning influences.

Consider next, then, Wright's recollections of what three books did for him in the years from 1912 to 1914. Before this he had accepted that science requires a rigid determinism of the Laplacean kind, but he had worried about where consciousness could find its place. This determinism was disturbed by reading Bergson in 1912, "but not for long", in that Wright was to return to determinism "in practice but with a radical revision of the philosophical implications of science".[70] How, then, was this return made along with this radical revision? A decisive step came on reading, in that same year, the biochemist Benjamin Moore's book, *The Origin and Nature of Life* of 1912, with its suggestion that cells and higher organisms and societies could be seen as extensions to the series running from atom to molecule and to colloid. This Wright accepted, but still a "dilemma" remained: "absolutely deterministic laws of physics at one end of the scale, consciousness and apparent freedom" at the other end.[71] The resolution of this dilemma followed a reading in 1914 of Karl Pearson's *Grammar of Science* (the second edition of 1899). This resolution involved two moves, one that was provided directly by Pearson, and another that was definitely not, as Wright himself emphasises. The first was the move to considering laws of nature not absolutely, as "part of the eternal structure of the world", as Wright had hitherto considered them, but "as merely condensed statistical descriptions of how things are observed to behave". They would be, then, no different in kind from "statistical laws of voluntary human behavior such as the law of supply and demand". Wright, moreover, accepted from Pearson that the causes of any individual event widen out, quite unmanageably, into the history of the whole universe. So a deterministic treatment of single events is impossible.[72]

For Wright, the full significance of Pearson's statistical viewpoint only came in going way beyond anything Pearson sanctioned. For the second move was to monistic dual-aspect panpsychism. The hierarchy conclusion prompted by Moore's book contributed here, in suggesting that a molecule or atom might be like a minute organism, in that it has structure and incessant activity. So the properties of organisms seemed projectible all the way down the hierarchical series. What is more, when Wright considered, among other things, how evolutionary biology suggests the projection back into our remote animal ancestry of mind, there seemed no resolution of the problem of where mind came from originally, and that it was least arbitrary to have mind constituting the underlying reality of everything. Thus did Wright come to a panpsychist philosophy, one he was only later to learn resembled what Pearson's friend W. K. Clifford had embraced and, before him, Fechner and others, but one that Pearson himself never came close to endorsing.

It was the dual aspect feature of Wright's panpsychism — inner mind

as reality, outer observable actions and appearances of mind as matter
— that left science proceeding much as before, within precise metaphysical limitations. Introspective human psychology apart, it is not part
of science to "make imaginative interpretations of the internal aspects
of reality — what it is like, for example, to be a lion, an ant or an anthill,
a liver cell, or a hydrogen ion". Science, accordingly, does not explain
the behavior of an amoeba "as due partly to surface and other physical
forces and partly to what the amoeba wants to do"; that would only
lead to duplication, and so science sticks to the physical forces. In this
program for science the "unique creative aspect of every event necessarily escapes" the scientist.[73]

On a statistical view of lawful causality, however, higher organisms,
with their very numerous component parts, should "simulate complete
determinism" in their actions. The reason they do not is that it is of the
"essence of an organism" to contain many "switch or trigger mechanisms which bypass purely statistical behavior". Accordingly, at least
since 1927 and Lindbergh's transatlantic flight, Wright illustrated this
view by arguing that the direction of an aircraft's flight could be "fully
accounted for deterministically except for the product of a succession
of infinitesimals", that are traceable to the pilot's brain and are decisive
because of the switch and trigger mechanisms, physical mechanisms,
there and in the rest of his body and in the aircraft. "A high degree of
freedom of choice by the whole is thus consistent with apparent
deterministic behavior of the parts".[74]

It was after invoking this thesis in 1931 that Wright wrote that this
view "implies considerable limitations on the synthetic phases of
science", but "quantum physics" had shown, in any case, that "prediction can be expressed only in terms of probabilities, decreasing with the
period of time". What then of evolution?

As to evolution, its entities, species and ecologic systems, are much less closely knit
than individual organisms, One may conceive of the process as involving freedom, most
readily traceable in the factor called here individual adaptability. This, however, is a
subjective interpretation and can have no place in the objective scientific analysis of the
problem.[75]

Forty five years later Wright reaffirmed this conclusion adding only
that he would now write "individual selection" and "selective diffusion"
in place of "individual adaptability", so as to refer to the "coefficients

which are actually used in the mathematical formulation, but which nevertheless represent processes which may involve choices made by individual organisms".[76]

So, although the subjective viewpoint of "choices" does not belong in the scientific analysis, the physical bases of their outward manifestations — the trigger and switch mechanisms in any organisms — very much do. We have, here, therefore, a real locus for the influence of Wright's philosophy upon his biology. For he has made a judgment that, above the level of the individual organism, there is no organisation sufficiently integrated, no physical system that is closely knit enough, for any bypassing of statistical behavior such as trigger and switch mechanisms allow in any individual organism. Species, even ecologic systems, may be loosely called organisms, but they are too loosely organised to be judged organisms in the full sense that individual organisms should be.[77]

That this judgment was influential in Wright's science is apparent when we notice the central place taken by individual adaptability in his analysis of the evolutionary process in relation to genetics, including developmental no less than transmission genetics. "Individual adaptability is, in fact, distinctly a factor of evolutionary poise". It is not only of "the greatest significance as a factor of evolution in damping the effects of selection" and keeping these from being excessive compared to the inverse of the population size and to gene mutation rates, "it is itself perhaps the chief object of selection", he says, continuing:

The evolution of complex organisms rests on the attainment of gene combinations which determine a varied repertoire of adaptive cell responses in relation to external conditions. The older writers on evolution were often staggered by the seeming necessity of accounting for the evolution of fine details, of an adaptive nature, for example, the fine structure of all the bones. From the view that structure is never inherited as such, but merely types of adaptive cell behavior which lead to particular structures under particular conditions, the difficulty to a considerable extent disappears. The present difficulty is rather in tracing the inheritance of highly localized structural details to the more immediate inheritance of certain types of cell behavior.[78]

This is a line of thought that has no equivalent in Fisher. When Fisher addressed the worries of the older writers, he insisted that they were all taken care of by combining two resources: the resolution of those worries offered by Darwin in the *Origin*, and the resolution, made possible by heredity turning out to be particulate not blending, of any remaining difficulties that Darwin had had.[79] It is not that Wright's

individualism — his emphasis on individual adaptability — traces solely
to his preoccupation with the hierarchy of organisms from ions to biota,
this tracing in turn to his philosophical debts to Moore; it is rather than
he had philosophical as well as scientific, especially physiological-
genetical, reasons for that emphasis.

It is easier, when we take these reasons into account, to see why
Wright is an organismic as well as a genic selectionist, but not a group
selectionist. What Wright calls intergroup selection is — remember the
bulls — a selection as to which groups within the species will alter the
whole species by exporting to the rest of the species successful genotypic combinations. That Wright has no level of selection higher than
this is because he finds no sufficiently persistent and integrated entities
above the level of individuals in the hierarchy of organisms. Conversely,
his requiring of a theory of evolution that it bring out the consequences
of persistence and variation at many levels of organisation, from genes
up to individuals, reflects his finding that below the species there are
several consequential levels of organismic integration. His inclination to
make such requirements belongs to his philosophical life no less than to
his scientific life, because the causal consequences of integrated organisation formed a subject that was common to his private philosophical
quest and his public scientific career.

15. FISHER AND WRIGHT: COMPARISONS AND CONTRASTS

Sometimes, in a comparative and contrastive study, one can draw on
those familiar distinctions that have often served well on other occasions: the distinctions made, to cite only a few examples, between
classical and romantic outlooks; or tenderminded and toughminded
philosophies; or this-worldly and other-worldly thinkers; or conservative and rationalist attitudes. Certainly those distinctions are relevant
and fruitful when brought to the case of Fisher and Wright. But it will
be evident already that no one of them fits well enough to be usefully
developed in a sustained way.

Equally, the distinctions often brought to the analysis of the history
and philosophy of biology do not provide obviously accurate bases for
the comparisons and contrasts one needs to make between Fisher and
Wright: reductionism versus holism, mechanism versus vitalism, and so
on. To be sure, it is true that much in Fisher does seem to assimilate his
science to the kind of atomistic, reductionist, determinist, mechanistic

physics that is traditionally associated with Laplace and the Laplacean school two centuries ago now. Garland Allen has accordingly taken Fisher to be a specimen mechanist materialist — working in a billiard ball universe or, at least, with analogies drawn from such a universe — to be contrasted with the evolutionary geneticist I. M. Lerner as a specimen holist materialist.[80] Likewise, Jean Gayon, in his comprehensive and authoritative treatise on the history of the theory of natural selection from Darwin to Kimura, contrasts Fisher's Fundamental Theorem with Wright's adaptive topography, seeing the first as Laplacean in character and inspiration, while the other draws on a hierarchical conception of reality.[81] Now, the view taken of Wright in the present paper agrees almost entirely with Gayon's presentation, and indeed is indebted to it; the only serious point of divergence being a strategic one; where Gayon concentrates on the adaptive topography in Wright, the focus here has been on the theory of heterogenetic and homogenetic balance. However, where Gayon finds Fisher's statistical mechanical analogies aligning Fisher with Laplace, they have been seen, here, following Fisher himself, as distancing Fisher from that program in physics and as making him post-Boltzmannian rather than post-Laplacean in his indeterminism.

But does there not remain a sense in which Fisher's science is, at least, more reductionist, less holistic than Wright's. Perhaps there does. However, rather than trying to identify that sense, it may be more valuable here to suggest that there is a decisive sense in which both Fisher and Wright are reductionistic in their theorising, but that Wright's reductions are much more complicated than Fisher's. Both are reductionist in one respect, surely. For both, the causal theory of evolution is to be reduced to Mendelian genetics through a statistical treatment of the consequences of that Mendelian genetics. The difference is that Wright's notions about the reducing science — Mendelian genetics — are more comprehensive and introduce more complications, most obviously physiological and developmental complications. Not coincidentally, his statistics takes him less of the way toward his evolutionary conclusions.

It is implausible to say, therefore, that Fisher did things differently from Wright because he was more reductionist; by itself that judgment would do very little for us in our efforts to understand why their evolutionary theories came out so divergently. We have, instead, to characterise their science less abstractly; saying, perhaps, that their integrations of Darwinism, Mendelism and biometry differed as they did because, among other things, one combined that integrative ambition with eugenics, Darwin family connections and statistical mechanics, while the other came to it through Castle, physiological genetics and livestock breeding.

If we turn to characterise the two men's universes rather than their science, there are difficulties, again, in ensuring that we are comparing what is comparable. For Fisher, the relationship between philosophy and science was such that he sought a unified account of the physical and social world that grounded his philosophical arguments in scientific foundations. For Wright, there is no such movement from science to philosophy to give a single synthesis. This difference between the two is reflected in Wright's response to Fisher's thermodynamic analogies. Writing in 1931, in the *Journal of the American Statistical Association* on 'Statistical Theory of Evolution', Wright opened appropriately by summarising the "interesting comparison" made by Fisher on "the position of the evolutionary principle in biology" — significantly Wright uses the word evolution here, throughout, where Fisher would have talked of natural selection — and the second law of thermodynamics in physics. Having given Fisher's positive analogies, Wright gives only one of Fisher's disanalogies: the second law entails disorganisation, "a passage from less probable to more probable states"; evolution, as usually described, involves, he says, the very reverse. Tellingly Wright does not signal any direct disagreement with any of these Fisherian analogies or disanalogies. Instead, in keeping with his own view of the relationship between science and philosophy, he says that the following three options are "philosophical questions that I shall not attempt to discuss": whether "evolution is a mere eddy" in the general running down of the universe, or whether "the developmental side of nature" so conspicuous in biology is an "aspect of reality more basic than increase in entropy in physical systems" or whether "time is essentially" directionless. The Fisherian comparison, he concludes, "brings out the difficulty of accounting for the evolutionary process on the same basis, statistical theory", as that leading to the entropy law in physics. Yet, says Wright, switching back from philosophy to science: "it seems the only course open to scientific analysis"; for, as he explains, the possibility of a "physiological rather than statistical interpretation" is not available with the demise of Lamarckian inheritance.[82]

That Wright never accepted that evolution was a counterentropic eddy in an otherwise entropic universe is only hinted at here or elsewhere. But years later he would end his piece on 'Organic Evolution' for the *Encyclopedia Britannica* by saying that, although each species has been "treated as if it evolved independently of the rest of the world except for a rather mechanical-seeming connection through the concept of selection pressure", in fact the "evolution of each species is merely an aspect in the evolving pattern of life as a whole and indeed of the world as a whole". Natural selection, he declares in closing, "is an abstraction of the complicated reciprocal process" whereby the "pressure of the species" to expand in the world and the pressure of the world "to keep the species in its place" results, through devious ways, "in a general trend toward progressive elaboration of the patterns of organization of both".[83]

Where Fisher restricted his optimism to the future of civilisation, provided intelligence was now allowed to take it under eugenic direction, Wright seems to have had a far wider cosmic hope for progress. However, if that tempts us to think that more of traditional religion lived on in Wright than in Fisher, we are checked by two reflections. The first is that the nearest equivalent for Wright of Fisher's college chapel was the Unitarian Church in Madison that Wright did indeed attend; but it is no full equivalent, being a far less demanding institution, metaphysically speaking, and one where Wright never went so far as to preach, it seems. The second is that when Wright did consider the metaphysics of theism the outcome was negative. For anything physical, such as a physical universe as a whole, to have a mind of its own, as a person has, it must, argued Wright, have a comparably tightly knit organisation; but this the universe, even with its ubiquitous fields of force, gravitational and otherwise, seems manifestly to lack. When it came to religion in general, Wright was no more a disbeliever than Fisher, but where Fisher had his Christianity to reconcile with his biology, Wright had only, it seems, some minimal Unitarian sympathies to preserve.[84]

When we turn to the political elements in the case, we have a procedural drawback to face. Fisher wore his politics on his sleeve, most overtly when writing as one eugenist for others. By contrast Wright never openly declared his political loyalties in print. There is, however, an indirect way of getting at a rough, intuitive comparison and contrast between the politics, in the broadest sense, of the two men. This is to look at their stand on those topics in genetics that are most ideologically telling: most especially, the nature-nurture controversy regarding human intelligence; human racial differences; the controversy over genetic damage from radiation; and eugenics itself.

Before doing this, however, we need to appreciate the difficulties in doing so. To some extent, it is possible to introduce a convenient simplifying assumption, by taking Fisher, as he is naturally viewed from a radical, socialist perspective; that is, as a pretty pure specimen of that conservative liberal ideology so manifestly adapted to the aspirations of the professional middle class in Edwardian England, as those aspirations were pursued through the eugenics movement. Fisher, as an extreme hereditarian and ultra-selectionist fits, almost perfectly, then, the historiographies that see Malthusianism and Social Darwinism as succeeded in the twentieth century by those other varieties of biological determinism that Fisher upholds. Nor shall we resist this assumption here. Perhaps the English eugenics movement was not so dominated by professional people among the middle class as was once thought, but for Fisher himself it was a movement especially appropriate to that social role.[85] Certainly Fisher rejects determinism in some senses, but he remains a biological determinist in holding that the properties in people decisive for the future prospects of civilisation are causally conditioned, and in that sense determined biologically, albeit probabilistically, in that their fate is only controllable for the best if the causation that biologists study — the causation of heredity and selection — is intelligently directed.

It is defensible, therefore, to ask how far Wright's thinking diverges from Fisher's in ways that might make him a less appropriate target for a radical, socialist critique of his thought. The answer, however, turns out once again to be less easy than one might think in advance.

On IQ and its heritability, Wright appears far too uncritical for radical, socialist comfort. Wright had taken up this topic, first, in a paper of 1931 and he returned to it at length in 1978 in the fourth volume of *Evolution and the Genetics of Populations*. He was certainly wary of drawing genetic comparisons among populations with different cultures; and he was certainly wary of saying how far IQ tests measured intelligence as that concept was generally understood. However, he saw these two cautions as consistent with taking IQ as a repeatable measure of some aspect of mental capacity, and with using data from studies of adopted children and from twin pairs to estimate its heritability within the population in question, and with siding more with Burt and Jensen than with their critics. Again, under the heading 'Temperament,' Wright joined those who had calculated a moderately strong heritability for such traits as responsibility, self-control and intellectual efficiency.[86]

One main conclusion from the following chapter, on 'Racial Differentiation in Mankind,' is that human evolution in the prehistoric and historic periods has consisted mainly of the "last phase of the shifting balance process": namely, "excess gene flow from a limited number of primary and secondary centers in which culture, and presumably genetic capabilities for it, have reached selective peaks". In working toward this conclusion, Wright avoids any assumption that there are genetic bases for this or that particular cultural trait. Earlier in that volume he had concluded that Fisherian individual selection and Hamiltonian "familial selection" are not entirely explanatorily adequate, and that "intergroup selection (Wynne-Edwards)" is also needed in explaining the evolution of "behavior". Now, he considers the "considerable but incomplete correlation" between the evolution of the genetic system and the evolution of culture. Stressing that if "the multiple genetic aspects of mental ability could be measured more independently of culture than is the case", then doubtless "each local race" would turn out to have "its own unique combination of favorable qualities", he urged also, however, that there have been "wide differences" among peoples "in average intellectual ability and cultural level from the standpoint of progress towards the situation in civilized man". The "capacity to anticipate and plan for the future" would be favored "under northern conditions and selected for insofar as it has a genetic basis".[87]

Such tentative conclusions and their lack of active policy implications would not have satisfied anyone who accepted Fisher's views on civilisation. When it came to radiation and genetics, Wright was, however, less reluctant to be normative. Wright's analysis rests on the assumption that "damage to society" can be assessed "more objectively than personal illness, pain, and frustration by classifying human phenotypes according to the ratio of social contribution to social cost", that is, cost of "rearing, education and maintaining standards of living". Thus, in an extreme case, a lethal mutation causing perinatal infant death is costly genetically, and distressing personally, but not very costly socially, because by contrast to a disabling but not fatal disease, there is no great burden to society. Not all "social damage" has genotypic bases; some is due to "reliance on inherited wealth", for instance; but there is more damage from those who "contribute negatively because of antisocial activities, and here genotype plays more role, though how much is a controversial question". Wright is hardly as unequivocal as the eugenists of old would have liked, but what he says is hardly grist for their opponents' mills either. Likewise, with the issues dividing Dobzhansky and Muller on the topic of genetic load. He distances himself from Muller, and sides silently with Dobzhansky, his onetime co-author, over the levels of heterozygosis it is proper to assume and over the advantages of those levels; but he concludes that Muller's campaign to reduce our exposure to radiation is "fully" justified.[88]

Muller was hardly less committed to eugenics than Fisher, but Wright kept his distance from their position. Although listed on its letterhead as one of 108 members of the advisory council of the American Eugenics Society, Wright never endorsed any eugenic policy. His reluctance stemmed, it seems, from the view that any positive eugenic measures required agreement as to what an ideal society would be, and no such agreement is achievable. "An industrial civilisation is a more complex organism" than an agricultural one and, therefore, he once wrote, it "requires development of many diverse types". He suspected that "the political and other choices" determining what sort of civilisation there is, tend "more or less automatically" to direct "the genetic character of the population", though perhaps with a considerable lag time. A "positive eugenic program" would then be directed at correcting the maladjustment, arising from this lag, either by changing the genetics to fit to social organisation, or by changing the social organisation "so as to fit types of individuals which seem admirable but have insufficient scope in the existing society". Needless to say, Fisher would never have agreed that an adjustment lag is all that needs correction, nor therefore that these corrective measures would be appropriate. But, although Wright's misgivings about anything like Fisher's eugenics are profound, they are insufficiently grounded in dissatisfaction with the existing social order to appeal to radical, socialist critics — or indeed, radical, conservative critics — of all eugenics.[89]

There is one impression that is hard to resist, but it does rather little for us historiographically. Fisher was rarely drawn to moderate, middle-of-the-road positions, while Wright often was; indeed his very language, as an evolutionary theorist, of balance and poise (meaning a very fine balancing) fits such a temperament all too obviously. Such impressions cannot take us, however, significantly beyond our earlier conclusions that, for instance, in evolutionary biology, we have in Fisher the most ardent, vindicative selectionist the world had ever — or perhaps has ever — seen, while in Wright we have an adjudicative theorist of evolution.

Quite generally, it does seem that the contrasts and comparisons one can draw between Fisher and Wright on political and ideological issues certainly contribute to the large picture, but more by way of confirming conclusions reached on other biographical grounds than by providing an ultimate resolution of anything that is otherwise inexplicable.

16. CONCLUSIONS

Returning to those themes announced at the opening of this paper, we may begin with the question of materialism, more precisely mechanistic materialism and its relations to Darwinian evolutionary theory. On the face of it, the case of Fisher and Wright must seem to make trouble for any view that Darwinian science has followed a succession from mechanistic to dialectical materialism. For consider where they stand, generationally speaking. Coming of age just before the first World War, they engaged a generation of earlier writers — such as Karl Pearson, Henri Bergson and Lloyd Morgan, to name only a few — who had refused to agree that evolution had to be interpreted mechanistically and materialistically. Fisher and Wright were to disagree with much that these authors proposed, either as science or as philosophy. However, each of the two men, in his different way, came to believe strongly that Darwinian biology does not require one to be mechanistic and materialistic in one's metaphysics.

Their cases suggest, therefore, that we have to be wary of constructing some historiographical abstraction — Darwinism, the Darwinian worldview or the Darwinian tradition or whatever — and then presuming that there is something that can be identified as the natural, dominant or inherent metaphysical character of this abstract construction. It may well be that Darwin himself — one thinks especially of his notebooks of the late 1830's — felt the need to make assumptions about consciousness and will that were deterministic (if not mechanistic) and materialistic, in order, as he saw it, to bring mind and so man within the causal, explanatory scheme he was developing for plant and animal structures and habits. It may well be that in assimilating Darwin's thinking some people, later, felt that they had to make similar assumptions. However, even to acknowledge this much is to leave open the possibility that yet others have felt differently, so that we will only be begging historical questions if we lay down an essentialist stipulation that insofar as these others did this then they were not really true Darwinians. Richards's history seems better fitted to avoid this kind of question begging than Allen's and is, therefore, confirmed to that extent at least. Quite generally, what we need to take in is the possibility that longstanding empirical convictions and metaphysical anxieties about mind and body, will and consciousness, creativity and causality and so on have continued to live on among professional twentieth-century biologists, rather than being left behind with the passing of the great nineteenth-century debates among men of science, men of letters and men of the cloth. The disagreements between Allen, Greene and Richards have done us a welcome favor in making us aware of that possibility.[90]

Turning now to the Bowler theses, one has to note that whether these are seen to be confirmed depends very much on how one interprets them. Sometimes, Bowler seems to be saying that it was idealistic philosophies of nature that fostered progressionist and developmentalist views of evolution, so that, conversely, the new Mendelian genetics of the twentieth century could be effective in discrediting such views because it challenged those philosophies of nature.[91] But this thesis is surely not supported by the case of Fisher and Wright. In Wright we have someone who saw his own philosophy as explicitly idealistic; and even Fisher can not be seen as confronting and repudiating idealism in the philosophy of nature. What does one need to concentrate on, instead, therefore, in understanding Fisher's and Wright's departures from the progressionist and developmentalist views so sidespread in the sixty years after Darwin's *Origin*? Here, surely, the obvious answer provides the right place to start, at least, and that is the conjunction in these two of Mendelism, biometry and Darwinism.

A further reflection is also pertinent. The biometricians, as led by Weldon and Pearson, had already upheld an interpretation of evolution that was both selectionist and nondevelopmentalist. One possible way to understand Fisher and Wright, is to see them as bringing Mendelism into conjunction with biometry in such a way that the synthesis retained this selectionism and lack of developmentalism. To follow up this interpretation, one would have to ask, further, how their Mendelism could be brought into such a synthesis in that way. Here, surely, it would be relevant to cite those features of heritable Mendelian variation — as Fisher and Wright conceived of it — that made it, in and by itself, quite incapable of directing the process of evolution: multiple factors, each of small effect, influencing quantitative traits; the random shuffling of factors in recombination without linkage; the rarity and fortuitousness of mutations and so on.

Bowler himself would not deny that these things are directly relevant, but he would rather, it seems, rest his case on a more general thesis: namely, that, with the new Mendelian conception of heredity, it became possible to distinguish transmission genetics from developmental genetics, and so to divorce evolution from development by integrating the theory of natural selection with Mendelian transmission genetics. This separatism thesis works well, it must be conceded, for Fisher. But it is hardly acceptable without qualification for Wright. It goes without saying, naturally, that the contrasts between mitosis and meiosis are fundamental for Wright's understanding of the contrast between development and evolution. It also goes without saying that, as Wright put it himself, "chance variation" and the "influence of the environment" are both "denied in the Aristotelian conception of evolution as analogous to individual development, the realization of an innate potentiality, irrespective of environment (except for provision of the necessary conditions for life)". But it does not follow from these familiar points that the problems of developmental genetics became entirely separable from those of evolutionary genetics. Indeed, Wright constantly relates to one another his views on these two subjects. Thus, in explaining the differences between Richard Goldschmudt's macro-mutationist views and his own and others' "neo-Darwinism", he stressed that there lay at the root of those differences divergent conceptions of the "physiological relations between germ plasm and organism". Goldschmidt seems to hold, he says, that "the conception of the organism as an integrated reaction system requires a corresponding *spatial* integration of the germ plasm" so that essential change in the reaction system can only come about "by repatterning of the chromosomes". To himself and others, Wright says, "a *temporal* integration is all that is necessary, or even possible, with the chain reaction as the simplified model". He goes on to explain how, with branching hierarchical systems of chain reactions, there can be many reaction systems based on the same set of genes and such systems can then evolve "more or less independently of one another".[92]

In Wright's case, therefore, it seems best not to say that evolutionary biology is tied to transmission genetics while sundered from developmental genetics, but to say that the integration of a new Mendelian genetics, of transmission and development, with Darwinism and biometry, was wrought in such a way that developmentalist evolutionary theory was undercut and an opportunistic, arboriform process was seen to result, as by Darwin himself, from the contingencies of variation interacting with the contingencies of environmental influences. In so far as Wright is a representative figure, therefore, the emergence within the Mendelian program of a new physiological and developmental genetics, must be included within the history of evolutionary biology in the early twentieth century.

So many probabilistic themes have proven relevant to the case of Fisher and Wright that the only conclusion to be confidently drawn is that there is no single conclusion. Even if we focus on the departure of science from some supposed Laplacean determinism — that may or may not have been a genuine historical reality — the story remains highly complicated.[93] Among the many reasons for this complexity four can usefully be stressed here. First, as several authors have insisted lately, there is more, far more, to the history of statistical thinking than the history of mathematical statistics. Were one to concentrate on Fisher's and Wright's mathematical consensus one would miss almost all the interesting and instructive comparisons and contrasts to be made between their lives and works as evolutionary theorists. Second, as these same authors have urged, since the mid-nineteenth century statistical thinking has ranged over a remarkable array of domains from social theory to astronomy, and statistics has itself developed as it has because of that wide range of applications. Wright and Fisher, in their different ways, vividly instantiate that theme. Fisher learned the style, habits and skills of his statistical thinking about genetics and evolution from Jeans and Boltzmann on gases as much as from Yule and Pearson on parental and fraternal correlations; while for Wright a decisive early statistical ambition was to move beyond Jennings's analyses of systems of livestock breeding.

Third, the scientific implications of statistical thinking have never been straightforward in their interpretation. Pearson and others saw statistics, positivistically, as leading science away from causation to correlations and laws alone. Wright and Fisher agreed in taking precisely the opposite view; for them statistical techniques were a way to further the traditional aim of science in finding the causes, often hidden causes, behind phenomena. Indeterministic or probabilistic causation was never reducible to statistical correlation. Fourth, the philosophical implications of statistical thinking have been even less straightforward in their interpretation. By the time of Fisher's and Wright's undergraduate years it was a commonplace that, in physics itself, mechanics was, arguably at least, no longer as mechanistic as it had been, nor matter theory as materialistic, nor causation as deterministic; that, in sum, the Laplacean intellect was no longer an unproblematic ideal. But, equally, it was evident that there was no agreement as to how to understand the ultimate philosophical significance of these developments. What the serious but amateurish philosophising of our two young evolutionary theorists shows is that there was no one simple and persuasive response that it was obviously correct to make to that rich, rewarding but confusing state of discussion. As on other occasions, therefore, a virtue of history, especially when it includes biography, is that we can learn how, in the long run, things are never likely to go as simply as either our philosophical predelictions or our scientific upbringing may incline us to presume.[93]

NOTES AND REFERENCES

[1] Almost everything in this paper is indebted to extensive discussions over the years with John Turner and to his papers on Fisher and Wright; so much so that large parts could appropriately be thought of as jointly authored. The paper also draws throughout on the publications and conversation of William Provine. I am grateful, too, for valuable talks with Garland Allen, John Beatty, James Crow, Anthony Edwards, Jonathan Harwood, Ernst Mayr, Robert Olby and David West.

[2] For a fuller discussion of these and other rationales, see the editors' introduction and other contributions to R. C. Olby, G. N. Cantor, J. R. R. Christie and M. J. S. Hodge, eds., *Companion to the History of Modern Science*, London and New York: Routledge, 1990.

[3] G. Allen, 'The Several Faces of Darwin: Materialism in Nineteenth and Twentieth-Century Evolutionary Theory', in D. S. Bendall, ed., *Evolution from Molecules to Men*, Cambridge: Cambridge University Press, 1983; pp. 81—102; J. C. Greene, *Science, Ideology and World View*, Berkeley: University of California Press, 1981; R. J. Richards, *Darwin and the Emergence of Evolutionary Theories of Mind and Behavior*, Chicago and London: The University of Chicago Press, 1987; pp. 405—407.

[4] P. J. Bowler, *The Non-Darwinian Revolution: Reinterpreting a Historical Myth*,

Baltimore: Johns Hopkins University Press, 1988; *The Mendelian Revolution: The Emergence of Hereditarian Concepts in Modern Science and Society*, London: Athlone Press, 1989; *Charles Darwin: The Man and His Influence*, Oxford and Cambridge, Mass.: Basil Blackwell, 1990.

[5] L. Krüger *et al.*, eds., *The Probabilistic Revolution*, 2 vols., Cambridge, Mass: MIT Press, 1987; T. M. Porter, *The Rise of Statistical Thinking, 1820—1900*, Princeton: Princeton University Press, 1986; S. Stigler, *The History of Statistics: The Measurement of Uncertainty before 1900*, Cambridge, Mass.: Harvard University Press, 1986; G. Gigerenzer *et al.*, *The Empire of Chance. How Probability Changed Science and Everyday Life*, Cambridge: Cambridge University Press, 1989; Ian Hacking, *The Taming of Chance*, Cambridge: Cambridge University Press, 1990.

[6] J. F. Box, *R. A. Fisher: The Life of a Scientist*, New York: John Wiley, 1978; J. H. Bennett, ed., *Natural Selection, Heredity and Eugenics. Including selected correspondence of R. A. Fisher with Leonard Darwin and others*, Oxford: Clarendon Press, 1983; J. H. Bennett, ed., *Statistical Inference and Analysis. Selected Correspondence of R. A. Fisher*, Clarendon Press, 1990; W. Provine, *Sewall Wright and Evolutionary Biology*, Chicago: University of Chicago Press, 1986.

[7] Provine, *Wright*, pp. 18 and 95—7; Wright, 'Panpsychism and Science', in J. E. Cobb and D. R. Griffin, eds., *Mind in Nature*, Washington: University Press of America, 1975, pp. 79—88; J. F. Crow, 'Sewall Wright's Place in Twentieth-Century Biology', *Journal of the History of Biology* **23** (1990) 57—89. Further evidence of Wright's philosophical views, see items 26, 38 and 104 in the full bibliography in Provine, *Wright* and also in *Evolution. Selected Papers* (see n. 14).

[8] Box, *Fisher, pp. 288—95; The Social Selection of Human Fertility*, Oxford: Clarendon Press, 1932; 'Indeterminism and Natural Selection', *Philosophy of Science* **1** (1934) 99—117; *Creative Aspects of Natural Law*, Cambridge: Cambridge University Press, 1950. All three pieces are reprinted in facsimile in J. H. Bennett, ed., *Collected Papers of R. A. Fisher*, 5 vols., Adelaide: The University of Adelaide, 1971—4. In citations of Fisher's writings that are in this invaluable collection, I shall refer to them by volume number together with the number of the item given it by Bennett. These three are, accordingly, at *CP, 3*: 99 and 121 and 5: 241. Box and the first volume of *CP* contain a full bibliography.

[9] Fisher, *Eugenics Review* **23** (1932) 88—90, in Bennett, ed., *Natural Selection*, pp. 287—8.

[10] Fisher, 'On the Dominance Ratio', *Proceedings of the Royal Society of Edinburgh* **42** (1922) 321—41; *CP, 1*: 24.

[11] Provine, *Wright*, pp. 232—276; *Origins of Theoretical Population Genetics*, Chicago: University of Chicago Press, 1971.

[12] Fisher, 'Darwinian Evolution of Mutations', *Eugenics Review* **14** (1922) 31—34, pp. 33—4; *CP, 1*: 26.

[13] Provine, *Wright*, pp. 232—276.

[14] Wright, 'The Genetical Theory of Natural Selection. A Review', *Journal of Heredity* **21** (1930) 349—59, pp. 352 and 355. Where, as in this case, the item is reprinted in facsimile in S. Wright, *Evolution. Selected Papers. Edited and with Introductory Materials by W. B. Provine*, Chicago: University of Chicago Press, 1988, I shall give the item number and the page references there (this collection does not reproduce the

original page numbers, although of course they can be reconstructed from what is given). Thus, in this case, *ESP, 9*: 83 and 86.

[15] In addition to Box, Bennett and Provine as already cited, see Provine, 'The R. A. Fisher—Sewall Wright Controversy and Its Influence upon Modern Evolutionary Biology', *Oxford Surveys in Evolutionary Biology* **2** (1985) 197—200, and J. R. G. Turner, 'Random Genetic Drift, R. A. Fisher, and the Oxford School of Ecological Genetics', in L. Krüger *et al.*, eds., *The Probabilistic Revolution* **2**, pp. 313—54.

[16] In addition to Box and Bennett, see B. J. Norton, 'Metaphysics and Population Genetics: Karl Pearson and the Background to Fisher's Multifactorial Theory of Inheritance', *Annals of Science* **32** (1975) 537—53; 'Fisher and the Neo-Darwinian Synthesis', in E. G. Forbes, ed., *The Human Implications of Scientific Advance. Proceedings of the 15th International Congress on the History of Science*, Edinburgh: Edinburgh University Press, 1978; 'Fisher's Entrance into Evolutionary Science: The Role of Eugenics', in M. Grene, ed., *Dimensions of Darwinosm*, Cambridge: Cambridge University Press, 1983, pp. 19—29; 'La Situation Intellectuelle au Moment des Débuts de Fisher en Génétique des Populations', *Revue de Synthése. IIIe serie: 103—4* (1981) 230—250. (A volume devoted to *R. A. Fisher et l'Histoire de la Génétique des Populations*).

[17] On mutations and their rates, see R. C. Olby, 'La théorie génétique de la sélection naturelle vue par un historien', *Revue de Synthése, IIIe serie: 103—4* (1981) 251—289. See the comments of Benett, *Natural Selection* p. 17, on the interpretation of Fisher given by various historians of science.

[18] The tradition of identifying natural selection as a known cause is discussed in M. J. S. Hodge, 'Natural Selection as a Causal, Empirical, and Probabilistic Theory', in L. Krüger *et al.*, eds., *The Probabilistic Revolution*, vol. 2, pp. 233—270 and 'Darwin's Theory and Darwin's Argument', in M. Ruse, ed., *What the Philosophy of Biology is. Essays dedicated to David Hull*, Dordrecht, Boston, London: Kluwer, 1989, pp. 163—182.

[19] Fisher, 'The Measurement of Selective Intensity', *Proceedings of the Royal Society of London, B* **121** (1936) 58—62, p. 58; *CP, 3*: 147.

[20] For a more extensive interpretation of Fisher's evolutionary theory along the lines taken here, see J. R. G. Turner, 'Fisher's Evolutionary Faith and the Challenge of Mimicry', *Oxford Surveys in Evolutionary Biology* **2** (1985) 159—196 and his *op.cit.* n. 15. For a comprehensive analysis of how Fisher's views stand today, see the two-part article, E. G. Leigh, 'Ronald Fisher and the Development of Evolutionary Theory', *Oxford Surveys in Evolutionary Biology* **3** (1986) 187—233 and **4** (1987) 212—263.

[21] See the articles of Turner, Olby and Leigh cited in notes 15, 17 and 20.

[22] Fisher, *The Genetical Theory of Natural Selection*, Oxford: Clarendon Press, 1930, and the second edition, published by Dover, New York, 1958.

[23] 'Paper on "Heredity" (comparing methods of Biometry and Mendelism)', in Bennett, ed., *Natural Selection*, pp. 51—8; the quoted words are at pp. 56—7.

[24] R. A. Fisher and C. S. Stock, 'Cuenot on Preadaptation. A Criticism', *Eugenics Review* **7** (1915) 46—61; p. 60. *CP, 1*: 5.

[25] *Ibid.*, pp. 60—1.

[26] 'On the Dominance Ratio' (see n. 10), p. 321.

[27] *Ibid.*, pp. 321—2. Provine, *Wright*, p. 241, has Fisher here making the investigation of selection comparable to the theory of gases. But this is surely an incorrect reading.

[28] *Ibid.*, p. 340; 'Croonian Lecture: Population Genetics', *Proceedings of the Royal Society of London, B* **141** (1953) 510—523; p. 515. *CP, 5*: 252.

[29] *Genetical Theory*, 1st ed., p. 11; 2nd ed., p. 11.

[30] *Ibid.*

[31] *Ibid.*, pp. 35 and 46 (1st ed.); pp. 37 and 50 (2nd ed.).

[32] G. R. Price, 'Fisher's "Fundamental Theorem" Made Clear', *Annals of Human Genetics, London* **36** (1972) 129—140, pp. 132 and 131. Price argues (p. 131) that since "the standard of 'fitness' changes from instant to instant, this constant improving tendency of natural selection does not necessarily get anywhere in terms of increasing 'fitness' as measured by any fixed standard," and mean fitness is "about as likely to decrease under selection as to increase".

[33] See Wright's review cited in n. 14.

[34] 1st ed., pp. 35—37; 2nd ed., pp. 38—40.

[35] 1st ed., pp. 36—7; 2nd ed., p. 39.

[36] 1st ed., p. 37; 2nd ed., pp. 39—40. Fisher seems never to have gone explicitly in quest of "some more general principle".

[37] 'The Social Selection of Human Fertility' (see n. 8), pp. 8—9.

[38] 'Retrospect of the Criticisms of the Theory of Natural Selection', in J. S. Huxley *et al.*, eds., *Evolution as a Process*, London: Allen and Unwin, 1953, pp. 84—98. *CP, 5*: 314. This article was evidently written in the early 1930's. See Bennett, ed., *Natural Selection*, p. 202.

[39] In 1934, in 'Indeterminism and Natural Selection' (n. 8), Fisher wrote (pp. 110—111) revealingly: "Two characteristics of Darwin's evolutionary thought which, in the writer's opinion, give it its supreme and lasting value are, first, that while an active theorist, willing to follow long chains of reasoning if he felt their foundations secure, he constantly brought his speculations into contact with a candid and thoughtful scrutiny of living things themselves; and, secondly, that he was never willing to curb or limit his thought by theoretical considerations brought in from other fields of study. Thus his theory of evolution made demands upon the age of the earth beyond what the physicists, for many years after his death, regarded as physically possible. We now know that the evolutionists were right, and that it was the physical theory that was faulty. Again, in regard to men and his mental and moral characteristics, (his Soul in all but the meaning given to the word by theological theory), Darwin did not shrink from the plain meaning of the facts that fighting cocks could be bred for courage, or dogs for fidelity, and that the amazing plumage and ornaments of many male birds indicated a discriminative choice on the part of the females, which only an anthropocentric prejudice would hesitate to call aesthetic appreciation."

[40] 'Student', *Annals of Eugenics* **9** (1939) 1—9. *CP, 4*: 165.

[41] 'Paper on "Evolution and Society"', in Bennett, ed., *Natural Selection*, pp. 58—62. See pp. 58, 61 and 62.

[42] 'The Evolution of Sexual Preference', *Eugenics Review* **7** (1915) 184—192. See pp. 191—2. *CP, 1*: 6. On Fisher's Nietzschean friends, see Box, *Fisher*, p. 18.

[43] 'Indeterminism and Natural Selection' (n. 8) p. 99.

[44] *Ibid.*

[45] *The Social Selection of Fertility* (n. 8).

[46] 'Indeterminism and Natural Selection' and *The Social Selection of Fertility*.

[47] *Creative Aspects of Natural Law* (n. 8), pp. 4—5.

XIII

[49] *Ibid.*, p. 7.

[50] *Ibid.* p. 13. Articles on Bergson and on Smuts by T. A. Goudge (in P. Edwards, *Encyclopedia of Philosophy*, 8 vols., New York and London, MacMillan, 1967) give good introductions to their evolutionary views.

[51] *Ibid.*, pp. 20—22. See also 'The Renaissance of Darwinism', *Listener* **37** (1947) 1001. *CP, 4*: 217, and John Turner's two articles (nn. 15 and 20).

[52] Fisher to C. S. Sherrington: 22 January 1947 in Bennett, ed., *Natural Selection*, pp. 261—2.

[53] Here, as always in discussing Wright's career, I am drawing directly on Provine, *Wright.*

[54] *Ibid.*, pp. 180—1 and p. 16.

[55] See the three Wright articles cited in n. 7, and Provine, *Wright*, p. 95.

[56] 'The Relation of Livestock Breeding to Theories of Evolution', *Journal of Animal Science* **46** (1978) 1192—1200. *ESP, 1.*

[57] J. F. Crow, 'Sewall Wright's Place in Twentieth-Century Biology', *Journal of the History of Biology* **23** (1990) 57—89. Cf. Provine, *Wright*, p. 235.

[58] *Ibid.*

[59] 'Evolution in Mendelian Populations', *Genetics* **16** (1931) 97—159; p. 142. *ESP, 11*: 143.

[60] 'The Shifting Balance Theory and Macroevolution', *Annual Review of Genetics* **16** (1982) 1—19, pp. 4—6.

[61] 'The Roles of Mutation, Inbreeding, Crossbreeding and Selection in Evolution', *Proceedings of the Sixth International Congress of Genetics* **1** (1982) 256—66. *ESP, 12*: 161—171.

[62] 'Evolution', pp. 142—3, 100 and 98.

[63] *Ibid.*, pp. 144—5.

[64] *Ibid.*, p. 145.

[65] *Ibid.*, pp. 145—6.

[66] *Ibid.*, p. 150.

[67] *Ibid.*, p. 151.

[68] *Ibid.*, pp.157—8.

[69] 'Gene and Organism', *American Naturalist* **87** (1953) 5—18; 'Biology and the Philosophy of Science', *Monist* **48** (1964) 265—90. 'Panpsychism and Science' (n. 7).

[70] 'Panpsychism and Science', p. 80.

[71] 'Biology and the Philosophy of Science', p. 281.

[72] *Ibid.*

[73] 'Gene and Organism", p. 18.

[74] 'Panpsychism and Science', p. 84.

[75] 'Evolution', p. 154.

[76] 'Panpsychism and Science', p. 87.

[77] 'Evolution', pp. 154—5. I see my analysis here as consistent with the rather different one given in Provine, *Wright*, pp. 95—7.

[78] *Ibid.*, p. 147

[79] See Fisher, 'Retrospect' (n. 38).

[80] See Allen's paper cited in n. 3.

[81] J. Gayon, *La Théorie de la Selection: Darwin et l'aprés Darwin*. Thése de Doctorat, Université de Paris I, 2 vols., 1989; vol. 2, pp. 646—7. This important study is expected to appear in a revised, published version in 1992.

[82] 'Statistical Theory of Evolution', *Journal of the American Statistical Association* **26** (1931) supplement: 201—8; p. 201. *ESP 10*: 89—96.

[83] 'Evolution, Organic', in *Encyclopedia Britannica*, 14th ed. revised, 1948, vol. 8, pp. 915—29; p. 929. *ESP 26*: 524—538.

[84] Provine, *Wright*, p. 460; 'Panpsychism and Science', p. 85.

[85] On the ideology and politics of English eugenics see D. A. Mackenzie, *Statistics in Britain 1865—1930: The Social Construction of Scientific Knowledge*, Edinburgh: Edinburgh, 1981; R. C. Olby, 'The Dimensions of Scientific Controversy: The Biometrician-Mendelian Debate', *British Journal for the History of Science* **22** (1989) 299—320; G. R. Searle, 'Eugenics and Class', in C. Webster, (ed.) *Biology, Medicine and Society 1840—1940*, Cambridge: Cambridge University Press, 1981, pp. 217—242; D. Kevles, *In the Name of Eugenics*, New York: Knopf, 1985: E. J. Larson, 'The Rhetoric of Eugenics: Expert Authority and the Mental Deficiency Bill', *British Journal for the History of Science* **24** (1991) 45—60. For a radical, socialist perspective, see R. Lewontin, S. Rose and L. Kamin, *Not in our Genes. Biology, Ideology and Human Nature*, New York: Pantheon, 1985. It is their notion of biological determinism that I am referring to here.

[86] 'Statistical Methods in Biology', *Journal of the American Statistical Association* **26** (1931) supplement: 155—163; *Evolution and the Genetics of Populations*, 4 vols., Chicago and London: University of Chicago Press, 1968—1978, vol. 4, pp. 296—420.

[87] *Ibid.*, pp. 53, 458 and 455—6.

[88] *Ibid.*, vol. 3, pp. 477—9. Wright's fullest treatment of these issues is in 'Discussion on Population Genetics and Radiation', *Journal of Cell and Comparative Physiology* **35** (1950) 187—210 and 'On the Appraisal of Genetic Effects of Radiation in Man', in *The Biological Effects of Atomic Radiation: Summary Reports from a Study by the National Academy of Sciences*, Washington: NAS-NRC, 1960, pp. 18—24.

[89] Provine, *Wright*, p. 180 and Wright to P. Popenoe, May 13, 1931, as printed there on p. 181.

[90] See n. 3.

[91] See n. 4.

[92] 'Evolution, Organic' (n. 83), p 924; 'The Material Basis of Evolution', *Scientific Monthly* **53** (1941) 165—70; pp. 168—9. *ESP, 27*: 389—94.

[93] (Added in proof). Jean Gayon's book (see n. 81) now interprets Fisher much as I have here: *Darwin et l'après-Darwin*, Paris: Editions Kimé, 1992. On Philosophy and the architects of the synthetic theory, see also J. Harwood, 'Metaphysical Foundations of the Evolutionary Synthesis', forthcoming.

XIV

Natural Selection as a Causal, Empirical, and Probabilistic Theory

Darwin's conforming of his theory to the old vera causa *ideal shows that the theory of natural selection is probabilistic not because it introduces a probabilistic law or principle, but because it invokes a probabilistic cause, natural selection, definable as nonfortuitous differential reproduction of hereditary variants.*

Chance features twice in this causal process. The generation of hereditary variants may be a matter of chance; but their subsequent populational fate is not; for their physical property differences are sources of causal bias giving them different chances of survival and reproduction. This distinguishes selection from any process of drift through fortuitous differential reproduction in the accumulation of random or indiscriminate errors of sampling. To confirm the theory of natural selection empirically is to confirm that this probabilistic causal process exists, is competent, and has been responsible for evolution. Such hypotheses are both falsifiable and verifiable, in principle, if not in practice.

Natural selection has been accepted and developed by biologists with very diverse attitudes toward chance and chances. But the theory and its acceptance have always involved probabilistic causal judgments that cannot be reduced to correlational ones. So, the theory has contributed to a probabilistic shift within the development of causal science, not to any probabilistic rebellion in favour of science without causes.

1

This chapter proposes a framework for integrating biologists' and philosophers' analyses of natural selection as a causal, probabilistic, and empirical theory of evolution. Throughout, the argument will be that the probabilistic character of the theory, whether in Darwin's day or ours, can only be properly understood when its distinctively causal and empirical character is kept in view.

Until fairly recently, perhaps only a decade and a half ago, philosophical commentary on natural selection rather rarely interested biologists, who were understandably impatient, for example, with endless variations on the old tautology complaint. Equally, biologists' disagreements over genetic drift, the classical and balance theories, group selection, and so on attracted little attention from philosophers.

I am much indebted here to discussions with John Beatty, David Hull, Larry Laudan, Rachel Laudan, Ernst Mayr, Bernard Norton, Michael Ruse, Sam Schweber, and John Turner. For the excellent interdisciplinary opportunities provided by two Bielefeld conferences, I am very grateful indeed. The present chapter incorporates material from an earlier piece, where some historical points were treated somewhat more fully: "Law, Cause, Chance, Adaptation and Species in Darwinian Theory in the 1830's, with a Postscript on the 1930's," in M. Heidelberger, L. Krüger, and R. Rheinwald, eds., *Probability since 1800. Interdisciplinary Studies of Scientific Development* (Bielefeld: Universität Bielefeld, 1983), pp. 287–330.

XIV

Happily, this phase is now past, as anyone will know who reads in such journals as *American Naturalist, Annual Review of Ecology and Systematics, Biology and Philosophy, Journal of Theoretical Biology, Paleobiology, Philosophy of Science, Studies in History and Philosophy of Science, Synthese*, and *Systematic Zoology*. For there one finds the technical resources provided by both biology and philosophy often combined and applied to problems of common concern to the two disciplines.[1]

This chapter aims to contribute to this most welcome trend. But it does not set out to do so directly. Rather the hope is to clarify the probabilistic character of the theory of natural selection, as it concerns biologists and philosophers alike, by beginning from a point of departure that lies within the discipline of neither party.

This point of departure is an historical one, namely, Darwin's original understanding of the theory of natural selection as a causal, empirical, and so explanatory theory. There will be, however, no concern with history for its own sake, nor any attempt to settle current disputes by invoking venerable authority. Instead, it will be argued that Darwin's conception of the character of the theory is still appropriate today, because it conforms to what is common to the best explications of the theory given of late by biologists and philosophers. Accordingly, the proposal will be that those developments in this century—most notably in Mendelian population genetics and molecular biology—that have made the current versions of the theory no less causal, no less empirical, and no less probabilistic have also made Darwin's original conception of what kind of theory it is more, and not less, instructive.

Although this is a historical proposal, albeit an overtly normative one, it will not be defended by offering a narrative analysis of developments from Darwin to Dobzhansky and beyond. For it will be proposed that we can abstract and generalize, from the century and a quarter since 1859, and insist that answering certain enduring clusters of questions in certain ways has always been characteristic of any thoroughgoing commitment to natural selection as a theory of evolution.[2] The principal thesis will be, accordingly, a very simple, even simple-minded, one, namely, that in trying to understand the theory of natural selection, whether in the original Darwinian or in any subsequent neo-Darwinian context, it is always best to follow Darwin's own strategy and concentrate on distinguishing some four clusters of questions:

1. *The definition question*: What is natural selection? How is this process (or agency or force or whatever) to be defined?

2. *Existence, that is, occurrence and prevalence, questions*: Does it exist, is it going on anywhere? How widespread, how prevalent is it? Among what units and at what levels—organisms, colonies, species and so on—is it occurring?

3. *Competence, that is, consequence and adequacy questions*: What sorts, sizes, and speeds of change does it suffice presently to produce? What are its possible and actual consequences?

4. *Responsibility, that is, past achievement questions*: What has it done? For how much of past evolution has it been responsible? What is to be explained as resulting from it?

Obviously, such question clusters might be distinguished in other ways. But this way will serve our purpose here. For it will be argued throughout this chapter that any correct answers to the definition question will have as a corollary that all the other three clusters concern empirical questions, albeit very diverse questions, about probabilistic causal processes, questions that can be given answers that are testable in principle if not in practice.

2

Before proceeding to argue for this thesis, however, mention should be made of three analytic resources that will be deployed in due course.

First, I have not assumed the privileged correctness of any particular account of testability, whether Carnapian, Popperian, Duhemian, or whatever. But I have assumed what I take to be accepted by most accounts of theory testing, namely, (a) the elementary logical point that "denying the consequent" is valid or deductively correct while "affirming the consequent" is not, so that a statement's truth is not validly inferred from the truth of its consequences, while its falsity is validly inferred from their falsity; (b) the familiar epistemological point that theories can often only be made to have testable predictions as deductive consequences by conjoining them with additional auxiliary hypotheses, so that the elementary logical contrast does not in itself sanction any methodological imperative whereby predictive errors should always be instantly accepted as conclusively falsifying the theory rather than the auxiliaries; (c) the obvious methodological point that, in judging the respective merits of two or more theories confronted with some observational reports accepted as facts, we have to evaluate, among other things, the different sets of auxiliary hypotheses that need to be conjoined with those respective theories if they are to predict those facts.[3]

Second, I sometimes distinguish between something being true as a matter of definition or meaning and something being true as a matter of fact or experience. Now, thanks especially to the teachings of Quine (whose own position has shifted more than is usually appreciated), all such distinctions are often held to be difficult to defend.[4] However, my uses of them do respect these difficulties. For it is not assumed here that something is definitionally rather than factually true independently of all contextual considerations; with changes in the theoretical context, especially, one might well want to revise earlier judgments as to what is best taken as definitional and what as factual. But that possibility only confirms that one can often clarify the assumptions constituting the theoretical framework current at any time by seeing how they make it reasonable to draw the line here rather than there. To drop all such distinctions would thus be to forgo needlessly a very useful source of light on those assumptions themselves.

Third, in conformity with Salmon's developments of Reichenbach's views, I take a causal process to be a physical process, one wherein energy is transformed and transmitted, and one whose later temporal stages can be altered by interfering with earlier stages.[5] Thus, by contrast, an abstract, mathematical "process," whereby successive numerical values are "generated" by an algorithmic "operation," is not a causal process because not physical, while the successive positions of a shadow passing over the ground do not form a causal process because no alterations of later positions can be wrought by actions at earlier ones. Among the successive stages of a causal process there hold relations of causal relevance. And, in conformity with recent explications of what it is to be a causal factor, Giere's, for instance, I accept that relations of statistical relevance cannot in and of themselves constitute causal relevance.[6] Accordingly, I assume that insofar as a demand for explanation is construed as a demand for an identification of what is causally relevant to what is being explained, then neither mere mathematical representations nor purely statistical description can by themselves meet this demand.

Finally, I should emphasize that I have not brought to this analysis of natural selection theory any one philosophical account of what a scientific theory is: the "received view," for instance, dominant in the heyday of logical empiricism, or its more recent rival, the "semantic view." Both of these have been shown to clarify some leading features of Darwin's own and later evolutionary theorizing.[7] But no one such view is uniquely helpful in exhibiting the most instructive continuities in the roles of probabilistic notions in natural selection theory as those roles have developed over the last century and a quarter.

3

On going back now to our starting point in Darwin's work, it will be evident that for my purposes here it is not necessary to read into his writings any philosophical resources developed only in our own times. On the contrary, if our understanding of natural selection theory is eventually to benefit from those resources, we need to begin by taking Darwin on his own terms. We need an analysis of the problems he saw natural selection as solving, an analysis that brings out why he deliberately gave his argument for natural selection a very distinctive structure.

The structure is most easily discerned by comparing *On the Origin of Species* (1859) with its predecessors, the manuscript *Sketch* of 1842 and *Essay* of 1844. Then it is apparent that Darwin knowingly conformed his argument to the *vera causa* ideal for a scientific theory.[8]

The phrase *vera causa* meant not the true cause but a true cause, that is, a real cause, one known to exist, and not a purely hypothetical cause, merely conjectured to exist. So, the *vera causa* ideal, as Darwin sought to conform to it, required that any cause introduced in a scientific theory should be not merely adequate to produce the facts it is to explain on the supposition that it exists. For the existence of the cause is not to be accepted on the grounds of this adequacy. Its existence should be known from direct independent evidence, from observational acquaintance with its active presence in nature, and so from facts other than those it is to explain.

This *vera causa* ideal had been first given this formulation in the Scots moral and natural philosopher Thomas Reid's teachings on the import of Newton's first rule of philosophizing, the one that specified that no causes are to be admitted except such as are both true and sufficient to explain the phenomena.[9]

Reid, going beyond Newton's own understanding of this principle, used his novel explication of it to argue for the conclusive epistemic superiority of the Newtonian gravitational force over the Cartesian ethereal vortices as explanatory of the planetary orbits. That force with its determinate law, unlike those vortices, was a well-evidenced cause for the orbits up there, because the orbits themselves were not the sole evidence for its existence. It was a real and true cause, not a hypothetical and conjectural one; for it was known to exist from our direct and familiar experience of swinging pendulums and falling stones down here on earth.

It was the teachings of Charles Lyell, in the three volumes (1830, 1832, and 1833) of his *Principles of Geology*, that mediated between this Reidian *vera causa* tradition, in the epistemology of physics, and Darwin's understanding of what evidential demands would have to be met in solving the problem of the origin of species.

All the causes of change on the earth's surface were presumed, in Lyell's system, to persist undiminished in intensity, and so in efficacy, into the present, the human period, and on into the future. Now, as at all times, habitable dry land is being destroyed by subsidence and erosion in some regions, while it is being produced by sediment consolidation, lava eruption, and elevationary earthquake action in others. Likewise, there is a continual, one-by-one extinction and creation of animal and plant species, a constant exchange of new species for old, adequate to bring about a succession of faunas and floras in the long run.[10]

Lyell's most explicit rationale for this presumption of the persistence of all such causes of change into the present and future was the ideal of explanation by *verae causae*, causes known to exist from direct observational evidence independent of the facts they are to be invoked to explain. Accordingly, his system was to exemplify, no less explicitly, an epistemological analogy. In geology, only causes active in the present, human period are accessible in principle, although often not in practice, to direct observation. So, in this science, the brief human present is to the far vaster prehuman past as the terrestrial is to the celestial in Newtonian physical astronomy.

We have here, then, in this *vera causa* ideal the source for Darwin's structuring his argumentation in the *Origin* as he did. For, to conform to this ideal, the argumentation on behalf of a causal theory had to make three distinct evidential cases: for the existence of the cause, for its adequacy for facts such as those to be explained by it, and, finally, for its responsibility for those facts. Thus the first two cases, in establishing that the cause exists and can produce effects of the appropriate sort and size, would constitute argumentation showing what the theory is, what cause it introduces, that it is a true theory, one of a true cause, and an adequate theory, one of a true cause capable of such effects, and so no mere conjectured hypothesis, while the third would constitute its verification, as the true theory for the particular facts it is to explain.

As anyone who has consulted the *Origin* will recall, the one long argument of that book does indeed make three distinguishable evidential cases for three theses about natural selection: first, that it is a really existing process, one presently at work in nature; second, that, as it now exists, it is adequate for the adaptive formation of species and their adaptive diversification, in the very long run, into genera, families, and so on; third, that it has probably been the main agency in the production in the past of the species now extant and of those extinct ones commemorated in the rocks.

Thus the first case evidences both a tendency in wild species, as in domesticated ones, to vary heritably in changed conditions, and a struggle for life wherein variant individuals are surviving and so reproducing differentially, as when domesticated species are bred selectively by man. The second argues for the ability of this natural selective breeding—so much more sensitive and sustained, precise, and prolonged than man's as it is—to produce and diversify species in eons of time and changing conditions. Finally, the third adduces many facts of various kinds—geological, embryological, and so on—about extant and extinct species; and it is argued that these facts can best be explained—most intelligibly connected by referring them to unifying laws, that is—on the supposition that this cause was mainly responsible for producing and diversifying those species in irregularly ramified lines of descent diverging adaptively from more or less remote common ancestral stocks.

4

We have, then, to acknowledge that Darwin, in the *Origin*, was bringing to a problem in biology evidential and explanatory ideals first explicated, principally if not solely, in legitimating Newtonian celestial mechanics. But to acknowledge this is to raise at once an issue of immediate bearing on our efforts to understand the probabilistic character of Darwinian theory. How could Darwin have thought that such an ideal, originally legitimating the subsumption of the solar system under the laws of a deterministic classical mechanics, was at all appropriate to his arguments on behalf of a probabilistic causal process, natural selection as a cause for the adaptive formation and diversification of species? And, more generally, even if Darwin saw no inconsistency here, has it not turned out that Darwinian theory has developed since so as to make quite inappropriate his notion of the evidential and explanatory challenges raised by any commitment to natural selection as a theory of evolution?

These are questions that could be resolved by constructing and contrasting caricatures of "Laplacean" and "Darwinian" science, so as to secure a quick verdict, namely, that in its concern with what is "historical," "unique," "statistical," and so on, Darwinian science is utterly unlike any Laplacean program and must, therefore, have arisen in a repudiation of everything we associate with the French mathematician and physicist.

Not surprisingly, to anyone who has studied how Darwin came to natural selection, while he was filling his *Notebooks* B–E and M–N in the years from July 1837 to July 1839, two decades before publishing the theory, this response is much too quick.[11] These decisive sources reveal a far more complicated history than this response can allow. Moreover, to see why these sources should do so, is to be in a better position to decide how far such a response can be vindicated by what has happened to natural selection theory since Darwin.

In understanding how Darwin's notebook theorizing is related to the Laplacean heritage, or indeed to any prior precedents, it is indispensable to recognize that although Darwin had broken fundamentally with his mentor Lyell over the organic world, before *Notebook* B was opened in July 1837, he had not departed, nor would he ever, from Lyell's main teachings on the physical world of land, sea, and climate change.

On the physical side, then, Darwin remains in conformity with the Laplacean heritage. Lyell himself had revised but not repudiated the Huttonian theory of the earth, as it was upheld by John Playfair. A principal interpreter of Laplacean science to the British, Playfair, in his *Illustrations of the Huttonian Theory of the Earth* (1802), drew explicit analogies between Laplace's and Lagrange's conclusion as to stability in the solar system and the leading Huttonian thesis of a permanent stable balance in the actions of the igneous and aqueous agencies modifying the earth's surface.[12] John Herschel supported Lyell's neo-Huttonian position, in his *Preliminary Discourse in the Study of Natural Philosophy* (1830), and Darwin allied himself with Herschel in the late 1830s in efforts to construct a theory of crustal elevation and subsidence on the assumption that for the whole earth the forces arising from subterranean heating and cooling were balanced, and so subject only to reversible local fluctuations in the long run of the geological past, back at least to the time when the oldest known fossiliferous rocks were laid down.

Such commitments to stability and reversibility in the system of physical causation at work on the earth's surface were upheld in the face of explicit dissent from those, notably Sedgwick and Whewell, who thought these analogies between celestial mechanics and geology profoundly erroneous. As Lyell saw it, however, to give up this stability and reversibility was to forgo the possibility of meeting the *vera causa* ideal in geology. If it is to be possible to have direct evidence of the real existence of a cause deployed in explaining some past effects recorded in the rocks, then it must be at work still today. And if its adequacy for such past effects is to be properly evidenced, then we should be able to presume that it could, indeed must eventually, reproduce effects of that character and magnitude in the indefinitely long run of the future. So, this reproductive adequacy requirement presupposed limited variability in the conditions of working of these persistent causes. For, a terrestrial world without that limitation would not be one where the same causes continued to produce similar effects because working in similar conditions. Lyell held, therefore, that such a terrestrial world would not be one safe for analogical inferences from the short human period to the vast, past, prehuman periods, the inferences that allowed one to bring those periods within a science of geology conforming to the *vera causa* ideal essential to all inductive science.

For the inorganic world of land, sea, and climate changes, Darwin was always to uphold this teaching as he found it proposed by Lyell and seconded by Herschel. So we have to ask how far, in developing his own quite novel account of the organic world to complement this Lyellian one for the inorganic world, he was in intent and in effect departing from the presuppositions made by such presumptions of stability and reversibility.

XIV

5

Here, we need to consider Darwin's theorizing over the year and a half before he arrived at the theory of natural selection late in 1838. For, when he eventually constructed this theory, he drew on views about chance and chances and about *verae causae* that he had already been working with explicitly.

From summer 1837 on, his theorizing was dominated by two theses: first, that all structural change, and so all adaptive change, is ultimately due to the effects of changing conditions on the impressionable immature organization possessed by the offspring of sexual as distinct from asexual generation; second, that the outcome, in the long run, of the changes wrought by changing conditions is an irregularly branched tree of life, wherein species are multiplied when lines split without ending and become extinct when they end before splitting. So, sex is the immediate means and the tree of life is the eventual result of adaptation to an earth's surface everywhere changing à la Lyell.[13]

Direct experience indicates, according to Darwin, that sexual generation leads to adaptive heritable variation in altered conditions; it may then be a *vera causa* for unlimited adaptive diversification with unlimited time and changing circumstances. Equally, common ancestry is a known cause for similarities among relatives, especially similarities not creditable to common adaptations to common ends. So, in the tree of life, the resemblances among the species of some supraspecific group, a genus, family, or class, may be explained as due to a common inheritance from common ancestors, while their differences will be largely due to adaptive divergences.

Darwin was led to reflect on chance and chances in the propagation of the tree of life by the very considerations that led him to prefer branching descent over special, independent creations of species.[14]

Lyell had had each species created independently of any other. The character of a new species was not determined by the structure and instincts of any older species already in the area. Rather, its character and so its supraspecific type are entirely determined by the conditions where it is being created together with those conditions it is destined to meet subsequently on later spreading into other areas. Here, then, conditions determine character not only through present needs but also prospectively, in that the species, thanks to divine prevision, is provided in advance for future contingencies.

It was from this view that Darwin dissented directly. A new species, he argued, has the characters distinctive of its genus thanks to heredity, to inheritance from older congeneric species already present. And it owes the structures and instincts distinguishing it from its congeners not to any provisionary adaptation to conditions not yet encountered, but to those conditions encountered during and so determining its divergence from those ancestral stocks.

To see how this alternative to provisionary character determination by conditions led Darwin to reflect on chance and chances, consider how he used his new tree of life to reinterpret and extend various demographic analogies expounded by Lyell. For Lyell had developed an analogy between the births and deaths of human individuals and the coming and going of species as quasi individuals. The intermittent recording in any region of species births, lives, and deaths, by the fossilization process, was thus compared with the periodic visits of census commissioners to one place in the nation.[15]

Darwin continued such demographic analogies in his *Notebook* B, in elaborating his branching tree representation of species multiplications and extinctions. He dwelt, especially, on the quantitative implications of high average degrees of relatedness, many species, that is, descending from a few ancestral species, "father" species in his phrasing. With high degrees of relatedness and a constant total species population, the chances are small of any individual species having living descendant species a long time from now; and the causes determining which ones will succeed in doing so may be impossible to analyze, just as with two flourishing human families, today, where many causes, such as hereditary disease and dislike of marriage, may eventually determine that one family rather than another has living successors.

These same genealogical, demographic analogies are elaborated further in considering the wanderings and colonizations whereby species settle in fresh sites, such as new land emerging from the ocean. Here Darwin again opposes prevision and provision; for he argues that, in any generic or familial group, those species that colonize some area do so because they happen to be fitted for it by structures and habits already acquired in adapting to conditions previously encountered. One would expect, then, Darwin reflects, that by the law of chances, larger groups of species would supply more successful colonists than others.

It is likewise with the structural diversifications whereby an ancestral stock may have one or more aberrant species among its descendants, a ground-dwelling species, say, arising in a group, such as the woodpeckers, that was originally and is still predominantly arboreal. Here, too, Darwin opposes providential plans in favor of adaptive opportunities and successes that accord with circumstantial contingencies and so numerical chances; such an aberrant species has been formed when some one species among an ancestral group has succeeded in adapting to the aberrant way of life, successful aberrances happening, therefore, more often in larger than smaller groups of species.

Pursuing these quasi-demographic and quasi-genealogical concerns, Darwin considers two ways whereby the adaptation in a father species may be related to the adaptations in descendants. It could be, he reflects, that in adapting to its own circumstances the father species ensures the adaptation of the offspring species to its circumstances. Or, over the long run, it could be that a father species is adaptively influenced by a succession of changing circumstances and produces numerous varieties, among which the best adapted alone are preserved and diverge to become species.

So far, then, in his *Notebook* B, Darwin is considering adaptive variety and species formation, and supraspecific group proliferation, without relating these changes to the individual maturations that are distinctive of sexual reproduction, and that ultimately make adaptive variation possible. In *Notebook* C he does take up this challenge, and he starts with two possibilities.

On the first possibility, adaptive species formation would trace to the production of "chance offspring" characterized by some slight peculiarity and by exceptional vigor, so that, among the males, they would be eventually more successful than others in winning mates in competitive combat and in passing on their peculiarity. By contrast, on the second possibility, the first step is not a chance prenatal innovation in an offspring, but a postnatal change in habits in some parents—as when some jaguars are tempted to swim after fish prey, on their region turning swampy, the changes in structure thereby acquired then being transmitted to their offspring.[16]

Now, it is this second possibility that Darwin develops most fully in the next six months, from spring to autumn 1838.[17] And in developing it, he continues his opposition to prevision and provision in adaptive species formation. For, although the structural variations acquired in habit changes are not chance variations or chance productions, they are initiated by chance encounters with new conditions, and not in planned anticipations of changed circumstances.

Darwin makes these corollaries of his habit theorizing most explicit in his *Notebook* M, begun in July 1838, where he reflects on the human will and its apparent freedom.[18] For in higher animals, new habits may arise from willed responses to whatever conditions or circumstances are being encountered. So, if willings are uncaused and unlawful, then structural changes tracing to habit changes may be too. To avoid this conclusion, Darwin takes a resolutely deterministic line. Free will is to mind, he says, as chance is to matter; that is, in both there is a misleading appearance, but only an appearance, of a lack of lawfully determining causation. Accordingly, Darwin insists that all responses to whatever is encountered are determined by existing hereditary organization and prior education and so on.

This determinism is complemented by a no less explicit materialism whereby everything mental arises from determinate material causation. In particular, any mental traits can be transmitted without conscious awareness in the material organization passing from one generation to another. By being subsumed within the corporeal, mental changes, such as instinct changes, are brought within the general account given for all adaptive change.

Darwin's reluctance, at this time, to credit adaptive change to chance variation is not then due to any misgivings about chance productions in general and as such. For his determinism makes chance variations the products of lawful, albeit hidden, causation. And here he was in conformity with the customary view that chance congenital variants, polydactylous offspring, for example, were not uncaused but, in accord with the commonplace ignorance interpretation of chance, caused by unseen prenatal conditions. For Darwin at this time, the most evident drawback to chance variations as a contribution to adaptive species formation in the wild was their rarity. Breeders, he emphasized, could deliberately pair rare, chance, congenital variants together and so make and perpetuate a race distinguished by that peculiarity. But, in the wild, free crossing ensured that such rare, chance congenital variation is counteracted. This contrast between species, as formed adaptively in the wild, and the products of the selective breeder's art was, for

Darwin, reinforced by the reflection that the artificial, selected varieties were monstrous, not adaptive, being fitted not to natural ends or conditions but human purposes.

Nor did Darwin's thinking about chance variation in relation to species formation change on his assimilation of Malthus late in September 1838. For Darwin's immediate response was to see Malthusian superfecundity as complementing crossing in contributing to the adaptiveness of structural change in the wild. Crossing ensured that structure changes were adapted to the slow, permanent changes affecting a whole country in the long run of physical change studied by geologists. For, with crossing, any adaptive variation elicited by local fluctuating changes in conditions is eliminated through the blending in offspring of parental characteristics. Complementing this means of retention and elimination, the Malthusian crush of population ensures, further, that only variations that were adaptive for the whole life of the individual, from conception to adulthood, would be retained and so, in the course of many generations, embedded permanently in the hereditary constitution as structure is adapted to permanent long-run changes in conditions.

Such, then, was the drift of Darwin's thinking about chance and chances, in relation to sexual generation and the tree of life, before he came, most likely in late November 1838, to his positive analogy between artificial and natural selection as both means of adaptation, and so before he came to that deployment of probabilistic notions that is so characteristic of the theory of natural selection as known in the *Origin*.

Plainly, before the selection analogy was arrived at and before it transformed his thinking in decisive ways, he had already the conceptions of chance and chances that he would integrate in that analogy. Those conceptions had always been such as to rest on two assumptions for which there were ample precedents: first, that, in an individual, the future consequences of its properties may be determined by conditions that can arise independently of the conditions that produced those properties in it earlier in its life; second, that, in a population, physical property differences may be causally adequate to move outcomes away from what frequency differences, rarity and commonness, alone would determine.

To confirm that these assumptions were, in themselves, not innovative, consider how Darwin's conceptions of chance and chances relate to the old distinctions of chance, necessity, and design.

In accord with the ignorance interpretation of chance, Darwin is not ultimately working with a trichotomy of chance, necessity, and design, only a dichotomy between design and the rest. For, while he will judge a chance production unplanned, unintended, or undesigned, he will not call it uncaused. Although not known to be necessitated by discernible causal antecedents, it is presumed to be so, no less than is one of known necessitation.

As with chance, so with chances: only ignorance requires us to distinguish them from causes. Chances are what are determined by numbers, by quantitative differences, when as far as we know all else is qualitatively, physically, causally equal. When all we know of two families is that one is larger than the other, then we give it the better chance of having descendants centuries hence. As soon as we learn that it is less healthy, that bet is off, for we now know a cause sufficient to shift the chances away from where they would be were the numerical difference all that nature was working with in determining what future will be produced from the present.

A distinction between accidental and necessary adaptations can then be upheld, by Darwin, as a distinction among productions that are, as all productions are, causally necessitated. For in an accidental adaptation, its productive causes are independent of its adaptive consequences, in that the conditions producing it are not those causing it to have consequences such as to make it count as adaptive. The fetal conditions producing extra length in some puppy's legs may be arising independently of the presence of the hares that make such legs post-natally advantageous. By contrast, a variant is a necessary adaptation if it is produced by the very conditions that make it advantageous, as in the thicker fur grown by puppies moved to a cold climate or the webbed feet developed by the swimming jaguars. Unlike necessary adaptations, then, accidental adaptations are chance productions, because they are produced by hidden causes acting early enough in life to be independent of the conditions encountered later, the conditions determining their advantageous consequences for survival chances.

In his conceptions of chance and chances, Darwin was not innovative; his leading distinctions relate in traditional ways to venerable contrasts between the accidental, the necessary, and the designed. We need next to see how, nevertheless, natural selection made a novel deployment of these conceptions.

6

As it emerged in late 1838 and early 1839, in *Notebook* E, Darwin's theory of natural selection made a dual deployment of chance and chances. Heritable variants sometimes arise "by chance," and among them will be some that have a "better chance" than the normal individuals of surviving to reproduce. It is thus a matter of chance as to what variations are arising in the conditions the species is now living in, but it is not a matter of chance as to which are most successful in surviving to reproduce.

Darwin did not have this dual deployment of chance and chances until he was explicitly drawing analogies between the selective breeding practiced by man and that going on in nature. To appreciate his own understanding of the place of these probabilistic notions in his new theory, it is necessary to consider his original rationale for developing these analogies as he did.

He seems to have come to these analogies, late in November 1838, in reflecting on sporting dog varieties, such as greyhounds; in these cases, man had not only made and maintained varieties adapted to his own purposes but also, in doing so, had given them structures and instincts that would be adaptive in the wild, in preying on hares, for example.[19] Accordingly, Darwin was soon reflecting that the outward structural form distinctive of greyhounds could be produced by selective breeding away from all hunting and hares. So, he reasoned, the superior adaptive power of the selective breeding going on naturally in the wild was not due to variation arising there differently from the case of domestication; rather, this superiority was due to the greater persistence and precision of the selection that would arise in a species making its living in the wild by hunting. So, those conditions of life would not be necessary for the elicitation of the requisite variation, but they would be determining the selective retention of it required for such an adaptation to be produced.

Going further, Darwin soon concluded that the conditions a structure was adaptive for would sometimes not be sufficient. To take an example he would elaborate later, he could not see how in an area the presence of woolly animals would affect the growth of plant seeds so that they became hooked and fitted for attaching themselves to the animals.

What Darwin's new theory did, therefore, was not to make changes in conditions less determining of adaptive change, but to make them less directly so. He was content to drop the thesis that conditions always had the power to determine adaptive change directly by working heritable effects upon growth and maturation, because the analogy with the breeder's art convinced him that adequate determination would come from the way different conditions determined chances of survival and reproduction among chance variants.

Darwin was used to considering the conjunctive chances of two rare chance variants coming together to breed and so to perpetuate their peculiarity.[20] He had emphasized that, in nature, their rarity made the chances very slight, there being no inherent tendency of like variants to pair together.[21] So, on going over to selective breeding of chance variants in nature, Darwin was from the first concerned with the consequences of their rarity. He had long argued that, with the reproductive isolation of a few individuals, following migration to an island, say, the conservative effect of crossing could be circumvented as it was in the breeder's assortative matings. Selective breeding in nature was, then, for Darwin a cause that worked slowly because of the initial rarity of chance variants, and was effective despite the counteractive effect arising from the crossing of these new variants with commoner older ones.

In his new appeal to chance variation, Darwin was thus concerned not only with the lack of direct adaptive determination by conditions of variant growth in individuals but also with the determination, by frequency alone, of the populational fate of such variants, as long as no causal interactions were biasing their chances of survival and reproduction. So, natural selection was, from its very inception, a theory as to when and why frequency alone was not solely determining, because causes for bias in those chances were present.

As an agency working causally to bias population outcomes away from where frequencies alone would otherwise have them, nature as a selective breeder, in Darwin, may remind us of the demon in Maxwell.[22] However, the resemblance must not be allowed to mislead us as to the contrasting rationales motivating the two theorists' essentially different proposals. Maxwell was concerned to dramatize how utterly improbable in nature is anything like the outcome secured by the

246

demon; for under all natural conditions there will be no such quasi-purposive interference as the demon exerts. By contrast, Darwin was out to establish that a quasi-designing form of selective breeding is an inevitable consequence of the struggle for existence and superfecundity, tendencies so ubiquitous and reliable as not to be construed as interferences at all.

These distinctive features of the theory were manifested vividly in Darwin's own earliest metaphors: of thousands of trials, that is, individual variations and struggles for survival over many generations, and of a grain in the balance, that is, of a consistent slight causal bias effective in a very long run in conformity with the law of large numbers.[23] Darwin did not need to have been a youthful afficionado of the gaming table, and of chemical paraphernalia, to find in loaded dice and tipped scales illustrations of the probabilistic causation constituting the process of change through the natural means of selection.

Darwin took the reiterated process of chance variants tried in the wild for adaptation, and so for causal bias in their chances of survival and reproduction, to be a quasi-designing process. At this time, he held ontogeny to recapitulate phylogeny; so he was also prepared to say that a growing mollusk was able to make a hinge for itself because it has in its heredity the innate, unconscious equivalent of a human craftsman's skill in making hinges, the skill gained by a long sequence of conscious trials, rejections of failure, and retention of successes.[24]

One can say, then, that Darwin gave up having variation arise as "necessary" adaptations, as necessary effects of conditions, in favor of having it arise "accidentally" or "by chance," when and only when he came to see that its fate was under the quasi-designing control of a natural selection analogous to the skilled practice of the breeder's quasi-designing art.

Darwin's insistence on the analogy, ever thereafter, was no accident of expository tactics, but an essential component in the original construction of the theory. What he did give up later, before the *Origin* (1859), was his early thesis that the maturation distinctive of the products of sexual generation was what made possible all hereditary adaptive variation. He went over to the view that the production of the male and female elements, ovules and pollen in plants, eggs and sperm in animals, was a disruptible budding process that could provide chance variation with which natural selection could work perfectly well. So, instead of the variation that made adaptation possible arising in a proper, distinctive, functioning of sexual reproduction, it now arose in disruptions of the replicative function common to all generation, sexual or asexual.

There was thus no inconsistency between Darwin's commitment to the old ideal of known causes and his acceptance of the venerable ignorance interpretation of chance. It was only in the interpretation of the production of chance variation that Darwin invoked this ignorance; and in doing so he did not preclude arguing that the process of selection of that variation counted as a *vera causa*, a cause known— from the struggle for life—to be existent in nature, and known—from the breeders' results—to be efficacious in producing permanent adaptive change in structures and instincts.

7

We are now in a position to see more clearly how Darwin's theorizing, with its twofold invocation of chance and chances, stands in relation to the legacy of the physical sciences as found in Laplacean exemplars.

For a start, we can see the disadvantages in one tempting historiographic view of the probabilistic elements in Darwin's theorizing: the view that insofar as Darwin retained deterministic, nomological commitments to reversible, stable causation, he was still in the thrall of those Laplacean physics precedents, while, by contrast, insofar as he took seriously chance occurrences, accidental changes, statistical trends, circumstantial contingencies, and the like, he was being liberated from this thralldom by those influences on him that came from the sciences of man and society, from Fergusonian conjectural history, Hartleyan associationist psychology, Queteletian societal arithmetic, Malthusian demographic theodicy, and the like.

This view is attractive because social theory obviously reflects social practice, so these influences could provide the mediation whereby Darwin's science was conditioned by his society, in accord with presumptions long made by many sociologies of knowledge.

The trouble with this view is that it tends to reduce all the diverse biological concepts of the early nineteenth century to so many echoes, emulations, borrowings, projections, extrapolations, analogies, and metaphors from physics, on the one hand, and social science on the other. And such reductions are very difficult indeed to square with all the traces we have of Darwin's deep indebtedness to rich traditions in physiology and natural history concerning vital forces, generation, geography, and so on. Moreover, it is surely arbitrary, as well as half-hearted, for a sociology of knowledge to presuppose that scientific theories are conditioned by their social context principally insofar as they are indebted to overtly social theories. More consistent and confident presuppositions for a whole-hearted sociological historiography would not require reading biological concepts and contexts out of the intellectual narrative, but would take them, and the physical science sources, too, as all suitable subjects for social conditioning, social construction, or social relations analyses.

To see one way whereby biological considerations led Darwin's theorizing away from the precedents set by the physics of the day, we may consider how heredity, for Darwin, could ensure long-run irreversibility. In Lyell, species losses, extinctions, are not reversible, in that no individual species returns once extinct; but any supraspecific group losses are reversible, for such a group may come to be represented again on the earth after a period when it has been temporarily missing. For that group, a genus or family or whatever, will reappear when conditions have changed, so as to be once more fitting for the creation of species with the structure and instincts characteristic of that group.

Darwin added heredity to adaptation in explaining the spatial and temporal representation of supraspecific groups on the constantly changing Lyellian earth's surface. In doing so, he gave a genealogical interpretation of classifactory groups that denied the possibility of any reversibility such as Lyell upheld. For Darwin, each supraspecific group has a single father, a monophyletic ancestry; and the assumption that a particular species never returns is extended to wider and wider groups, through the assumption that no group could ever arise again from a different future father species. This contingent impossibility was, for Darwin, confirmed by considering the cumulative changes in the lines leading from the simplest, remotest ancestral animals to the highest: birds and mammals. Here, he argued that heredity worked so that the oldest characters were most deeply embedded in the constitution, and so least susceptible to loss or alteration later. New characters would thus be added to these older ones and not substituted for them. The irreversibility of splitting and branching divergences likewise depends, for Darwin, on this same power of hereditary constitutional embedding, for inability or disinclination to interbreed depended on the accumulation and embedding of constitutional changes in two races descending from a common stock.

Heredity, although a principle of conservation, for Darwin, is therefore not at all closely analogous to the nearest equivalent principle in Newtonian physics: inertial mass. Nor, then, is the adaptive transformation and multiplication of species closely analogous to the outcome of successive actions of impressed forces on a body of constant mass. Selection, working with and complementing the powers of heredity, as Darwin judged it to do, was a force or cause with no close precedent in the exact sciences of the day.

However, in breaking with those precedents Darwin did not need, in his theory of natural selection, any new conceptions of chance and chances. As his heredity, adaptation, and selection theorizing developed, his twofold deployment of chance and chances could be made without breaking with available conceptions of these two.

Darwin could see natural selection as a *vera causa* because of his understanding of the relations between the short run and the long, especially as those relations were illumined by the analogy between artificial and natural selection. In the short run, in the hands of expert human practitioners, selective breeding was a causal, quasi-designing process with an approximately predictable outcome, in any instance, given a prior knowledge of the materials and the objectives. Superfecundity and the susceptibility of heredity to disruption in altered conditions entailed that an analogous process existed in nature, while comparison of that natural analogue with the achievement of the breeders indicated that, if they could produce marked racial differentiation in the short run, nature could produce distinct and diversely adapted species in her longer run.

There was the implication that the changes wrought by any selective breeding of hereditary variants would become less and less predictable, as one moved away from the shortest of artificial runs to the longest natural ones, but that loss of predictability was entirely tolerable, because Darwin was not seeking a theory as to why organisms of one particular group would tend to have descendants of another particular group: why fishes tend to have mammal descendants, or mastodons elephant descendants. There was for him no tendency in such descents, as so described. They had only happened once and were not expected to be repeated.

The ramifying, diversifying, complexifying tendencies for which Darwin sought adequate causation were general and so could be introduced with referential anonymity. His tree diagram, in the *Origin*, is labeled abstractly with letters and numbers, not proper names, so that it can represent generalizations about types of trends. To fail to appreciate the implications of this generality and abstraction is to misunderstand the entire Darwinian enterprise in the *Origin* and ever since. For their most general explanatory purposes, the generalizations Darwin and his latter-day successors have needed to establish have had to indicate, for example, when adaptive divergence rather than convergence was more, rather than less, likely to occur, or when extinctions rather than species splittings are to be expected.[25] And, in this generality and abstractness of its predictive and explanatory functions, Darwin's natural selection was in conformity with such exemplary *verae causae* as the gravitational force.

There was a break with the gravitational theory precedent, however, in that there was no law that was to natural selection as the inverse square law with proportionality to mass products was to that force. For gravitation, this law enabled the exact consequences of that force to be deduced for certain simple, suitably specified cases of one mass in motion around another in an otherwise empty universe. Natural selection has no equivalent law because its very existence requires, causally, processes of reproduction, heredity, and variation; and while these processes may be and were presumed to be conforming to laws of their own, they cannot exist and conform to those laws in an empty universe void of complex interactions between what is changed and the conditions determining how it is changed.

These sources for the lack of any equivalent for the gravitational law do not entail that selective breeding is on all occasions as likely to produce one outcome as another. Indeed, Darwin does not have to deny that, if on some occasion exactly the same heredity and variation were to be subject to exactly the same selection as on some earlier occasion, then the outcome would exactly resemble the earlier outcome. However, it is not a possibility that will ever be approximately instantiated. What Darwin needs, for his explanatory purposes, is the presumption that the departures from the impossible sequence specified by that assumption are not capricious, but are occurring because of causation similar in kind, although different in degree, from those producing the controllable, approximately predictable results of the animal and plant breeders.

That the existence of natural selection requires the causal processes of heredity and variation shows that there is no Anselmian demonstration possible for its existence from its essence, no demonstration in the style of Anselm's ontological proof for God's existence, which argued that His essence, as completely perfect, must necessarily include the perfection of actually existing. In the *Origin*, Darwin introduced the term natural selection, definitionally, after he had argued for the existence of the process in nature. And so he proposed natural selection as an appropriate name because the process was analogous to artificial selection. But this analogical definitional procedure presupposed no Anselmian aim of making selection exist in nature by virtue of its very essence as stated in its definition. On the contrary, Darwin argued for the appropriateness of the name because of the character of the process—arising from heredity, variation, and the struggle for existence—as already shown to exist in nature. He did not argue for its existence on the ground that its definition was such as to make that name appropriate.

We have seen how Darwin deployed traditional conceptions of chance and chances in a *vera causa* solution for his often unprecedented problem situation. And in doing so we are brought to see that it was intrinsic to his whole enterprise that he take up in certain distinctive ways various clusters of questions about the definition, the existence, the competence, and the responsibility of natural selection as a cause of evolution. We do well to work with such question clusters even after leaving Darwin behind.

8

On considering the first of these four clusters of questions, it will be evident that there are many ways to construe the definitional task it sets us. But that is only to be expected. Biologists call natural selection various things: an agency, a process, a factor, a cause, a force, and so on. Philosophers, meanwhile, always insist that definitions themselves come in several genres, although they rarely agree on what those genres are. The possible permutations are, therefore, plentiful for anyone proffering a definition of natural selection.

For our purposes here, however, one construal of this definitional task makes the best point of departure, for it introduces us directly to the issues of chance and causality raised by Darwin's and later versions of natural selection as a theory of evolution. On this construal, we confine ourselves, for a start, to intrapopulational selection of heritable traits distinguishing individual organisms. We take this selection to be a process, and we do what is always instructive for any process, namely, define it by stating the conditions necessary and sufficient for the process to occur. Such a definition allows us to integrate current textbook expository practices with longstanding themes about fundamental principles.[26]

In textbook presentations of definitions for natural selection, it is instructive that everything goes along standard lines only up to a certain stage, and then different authors tend to go in one of two directions. Thus, as to necessary conditions, there is agreement that these include variation, heritability of variation, and differential reproduction of heritable variation. So we quickly reach differential reproduction of hereditary variants as indispensable to the definition, because necessary for the process. Moreover, there is almost always explicit recognition that some further condition is necessary, so that once this is given there will be a set of necessary conditions that are jointly sufficient. The need for the further condition is apparent, because in a finite population of hereditary variants, even without selection, there will be differential reproduction in genetic drift, that is, in the accumulation of any successive indiscriminate or random sampling errors in the same direction. And such drift must not be allowed to count as natural selection.

However, in proposing a further condition to distinguish selection from drift, some framers of definitions lay down that the differential reproduction in selection be "consistent" or "systematic" or "nonrandom," all terms with no peculiarly biological content and drawn often from the terminology of statistics, while other authors insist that the differential reproduction must be due to differences in "fitness" or "adaptation," terms characteristic of the biologists', even formerly indeed the theologians', lexicon, terms with an apparent teleological import.

Any diversity of definitional proposals reminds us that it is rarely easy to decide on what grounds one should be preferred over another. At a minimum, however, we have surely to take two sorts of considerations into account. First, there are judgments already being made, independently of particular definitional analyses, as to which real or imaginary processes count or would count as cases of natural selection. So, we need a definition that respects these judgments without merely reflecting them uncritically. Second, we should be guided by what has motivated the development of the theory of natural selection, and by what has influenced the way this concept is related to others, just as, to take another example, with the concept of mass in physics. For, with mass, awareness of the different presuppositions about space, time, motion, matter, and force, in Cartesian, Newtonian, and Einsteinian physics, has motived decisions as to how to define mass itself and associated concepts such as weight, length, and so on.

Bringing these considerations to bear on natural selection, we are led, I shall argue, to one preferred resolution of the disagreements over what is needed, in defining selection, beyond differential reproduction of heritable variants. For we have to avoid two errors. On the one hand, we may be tempted to have a purely formal or mathematical restriction on differential reproduction in selection as distinct from drift: choosing a term such as "nonrandom" and then seeking for that term a purely formal or mathematical explication. On the other hand, we may be tempted to be finalistically biological rather than formalistically mathematical, and to require that the differential reproduction be due to differences in fitness or adaptation, with these terms explicated by reference to standards of design.

These two moves are errors, I submit, because their formalistic and finalistic quests lead us away from what is manifestly desirable: an explicit definitional insistence on causation itself, on, that is, its physical ingredients rather than on mathematical representations or teleological interpretations of its inputs and outputs. For, obviously enough, differential reproduction in selection is distinguished from any in drift by its causation; by contrast with drift, it is occurring because the physical property differences constituting the hereditary variation that is being differentially reproduced are not merely correlated with differences in reproduction—they are causally relevant to them.

When such causal relevance is present, we may call the differential reproduction nonfortuitous, a term better suited to biologists' conceptions than nonrandom precisely on account of its connotation of causation rather than mere correlation. So, intrapopulational selection may be defined as what is occurring when and only when there is the nonfortuitous differential reproduction of hereditary variants.

9

To bring out the advantages in such a physicalist, causalist explication of the concept of natural selection, we may start with those corollaries of it that can be clarified through elementary imaginary exemplars; then we can proceed to the issues raised by more complicated cases of intrapopulational selection, and by the extension of the explication to other levels, such as the interspecific. Elementary, imaginary scenarios are not to be dismissed as conceptually uninstructive merely because they are unrealistic. The interactions that make up any process that counts as a cyclone are bafflingly complex, obviously; but to acknowledge this is quite consistent with holding that the concept of a cyclone is such as to be usefully explicated through its illustrative exemplification in elementary, imaginary scenarios.

Consider two small populations of butterflies. In both, the only variation is in color: half are red, half are green, and this difference is inherited. Both populations are living and breeding in green environments and only die from predation by birds. However, one population is preyed on by birds that are color-blind, the other by birds that are color-sighted. Now consider four particular runs of breeding and predation over several generations of the butterflies. In each population, it is found that there is one run in which the proportion of red butterflies goes down from a half to about a quarter, and one run where it rises to about three-quarters.

Presumptively, only one of these four runs involves selection rather than drift, because only in one population is the heritable color difference causally relevant to survival and so to differences in reproduction. Where the predators are color-blind, increases in red or in green are equally to be expected. In the other population the increase in red is presumably an unusual outcome, as the increase in green is not. However, even in this population, we have to admit the possibility that in a particular run the red butterflies may sometimes have been picked off at a higher rate by the color-sighted birds through the bad luck of landing up near the birds, perhaps at night, rather than as a result of being spotted on account of their color.

So, to presume that the differential reproduction in such a cause is nonfortuitous, rather than fortuitous, is to presume that in this environment this physical property difference is causally relevant to reproductive success.

In making this contrast between selection and drift, we maintain an obvious analogy with paradigm cases of discriminate versus indiscriminate samplings, when those samplings are considered as physical, causal processes. If a person, without looking, is picking balls out of a bag, some of them red, some white, and differing in no other property, there is indiscriminate sampling with respect to color differences, whereas if he is looking, and going for red ones deliberately, the sampling is discriminate. Even with indiscriminate sampling no ball jumps spontaneously out of the bag into the sampler's hand, out of the population and into the sample. So, likewise with the birds and the butterflies: no spontaneous uncaused deaths are occurring whether predatory sampling is indiscriminate or otherwise. What the balls, picked by the indiscriminate sampler from the bag, have in common is that they were in the right place at the right time to come to hand. And likewise, in indiscriminate predatory sampling, a disproportionate number of red or green butterflies may happen to be in the right place at the right time—or wrong place from the prey's viewpoint.

All these samplings, indiscriminate or otherwise, are physical, causal processes with energy transformed and transmitted. So, it would be fallacious to think that because drift is a corollary of a mathematical property of the population, its finite size, while selection is a consequence of physical property differences among the individuals, it follows that drift is somehow a mathematical rather than a physical process. For selection in a finite population is sampling error, too, a discriminate sampling error. So, here, drift and selection are not to be contrasted as sampling with and without error, but as causally discriminate rather than causally indiscriminate erroneous sampling. An explanation that invokes drift invokes causation no less than a selection explanation does, but it invokes indiscriminate causation and so no causes of discrimination.

To explicate selection by contrast with drift allows for the indispensable distinction between selection for a property or trait and selection of a property or trait.[27] A property that is selected for is one that is itself causally relevant to its own differential reproduction. However, there will be selection of any properties that are correlated with one that is causally relevant, even though they are themselves not relevant causally. Thus if size differences are correlated with color differences and there is selection for color differences, then there will be selection of size differences, even if these size differences are not themselves of causal relevance, being only statistically relevant to this differential reproduction thanks to their correlation with those for which there is causal discrimination and so selection.

Defining selection as nonfortuitous differential reproduction is consistent with accepting that the instantiation of fortuitousness is description relative, so that, in our drift scenario, the deaths of the red butterflies were not fortuitous events qua deaths of butterflies, but were qua deaths of red butterflies in a population also including green ones preyed on by color-blind predators in green surroundings. For, even with such description relativity, it is still an empirical, causal matter whether a particular process of differential reproduction instantiates some description under which the process is not fortuitous.

Again, explicating selection by contrast with drift allows for—indeed encourages—the admission that in real life the ecology and genetics of butterflies and their predators may include some changes that are equally plausibly categorized as selection or drift.[28] Once one considers such familiar complications as linked genes, correlated responses to selection, patchy environments, frequency dependent effects, habitat preferences, and so on, it is possible to think of scenarios for which it is impossible to draw a sharp line between fortuitous and nonfortuitous differential reproduction. But, here, as always, it is to be emphasized that such impossibilities do not nullify the rationales for making the conceptual distinction.

To see why the distinction is indispensable, consider what sorts of causal theories of evolution are left once we have set aside orthogenetic and saltationist views that deny gradual change under the control of environmental conditions. To be schematic, we could draw up four options: two selectionist positions and two neutralist positions, where neutralism is the view that ascribes evolution to the populational fixation by drift of mutations that are adaptively neutral. For there would be Lamarckian selectionism and Weismannian selectionism, and likewise two versions of neutralism. Thus, in Lamarckian selectionism neither individual generation of hereditary variation nor its populational fate would be a matter of chance in relation to conditions. By contrast, in Lamarckian neutralism the fate but not the generation of the variants would be a matter of chance.

Now, in the last half-century, Lamarckian views have declined greatly, the relation between phenotype and genotype implied by the molecular biology of protein synthesis making them very implausible in most eyes. So, to seek to define selection in contrast to drift, as one of two options to go with Weismannist presuppositions about hereditary variation, is now the more appropriate.

10

Such a definition is implicit in Wright's familiar classification of factors of evolution. He starts from the point that in any selection, by contrast with drift, the direction and amount of the gene frequency change increment may be determinate in principle; and he concludes that selection is a "wastebasket category" including all causes of directed change in gene frequency not involving mutation or introduction of hereditary material from outside. Biologically speaking, he stresses, it will include factors as diverse as differential viability, dispersal beyond the breeding range, fertility differences, and so on.[29]

One advantage of this classificatory characterization of all selection is that we can circumvent the difficulties inherent in drawing lines between natural versus artificial selection or between natural versus sexual selection. This is an advantage, because insofar as all selection is nonfortuitous, drawing such lines is not desirable definitionally.

Another line, drawn within the category of selection, is inappropriate if our definitional proposal is acceptable; and that is Lewontin's suggested contrast between "tautological selection" and "functional selection."[30] In cases of the first, he says, we do not know why some genotype difference is subject to selection, but we think that it is and hence have to presume, tautologically as it were, that there must be some cause for its being so, while in the second the selection is arising from the known relation of a trait, such as color, to a known function, such as concealment from predation. Now, the very use of the unfortunate term tautological here may carry an unwelcome suggestion, albert presumably unintended by Lewontin, namely, that if the causal workings constituting some selection are not manifest, then we are somehow reduced to asserting even the existence of this selection, Anselm style, for reasons to do with the way selection as such is defined. But by any reasonable definition this will not be so. For a process to be nonfortuitous rather than fortuitous, in the requisite sense, is for it to be so quite independently of how far we have succeeded in observing the causal bases for the causal relevance. There have been many cases—some chromosomal polymorphisms in *Drosophila* being once a famous one—where it was accepted that some genotypic differences are causally relevant to survival and reproduction differences, and so subject to selection rather than merely to drift; and yet it remained unknown as to how and why these genotypic differences are making for this reproductive success difference.

To conclude that there is causal relevance does not require knowing, much less observing, how it works. Many experimental routines with caged populations of *Drosophila* establish that, under some range of conditions, one genotype or karyotype will improve its frequency far too consistently for drift to be a credible explanation; and yet the causal basis for this consistent superiority in reproductive performance is undisclosed.

Consider next how this definitional explication of selection can be extended to levels other than the intraspecific. Here, one needs to keep constantly in mind the conditions that must be met if the different contrasts of Lamarckian versus Weismannian and selectionist versus neutralist are to work. Failure to keep them in mind has led to confusions. Thus, for instance, it is sometimes suggested that evolution in some prebiotic molecular systems is somehow Lamarckian. But in the absence of any distinction between analogues of germ plasm and somatic tissue, or phenotype and genotype, the contrast between Lamarckian and Weismannian modes of change loses all force. Again, we have been offered various games as models for selectional processes whereby life might originate. But on inspection, it often turns out that the games are equally good models for neutralist drift scenarios, in that there is differential reproduction but it is left open as to whether it is fortuitous or not.[31]

Going the other way, to higher levels of organization than the species, there are proposals current as to "species selection." The proposals appear to presuppose a quasi-Weismannist rather than quasi-Lamarckian view of the relation between changes in conditions and the generation of interspecific differences; and the term "species selection" indicates an implicit contrast with a species level analogue of drift. But the proposals would often be clearer than they are if the quasi-Lamarckian and quasi-neutralist analogues had been argued against more explicitly.[32]

Hull has suggested that any selectionist proposal for any level can be clarified by insisting that it distinguish between replicators and interactors, examples of these being, respectively, genes and organisms at the familiar level of intraspecific selection. He emphasizes, what is more, that causal interactions, as a basis for consistent differential reproduction, are required, definitionally, if the process is to be properly called selectional. However, even Hull does not always sustain successfully the contrast between all selectional processes and any drift processes. Thus he defines a replicator as "an entity that passes on its structure directly in replication" and an interactor as "an entity that directly interacts as a cohesive whole with its environment in such a way that replication is differential." And he defines the selection process as one "in which the differential extinction and proliferation of interactors cause the differential perpetuation of the replicators that produced them."[33] Hull makes it clear, then, that he is as resolved as anyone to distinguish, causally, selectional from drift processes. However, the definition may not do this as it stands. After all, in our color-blind predation scenario there was differential extinction and proliferation of interactors (organisms) causing differential perpetuation of the replicators (genes) that produced them. So, once again, we need to insist that the differential perpetuation be caused nonfortuitously, that is, as a result of the causal relevance for reproductive success differences of those particular physical property differences.

For, consider how a difference can fail to be causally relevant, how it may be causally neutral. A genotype difference may make no phenotypic difference, or it may make a phenotype difference that the environment cannot detect and causally discriminate. By contrast, then, we should count as selection any differential reproduction of any genotypic variation that is occurring because of the causal effects of that genotypic variation. This will include, therefore, Lewontin's "tautological selection." All that some gene may do for its possessors is to enhance their enzymatic heat stability, and it may do this independently of any particular environmental conditions; but if its doing that makes a difference causally to reproductive survival, then its frequency must be credited to selection, not drift.

It will now be evident why the mathematics associated with the theory of natural selection may mislead us as to how to demarcate selection definitionally from drift. When the so-called coefficient of selection is introduced in textbooks, it is introduced merely as a coefficient of differential reproduction. So the manner of its introduction makes it equally applicable to differential reproduction in drift, provided that the drift has already occurred. Indeed, we can imagine a teacher of theoretical population genetics setting an instructive exam question. Here the student is told how some consistent differential reproduction in each generation has happened by luck, by drift, in some small population, and he or she is asked to use the mathematics introduced for selection to work out the resultant change in genotype frequencies. In an infinite population, granted, there is no random sampling error, and so any differential reproduction is selective, not drifting; and if the coefficients of selection are constant and all other sources of uncertainties are suppressed *ex hypothesi*, then the selection equations can be deterministic, not stochastic. It is only under such conditions that the so-called selection coefficient is appropriately termed; otherwise, it can mislead us in our efforts to understand selection as nonfortuitous by contrast with drift. It is on grounds of metaphysical presuppositions and empirical findings about causation, not for purely mathematical reasons alone, that we accept that deterministic equations become better representations of selective differential reproductions as population size gets larger.

Just as the implications of the mathematics of selection must be supplemented with physicalist and causalist notions as to their applicability, if they are not to mislead us in explicating the concept of selection, so with considerations of adaptation and fitness; these too will mislead us unless supplemented similarly.

11

There is no canonical explication of adaptation as it is understood in evolutionary biology. But there seems widespread agreement that adaptation is relative, in that an organism or an organ is not judged to be adapted unqualifiedly, but to be adapted or to be an adaptation, for example, to low rainfall or a herbivorous diet or for heat regulation or defense. Moreover, it seems accepted that such judgments often presuppose decisions as to how to distinguish one feature from another in the organism, and also decisions about the problems, resources, and constraints involved in its way of life. Here, the application of what have been called engineering criteria of structural and functional efficiency are often thought to have heuristic value.

On all such proposals for the explication of adaptation, we would have two ways of relating adaptation to selection. We might define adaptation narrowly enough so that some variation not counting as adaptive could nevertheless be subject to selection. Thus we might decide that a heritable prolongation of the period of fertility is not an adaptive difference even though it has selectional consequences. On the other hand, we might make having selectional consequences definitionally part of what is meant by adaptation. Either way, however, what is indispensable to the concept of selection remains the same to this extent: whether some adaptive differences are judged to be selectional differences as a matter of fact or whether all are to be so as a matter of definition, a selectional difference may still be, by definition, only one that is causally relevant to reproductive success. Putting the point another way: we should define selection by contrast with fortuitous differential reproduction. And if we do that, we may—although I see no strong reason to do so—define adaptation in terms of selection so defined. And even if we did this, selectional explanations would not lose their force when given for structures judged to be adaptations. It would be an empirical question whether a particular structure was produced by selection or not. And that question could not be justifiably settled merely by declaring in advance of inquiry into its history that it is adaptive; because for that judgment to be justified one would have had to make that inquiry. So, whatever empirical or definitional relations we decide do hold between adaptation and selection, the appropriateness of a physicalist, causalist explication of selection is not compromised.[34]

Turning now to fitness, we should note that many authors use this term more or less synonymously with adaptedness. But the concept of fitness that is of concern here is the one standard in population genetics, where the fitness of a genotype is defined in terms of the expected contribution of offspring made to the next generation by individuals of that genotype.

That fitness so defined is an expected quantity makes it appropriate to see it as a reproductive expectancy analogous to a life expectancy. Now, expectancies are in themselves not causal and so without explanatory content. If Jones has outlived Smith we cannot explain this difference by establishing that he earlier had the higher life expectancy and arguing that this duly caused him to live longer. It is physical differences that can make a causal difference and provide explanations; perhaps Smith smoked while Jones jogged. But expectancies estimated from physical differences can neither constitute causes nor provide explanations. For these reasons, it is misleading to conceive of fitness differences as causally mediating between the causes of a difference in reproductive performance and the difference in performance itself. We should not suppose that physical property differences somehow make organisms differ in fitness and that those fitness differences then somehow make them reproduce differentially. The contrasts between an expectancy and a performance or outcome must never be overlooked. According to the reference class to which an individual is assigned, he may have a high life expectancy or a low one. A smoking jogger has one qua smoker, another qua jogger, and a third—not algorithmically computable from the other two—qua smoking jogger. Likewise, then, for the genotype of an individual organism when referred to its various allelic classes, it has many reproductive expectancies. But the organism eventually makes only one reproductive contribution performance, and that is free of all reference class relativity.[35]

Because the relationship between fitness differences and reproductive performance differences is not causal and explanatory, we do not need one common but mistaken strategem in defending natural selection against the tautology charge. It is often thought that a successful defense depends on showing that fitness differences are contingently and fallibly rather than definitionally and infallibly related to performance differences. And it is often thought that to show this requires establishing criteria of fitness independent of reproductive success.[36] But this whole line of defense is mistaken because no such defense is appropriate. All that is needed if selection is to be saved from charges of tautology is that it be an empirical question as to whether any differential reproductive performance be nonfortuitously or fortuitously caused. Suppose in some case that the expectancy judgments are turning out false. As the tautology objector insists, they can always be revised retrospectively so as to be made true retroactively. But where the tautology objector and his opponent are both mistaken is in thinking that this point about fitnesses is decisive for the empirical rather than tautologous status of selection theory. What fitness estimates are being made ahead of time, and what revisions of them are made later, does not determine whether the physical differences the estimates are based on are or are not causally relevant to the reproductive performance differences. We now know that smoking is causally relevant to lung cancer, and to know this is to know that it is so whatever odds the life insurance companies are offering the tobacco companies' best customers.

In mathematical deductions of the consequences of a fitness having a given value, understood as an expected value, no epistemic gap, so to speak, between expectation and outcome ever arises. But this does not mean that the presence of such an epistemic gap is all that distinguishes real selection processes in nature from mathematical representations of them on paper. When we ask what it is that allows us to put some empirical content with the otherwise empty formalisms of the mathematics of selection theory, the answer should not refer to the riskiness as predictions of prospective estimates of fitness differences, but to the empirical status of our conclusions about causally relevant physical property differences.

Such an emphasis on these property differences can accommodate frequency dependent effects, as when a color variant in a prey population is at an advantage in survival and reproduction only as long as it is rare. In specifying the environmental conditions wherein a physical property difference has causal relevance to survival and reproduction, we may have to specify what the populational proportions of the variants are. But to do this is not to substitute mathematical for physical factors, for the frequency differences only make a causal difference because they have different physical consequences for such processes as predator discrimination learning.[37]

Considerations of fitness, then, its probabilistic and frequency dependent aspects included, bring us back to causation rather than away from it in any explication of what selection is, in the sense of how it is to be definitionally demarcated.

These considerations also confirm that it is fundamentally mistaken to answer the question as to what selection is by proposing that it is a law of nature, like a gas law, or, alternatively, that it is a framework principle, perhaps like the principle of the uniformity of nature.[38] In answering the question as to what natural selection is, we need only to insist that there is the term "natural selection," that the term's meaning can be given in a definition, that a concept is thereby explicated, and that when that is done natural selection is, definitionally, a kind of causal process, and so neither a law nor a principle. It is these reasons for this conclusion, as much as the conclusion itself, that are useful when we turn to those further question clusters concerning the existence and the competence and responsibility of natural selection for evolution.

12

The existence, competence, and responsibility questions are not independent of one another in every way. If selection does not really exist, it cannot be competent, nor therefore responsible for evolution. Again, if selection is judged to exist and to be competent for the job and is also thought to be the only conceivable cause competent, then it must be thought responsible. Thus Dawkins has argued along these lines for this last conclusion. His argument is not Anselmian, that is, solely from the essence of a cause, as expressed in a definition, to its existence; rather, as he acknowledges, it is in the Paleyan manner of the old arguments from and to design. For he argues that selection must have been responsible for evolution because no other cause could have done it. And his reasoning explicitly depends not only on premises concerning what selection is, definitionally, deemed to be, but also on premises concerning what anything adapted and complex enough to count as organisms will be like, whether here on earth or anywhere else where they have been produced.[39]

What I ask to be conceded is thus merely that some answers to these questions leave the answers to the others open. If selection is judged to exist but not to be competent, then that rules out responsibility; but if it is judged to exist, that leaves competence as an open question, and if it is judged competent, that leaves responsibility open. For, by competence is meant competence for the sorts and sizes of change that the whole past course of evolution includes; and by responsibility is meant responsibility for these past changes. Obviously, if any selection exists in nature, it is having some consequences and so is competent and responsible for those consequences, but this does not imply competence and responsibility for evolution in the senses needed here.

It should be stressed that in this century, as in the last, the disagreements have often been over matters of more or less rather than all or none. Neutralists, today, agree with their selectionist opponents that selection exists; but they think that there is less of it. They agree that it can produce and has produced some major changes. They agree with selectionists that manifestly adaptive structures such as vertebrate limbs should be credited to evolution by selection. They doubt, however, that selection can bring about changes as quickly as drift sometimes can; they also think drift responsible for much past evolution of molecular structures. Equally, for their part, selectionists admit some drift as existing and even as competent and responsible for some evolutionary changes.[40]

The controversies between selectionists and neutralists have thrown into a clear light the relations among various answers to existence, competence, and responsibility questions. The controversies themselves, and the acceptance on both sides that empirical findings bear decisively if not conclusively on their resolution, undermine entirely any notion that natural selection as a theory of evolution is somehow forced on us a priori because it is merely tautological in the truth-functional sense of a logic text. If any such version of the tautological view were correct, selectionists and neutralists should have sought a verdict on their dispute using only a lexicon of definitions and a table of truth functions. As has rightly been observed, if selectionism is tautologous, then neutralism is contradictory.[41] Any historical and philosophical analysis of natural selection as a theory of evolution must make plain why these recent controversies have not been about whether to prefer a tautology over a contradiction.

We may start by considering the existential issue. Here we should distinguish modest claims from sweeping ones, specific claims from general ones, and claims for one level from those for others. Thus someone might hold not merely that there is some selection going on, but that for most heritable traits there exist organism-environment interactions making those traits causally relevant to reproductive success. And such a general claim for intrapopulational selection might be matched by equivalent ones for interspecific selection, for example.

The generality of such a claim allows for it to be supported empirically in many ways. One might point to particular instances where selectional causation is established, and then argue that these instances are typical.[42] Or, more in the manner of Darwin, one might argue that, from what is known generally concerning hereditary variation and concerning life, breeding, and death in natural conditions, it is probable that there is plenty of selection going on. And both lines have been taken, as can be seen from any recent survey of selectionist work.

The quantificational form of any claim may have implications for its testability: its verifiability or its falsifiability. And two points, familiar especially from Popper's teachings, should be introduced here.

The first, an uncontroversial corollary of what entailment includes, is a quite general one about falsification and verification. It is that one can falsify any statement deductively by falsifying what it entails, that is, by falsifying what can be deduced from it, but not verify it deductively by verifying what it entails, while, conversely, one can verify a statement deductively by verifying what entails it, but not falsify it by falsifying what entails it.

The second point, far from uncontroversial in its entirety, comes when Popper applies this first one to law statements. Law statements, he holds, have unrestricted universal form; the law that all metals expand when heated is universal and unrestricted in that it is about all metals and about them wherever and whenever they are, here today or on Mars a million years hence. Now, Popper insists that we cannot learn the truth of such a law statement through deductively valid inferences from any statements that we can verify. For we can verify only statements about the existence of finite numbers of instances of such a law, as we only have experiential access to these instances. However, we may have experiential access to a counterinstance, and so can learn the falsity of the law through a valid deductive inference from a statement of the existence of a counterinstance.[43]

For law statements there is thus, Popper insists, an asymmetry; they are deductively falsifiable but not deductively verifiable. Moreover, for unrestricted existentially qualified statements the reverse holds. The unrestricted statement that there exist centaurs, somewhere at some time, is deductively verifiable by finding some centaurs at a particular time and place, but not falsified by not finding some at any particular time and place. And human experience is always limited, Popper holds, to finite findings at particular times and places.

For our purposes what is decisive, as Popper himself emphasizes, is that for statements that are not unrestricted these asymmetries do not hold. A restricted statement of universal form may be both falsifiable and verifiable, as the statement is that all the finite number of snails in my finite garden during this finite summer day are hermaphrodite. This can be falsified or verified by checking its finite instances. Equally, the statement that there exist now some fish in some particular pond is an existentially quantified statement that, thanks to its restrictedness, is both falsifiable and verifiable through finite observational findings.

We see, then, that if universal or existential statements are restricted in these ways, how easily falsifiable or verifiable they are depends on their content, it being much easier to verify or falsify a claim that there is an elephant in the room now than a claim that there is a gravity wave in the room now, or the claim that there was an elephant in Scotland in Roman times.

Now, as Popper seems to appreciate, testing natural selection as a theory of evolution involves testing claims that are restricted.[44] Consider even the very general selectionist claim that for most heritable traits there exist some organism-environment interactions making them causally relevant to reproductive success. This claim is of mixed quantificational form; the logic of *most* (very different from that of *all*, obviously) and of *exists* are both in play. But also we have to appreciate that there are several restrictions implicit in the content of the claim. Organisms cannot interact now with past or future environments, only with present ones; they cannot interact instantaneously with distant present ones. That the interactions are causal means that they are subject to restrictions that hold for all causation; further, as organism-environment interactions they are implicitly restricted even more.

Nothing I have said suggests that testing existence claims on behalf of natural selection is possible in practice, much less easy. The selectionist-neutralist controversy has highlighted the difficulties. But testability in principle cannot be denied.

There are many diverse issues involved in the competence question cluster. We have to distinguish between real competence, that is, the abilities of natural selection as it is found to exist in nature, and the hypothetical competence of selection as specified in abstract idealized models or imaginary scenarios. There is the obvious distinction between the short-run abilities of selection as experimentally accessible to us and its long-run powers. Again, very diverse questions can be asked about the ability of selection to cause equilibrium states or to cause change, to produce a particular sort of structure, such as the python's jaw, or to produce very general trends in adaptive diversification.

What is most instructive about the answering of all such questions is that nothing is thought to be true of all the workings of natural selection—except the definitional truth that they all include the conditions necessary and sufficient for selection to occur. In that sense there is no law of natural selection, no law of action for this cause.

One should not be surprised by this lack of law. For we would not expect all the cases of causal bias in the differential reproduction of hereditary variants to have anything in common that could be summarized in a statement of law, any more than we would expect all the cases of causal bias in any other sort of process to do so. A famous attempt to legislate for natural selection, Fisher's so-called fundamental theorem, only confirms this point. The proportionality it asserts between genetic variance in fitness and rate of increase in fitness only holds under some conditions, and so is not equivalent to the proportionalities asserted for gravitational attraction by Newton. As for the comparison Fisher made between his theorem and the second law of thermodynamics, this comparison is weakened on the same account. The reasons for his theorem not holding when it does not are not matched by any equivalent limitation on the applicability of the second law of thermodynamics.[45]

It is tempting to say that there is no nondefinitional general statement of law to be made about all the workings of natural selection, because what its consequences are depends on what materials it is working with and what conditions it is working in. But this would be misleading in that it would suggest that natural selection is somehow an agency distinguishable from the hereditary materials and their interaction with their environmental conditions. And, of course, that is just what natural selection consists of—the processes of interaction—and that is why there can be no such general statement of law.

This lack of law does not make it impossible in principle to have well-confirmed generalizations about the workings of selection in specified ranges of conditions and with genetic systems of specified properties. And the contribution here of mathematical theory and experimental inquiry seems to be much as it is in most physical sciences.

Mathematics can provide the means for consequence extractions. And in this role it leaves open the status of the premises, assumptions, or axioms whose consequences it extracts. These premises can be completely conjectural or they can be statements of more or less well-confirmed generalizations about what is found in nature. In either case, the issue of responsibility for past evolution may be left untouched. Fisher knowingly invoked all the phrases associated with the old *vera causa* ideal when he explained, in the preface to his *Genetical Theory of Natural Selection*, what his book was intended to do. For it was to develop, mathematically, the genetical theory of natural selection without reference to the responsibility issue. Indeed, he emphasized in correspondence that most of his book would stand if the world turned out to be created a few thousand years ago as in the Bible account.[46]

Many writers have dwelt on the deductive structures that can be given to the theory of natural selection. But what is meant and not meant by this has to be kept in mind. One may deduce a consistent differential reproduction in a postulated system of suitably variable, reproductive entities. In that sense a formal modeling of natural selection can be exhibited as a theorem derivable from an appropriate set of axioms.[47] Again, one can postulate that the conditions of Hardy-Weinberg equilibrium are satisfied by some Mendelian system, and one can then add supplementary specifications as to a coefficient of consistent differential reproduction, and so, much as in classical rational mechanics, deduce the equations for the resultant change.[48]

The fact that deductive techniques of consequence derivation have such application to formal representations of selective processes does not mean that in the defining of natural selection, and in the asking and answering of existence, competence, and responsibility questions about it, biologists are matching closely what is done by physicists working with a theory, such as Newtonian gravitation theory, that has a force law intrinsic to its very formulation. However, framing and testing generalizations about the competence of natural selection is not made impossible by the lack of such a force law analogue. A causal process can have testable causal consequences, and the theory of such a process can have derivable logical consequences, notwithstanding that lack.

Most natural selection theorists have, of course, wanted to go beyond questions about consequences, causal or logical, to the responsibility issue. They have sought to establish whether or not natural selection really did it, or at least most of it.

To establish this, more is required than merely making it plausible that selection could have caused evolution. Reconciling the supposition that it did with the fossil record or biogeographical data is not enough. Beyond that, from Darwin on, upholders of the theory have sought to show the explanatory superiority of that supposition over rival theories, in relation to such facts, and thereby to confirm that supposition itself. Insistence on some such link between explanation and confirmation has been a commonplace for centuries, and is not distinctive of biology, much less evolutionary biology. Nor is there a source of difficulty here that is sometimes thought to be peculiarly damaging to Darwinian explanations. For most structural or functional characters, it is easy to construct, imaginatively, some selectionist scenario for its production as an adaptation. Such scenarios have been deprecatingly dubbed "Just So" stories in recollection of Kipling's tale as to how, for instance, the leopard got his spots.[49] But, as the deprecators have appreciated, there is little that is surprising or distinctive about natural selection in the way it lends itself to such exercises of imaginative ingenuity. To have any theory supply a reasonably confirmed rather than merely a conjectured explanation, it has to be supplemented with auxiliary assumptions, if only assumptions as to initial conditions, and these assumptions must themselves be independently confirmed from their own evidence.

In the standard cases, such as industrial melanism in moths or sickle cell anemia in man, the requisite auxiliary assumptions about the genetics of the variation and ecology of the interactions with the environment are evidentially pretty well secured. Securing such assumptions is never easy, and it is conspicuous that in very few cases indeed has this been done. There is a vast way to go if biologists are ever to have for the evolution of reptiles from their fish ancestors any selectionist explanation that is on a par, evidentially, with what they have for the melanism and anemia cases. But moving further toward such a goal is possible in principle, if often impossibly difficult in practice; and this shows that responsibility theses on behalf of natural selection are empirical theses, testable in principle.

14

In taking natural selection as a theory of evolution conforming to the old *vera causa* ideal, we are led, therefore, to appreciate the tremendous complexities inherent in any sophisticated articulation and thoroughgoing acceptance of the theory. Simple-minded though it may be, the approach suggested by that old *vera causa* ideal can show us clearly why such acceptance cannot be reduced to anything at all simple, such as "belief in natural selection or the survival of the fittest." The answers to the existential, competence, and responsibility questions involved in accepting the theory have always been highly complex answers, conceptually and evidentially, for reasons arising from the distinctive probabilistic, causal, and empirical character of the theory itself, as it has descended historically from Darwin's own exposition. Inevitably, many writers on the theory, even today, may think their analytical obligations to the theory can be largely met by formulating some single propositional expression of a "law" or "principle" of natural selection, exhibiting its "structure" and then arguing for its "status" as "falsifiable" or "nontautologous" or whatever. But such exercises canot be counted as seriously engaging the theory as it has been developed, conceptually and evidentially, over the last century and a half. Equally, anyone working in some domain beyond evolutionary biology, whether linguistics, epistemology, or immunology, who wishes to represent his theorizing as Darwinian or selectionist, in some serious sense, should always be asking whether there are in his proposals any significant and not merely superficial analogues to the definitional, existential, competence, and responsibility issues that biologists have always had to face in making up their minds about natural selection.

This last declaration may seem to presume that there has always been a consensus among selectionists in biology over all fundamental matters. But, on the contrary, one virtue of approaching the theory from the old *vera causa* ideal is that we can see how the theory could be embraced and developed by people with strikingly contrasting metaphysical and ideological orientations.

Consider for a moment what diverse metaphysical and ideological attitudes toward chance itself are found within the Darwinian tradition. Darwin's own attitude was ultimately negative, in that he never saw the chanciness of chance variation as itself making a positive contribution to adaptation or progress. For him the decisive contrast was between the chanciness in the generation of chance variation and the determinacy in its populational fate insofar as that was directed by selection. Here, then, were echoes of all those schemes, such as Plato's in the *Timaeus*, with Manichaean affinities. Variation, as chancy, is wild, chaos, error, so many failures of orderly replication, adding up to nothing when left to themselves, while selection is the cosmogonic, quasi-designing shaper of this otherwise helpless material.

Among dominant figures in our century, Fisher can be seen as largely perpetuating similar attitudes. Moved by Christian sympathies and thermodynamical preoccupations as well as eugenic zeal, he gave us a view of the living world as subject not only, as the inanimate world is, to the degenerative tendency entailed by the second law of thermodynamics but also to another, counterdegenerative agency, natural selection, so that this natural selection, although formally comparable, according to Fisher, with the second law, is, cosmologically, to be contrasted with it, as he emphasizes quite explicitly. Thus, for Fisher, gene mutations, as errors in gene replication, and drift, as erroneous genetic wanderings in populations, can never add up to anything by themselves except bigger errors and wider wanderings. So, on the variation side, where there rule such disordering tendencies as the living world shares with the inanimate world, small is beautiful; for small variations can be most precisely shaped by selection and made into what they cannot make themselves. To have small mutations in large populations is thus the optimal combination, for then drift is minimal and variation is plentiful, while even small selection pressures can control the outcome. Thus, too, was Fisher attracted to indeterminism, as required by free will and creativity in man. For, to exercise the possibility of free will and creativity is to master and direct, as natural selection does, what is, insofar as it is indeterminate, merely a material contribution to any adaptive progress, such as human eugenic advances.[50]

By contrast, Wright, drawing on his Wundtian panpsychism and his shorthorn cattle breeding strategies, has seen chance processes as able, by virtue of their very chanciness, to contribute to the overcoming of the limitations arising from the determinacy of selection. For Wright, the stochastic exploration, by drift, of the field of variational possibilities can lead to a population moving to new adaptive peaks through selectional forces that would not otherwise have been brought into play.[51] So, a population structure that entails such drift is in the long run more optimal than one that does not. One can even think of such a population structure as a quasi-designed error machine.

Likewise, sexual reproduction itself, when it includes meiosis, with its randomized combinatorial consequences, can be conceived of as a well-designed mechanism for making replicative mistakes at the individual level. So, from such reflections, a much more positive attitude toward chance can arise, one that allows for a teleology of chance. For, if the mechanisms, such as meiosis, that now supply natural selection with its material, have themselves evolved by earlier selection, then the benefits of chanciness in variation, when changing through natural selection in an unpredictably changing world, are decisive for any understanding of how that evolution took place.[52]

Finally, an approach to natural selection through the old *vera causa* ideal does not have to deny that the theory has often been championed and elaborated by people, such as Pearson, who have followed Comtean and other positivist precedents in repudiating the very notion that science can and should construct causal rather than correlational theories and laws. The contribution of Pearson and other biometrical correlationists included new quantitative analyses of variation and differential reproduction. However, such measurement and computation analyses could proceed purely correlationally, avoiding issues of causal discrimination and relevance, only as long as questions about adaptation and population size in nature were not confronted. For, when biologists of the Pearsonian persuasion, as Weldon was, sought to instantiate empirically their claim for the effectiveness of selection in changing structures gradually and adaptively, they appealed to the sensitivity and intricacy of organism-environment interactions, as providing adequate causes, and so explanations, for the preservation of particular structures in natural populations; and in making that appeal they never claimed to reduce all their causal judgments to correlational ones.[53] It should now be plain why such a reductive claim was not made. Statistics may be a science of probabilities and correlation may be a statistical measure; but it does not follow that natural selection insofar as it is probabilistic is correlational rather than causal. A historical view of the theory confirms what philosophical examinations and scientific expositions suggest; the theory arose as a probabilistic contribution to causal science, not as a rebellious rejection of causation in favor of science without causes.[54]

A proper appreciation of Darwin's place in the probabilistic revolution can be enhanced, therefore, by recalling our own historical location. From Comte, in Darwin's day, to Hempel, in our own, empiricism in its more positivistic forms has construed the main questions about evidence and explanation in science as questions about universal statements of law rather than existential claims for causation. Darwin worked in an older, *vera causa*, tradition of empiricism than the Comtean one. We have had now two decades of philosophers priding themselves on seeing the limitations in any positivistic form of empiricism. So, perhaps, the present is an appropriate time to develop further the original interpretation of natural selection as a causal and empirical theory.

Notes

1. Two books by Michael Ruse introduce many of the issues and much of the literature: *Is Science Sexist? And Other Problems in the Biomedical Sciences* (Dordrecht and Boston: D. Reidel, 1981) and *Darwinism Defended. A Guide to the Evolution Controversies* (Reading, MA: Addison-Wesley, 1982). An invaluable collection of papers is now available: Elliott Sober, ed., *Conceptual Issues in Evolutionary Biology. An Anthology* (Cambridge, MA: The MIT Press, 1984). See also Sober's *The Nature of Selection* (Cambridge, MA: MIT Press, 1984), a masterly analysis of natural selection theory.

2. Peter Bowler brings out well the usefulness of analyzing enduring clusters of questions regarding evolution in his excellent study, *The Eclipse of Darwinism: Anti-Darwinian Evolution Theories in the Decades around 1900* (Baltimore: Johns Hopkins University Press, 1983). In doing so he acknowledges direct debts to Stephen Gould.

3. For a lucid introduction, see R. Giere, *Understanding Scientific Reasoning* (New York: Holt, Rinehart and Winston, 1979).

4. S. Haack, *Philosophy of Logics* (Cambridge: Cambridge University Press, 1978), introduces the complexities in Quine's position.

5. See, especially, W. Salmon, "Causality: Production and Propagation," *PSA 1980: Proceedings of the Biennial Meetings of the Philosophy of Science Association* (Philosophy of Science Association: East Lansing, Michigan, 1981) vol. 2, pp. 49–69, and earlier papers cited there.

6. Giere, *Understanding Scientific Reasoning* and "Causal Systems and Statistical Hypotheses," in L. J. Cohen, ed., *Applications of Inductive Logic* (Oxford: Oxford University Press, 1980). Giere's views are discussed in relation to natural selection theory in an important trio of articles: E. Sober, "Frequency-Dependent Causation," *The Journal of Philosophy*, 79 (1982), 247–253; John Collier, "Frequency-Dependent Causation: A defense of Giere," *Philosophy of Science*, 50 (1983), 618–625; and R. Giere, "Causal Models with Frequency Dependence," *The Journal of Philosophy*, 81 (1984), 384–391. The contrast between mathematical and causal considerations is insisted on in E. Sober and R. Lewontin, "Artifact, Cause, and Genic Selection," *Philosophy of Science*, 49 (1982), 147–176, an article reprinted in Sober, ed., *Conceptual Issues in Evolutionary Biology* (note 1).

7. Ruse prefers the received view, for example, in the essay "The Structure of Evolutionary Theory" in his collection *Is Science Sexist?* (note 1). J. Beatty argues for the semantic view in "What's Wrong with the Received View of Evolutionary Theory?" *PSA 1980: Proceedings of the Biennial Meetings of the Philosophy of Science Association* (Philosophy of Science Association: East Lansing, Michigan, 1981) vol. 2, pp. 397–426.

8. M. J. S. Hodge, "The Structure and Strategy of Darwin's 'Long Argument,'" *The British Journal for the History of Science*, 10 (1977), 237–246.

9. L. Laudan, "Thomas Reid and the Newtonian Turn of British Methodological Thought," in his *Science and Hypothesis: Historical Essays on Methodology* (Dordrecht and Boston: D. Reidel, 1981), pp. 86–110.

10. R. Laudan, "The Role of Methodology in Lyell's Science," *Studies in History and Philosophy of Science*, 13 (1982), 215–249; M. J. S. Hodge, "Darwin and the Laws of the Animate Part of the Terrestrial System (1835–1837): On the Lyellian Origins of His Zoonomical Explanatory Program," *Studies in History of Biology*, 6 (1982), 1–106.

11. *Notebooks* B–E were published in G. De Beer, M. J. Rowlands, and B. M. Skramovsky, "Darwin's Notebooks on the Transmutation of Species," *Bulletin of the British Museum (Natural History). Historical Series*, 2 (1960), 27–200, and 3 (1967), 129–176; and *Notebooks* M–N were published by Paul Barrett in H. E. Gruber, *Darwin on Man* (New York: E. P. Dutton, 1974).

12. On matters concerning reversibility in this period, see S. G. Brush, *The Kind of Motion We Call Heat: A History of the Kinetic Theory of Gases in the 19th Century* (New York: North-Holland, 1976) and "Irreversibility and Indeterminism: Fourier to Heisenberg," *The Journal of the History of Ideas*, 37 (1976), 603–630.

13. For a more detailed account of Darwin's pursuit of these theories, see M. J. S. Hodge and D. Kohn, "The Immediate Origins of Natural Selection," in D. Kohn, ed., *The Darwinian Heritage. A Centennial Retrospect* (Princeton: Princeton University Press, 1986).

14. Darwin's early thinking on chance and chances has been discussed in S. S. Schweber, "The Origin of the *Origin* revisited," *The Journal of the History of Biology*, 10 (1977), 229–316, and in E. Manier, *The Young Darwin and His Cultural Circle* (Dordrecht and Boston: D.

Reidel, 1978), pp. 117–122. For a comprehensive survey of notions of chance and chances in the biology of this period, see O. B. Sheynin, "On the History of the Statistical Method in Biology," *Archive for the History of Exact Sciences*, 22 (1980), 323–371.

15. M. J. S. Rudwick, "Charles Lyell's Dream of a Statistical Paleontology," *Paleontology*, 21 (1978), 225–244, is an invaluable study.

16. *Notebook* C, MS pp. 61–63 (note 11).

17. See, further, Hodge and Kohn, "The Immediate Origins of Natural Selection" (note 13).

18. See *Notebook* M in Gruber, *Darwin on Man* (note 11).

19. For the details, again, see Hodge and Kohn, "The Immediate Origins of Natural Selection" (note 13).

20. See his *Notebook* D (note 11).

21. Compare, for example, the discussion of congenital variation in Darwin's *Notebook* C (note 11), with the view of chance varieties in J. C. Prichard, *Researches into the Physical History of Mankind*, 2nd ed., 2 Vols. (London: T. Hughes, 1826), Vol. 2, p. 548. Whether Darwin had read Prichard by this time is not clear.

22. S. S. Schweber, "Aspects of Probabilistic Thought in Great Britain: Darwin and Maxwell," in Heidelberger, Krüger, and Rheinwald, eds., *Probability since 1800: Interdisciplinary Studies of Scientific Development* (Bielefeld: Universität Bielefeld, 1983), pp. 41–97.

23. Hodge and Kohn, "The Immediate Origins of Natural Selection" (note 13), gives a fuller account of these metaphors and analogies.

24. See, for example, Darwin's notes on this, probably from March, 1839, published in Gruber, *Darwin on Man*, p. 420 (note 11).

25. On the themes of this paragraph, see D. Hull, *Philosophy of Biological Science* (Englewood Cliffs, NJ: Prentice Hall, 1974). The view that evolutionary theory is not predictive is effectively countered in two articles by M. B. Williams: "Falsifiable Predictions of Evolutionary Theory," *Philosophy of Science*, 40 (1974), 518–537, and "The Importance of Prediction Testing in Evolutionary Biology," *Erkenntnis*, 17 (1982), 291–306.

26. See, for example, F. J. Ayala and J. A. Kiger, *Modern Genetics* (Menlo Park, CA: Benjamin-Cummings, 1980), pp. 657–658, and D. J. Futuyma, *Evolutionary Biology* (Sunderland, MA: Sinauer, 1979), pp. 300–301, for careful attention to the definitional issue. For complementary emphasis on physical properties and property differences, see Sober and Lewontin, "Artifact, Cause, and Genic Selection" (note 6).

27. On the importance of this distinction, see E. Sober, "Force and Disposition in Evolutionary Theory," in C. Hookway, ed., *Minds, Machines and Evolution. Philosophical Studies* (Cambridge: Cambridge University Press, 1984), pp. 43–62.

28. See the incisive analysis of discriminate and indiscriminate samplings in selection and drift in John Beatty, "Chance and Natural Selection," *Philosophy of Science*, 51 (1984), 183–211.

29. S. Wright, "Classification of the Factors of Evolution," *Cold Spring Harbor Symposium on Quantitative Biology*, 20 (1955), 16–24.

30. R. Lewontin, "Testing the Theory of Natural Selection," *Nature*, 236 (1972), 181–182.

31. See Beatty's critique, in his "Chance and Natural Selection" (note 28), of a game proposed by Eigen and Winkler.

32. J. Maynard Smith raises such questions about such proposals in "Current Controversies in Evolutionary Biology," in M. Grene, ed., *Dimensions of Darwinism. Themes and Counter-Themes in Twentieth-Century Evolutionary Theory* (Cambridge: Cambridge University Press, 1983), pp. 273–286.

33. D. Hull, "Individuality and Selection," *Annual Review of Ecology and Systematics*, 11 (1980), 311–332—see, especially, pp. 317–318.

34. There is now a sizable literature devoted to adaptation. Two influential contributions by biologists are G. C. Williams, *Adaptation and Natural Selection. A Critique of Some Current Evolutionary Thought* (Princeton: Princeton University Press, 1966), and R. Lewontin, "Adaptation," *Scientific American*, 239 (1978), 212–230. See, further, R. N. Brandon, "Adaptation and Evolutionary Theory," *Studies in History and Philosophy of Science*, 9 (1978), 181–206. Brandon's article is in Sober, ed., *Conceptual Issues in Evolutionary Biology* (note 1), together with other related discussions of adaptation, including one by Lewontin. Many recent proposals are discussed in valuable analyses by R. Burian, "Adaptation," in M. Grene, ed., *Dimensions of Darwinism* (note 32), pp. 241–272, and by C. B. Krimbas, "On Adaptation, Neo-Darwinian Tautology and Population Fitness," *Evolutionary Biology*, 17 (1984), 1–57.

35. The literature on fitness also grows apace. In relating my suggestions to others, it would be best to start, perhaps, with R. Brandon and J. Beatty, "Discussion: The Propensity Interpretation of 'Fitness'—No Interpretation Is No Substitute," *Philosophy of Science*, 51 (1984), 342–357. There is much in the propensity interpretation, with its application of the propensity view of probability associated with Popper, that I can agree with; but a confusion surely remains precisely where we need to be unequivocal: are fitnesses as propensities causal and so explanatory or are they not? The difficulties inherent in population-genetic fitness definitions and measurements are discussed in R. Lewontin, *The Genetic Basis of Evolutionary Change* (New York: Columbia University Press, 1974); R. Dawkins gives an instructive attempt to relate population-geneticists' and ecologists' notions of fitness in his book *The Extended Phenotype* (San Francisco:Freeman, 1982), Chapter 10. For a fuller discussion of fitness and the reference class relativity of probabilities, see S. Mills and J. Beatty, "The Propensity Interpretation of Fitness," *Philosophy of Science*, 46 (1979), 263–286, reprinted in E. Sober, ed., *Conceptual Issues in Evolutionary Biology* (note 1). See also A. Rosenberg, "Fitness," *Journal of Philosophy*, 80 (1983), 457–473, and E. Sober, "Fact, Fiction and Fitness: A Reply to Rosenberg," *Journal of Philosophy*, 81 (1984), 372–383.

36. See, for example, S. J. Gould, *Ever Since Darwin* (Harmondsworth: Penguin Books, 1978), pp. 39–48, reprinted in E. Sober, ed., *Conceptual Issues in Evolutionary Biology* (note 1).

37. See Collier, "Frequency-Dependent Causation: A Defense of Giere" (note 5).

38. I can agree, then, with many points made by H. C. Byerly against E. S. Reed's proposal that natural selection is a law of nature; but I cannot accept his own suggestion that it is a framework principle: E. S. Reed, "The Lawfulness of Natural Selection," *The American Naturalist*, 118 (1981), 61–71; and H. C. Byerly, "Natural Selection as a Law: Principles and Processes," *The American Naturalist*, 120 (1983), 739–745. I am in full agreement with Beatty when he insists that there is no law or principle of natural selection. See his discussion in "What's Wrong with the Received View of Evolutionary Theory" (note 7); see also on this question, in the same book, R. N. Brandon, "A Structural Description of Evolutionary Theory."

39. R. Dawkins, "Universal Darwinism," in D. S. Bendall, ed., *Evolution from Molecules to Men* (Cambridge: Cambridge University Press, 1983), pp. 403–425.

40. Futuyma, *Evolutionary Biology* (note 26), provides a good introduction here. See, also, R. Lewontin, The Genetic Basis of Evolutionary Change (note 35).

41. M. Ruse, *Darwinism Defended*, pp. 140–141 (note 1).

42. R. Lewontin, "Testing the Theory of Natural Selection" (note 30).

43. For these themes in Popper, the best introductions are his own books, *The Logic of Scientific Discovery* (London: Hutchinson, 1959) and *Conjectures and Refutations* (New York: Harper and Row, 1968).

44. Popper's views on natural selection are discussed in Ruse, *Is Science Sexist?* (note 1).

45. Fisher's original account is in his *Genetical Theory of Natural Selection* (Oxford: Clarendon Press, 1930), a revised edition of which was published by Dover in 1958. On the conditions

for the theorem holding, see J. R. G. Turner, "Changes in Mean Fitness under Natural Selection," in K. Kojima, ed., *Mathematical Topics in Populations Genetics* (New York: Springer-Verlag, 1970), pp. 32–78; G. R. Price, "Fisher's 'Fundamental Theorem' Made Clear," *Annals of Human Genetics (London)*, 36 (1972), 129–140; and R. C. Olby, "La Théorie Génétique de la Selection Naturelle Vue par un Historien," *Revue de Synthèse: Actes du Colloque R. A. Fisher et L'Histoire de la Génétique des Populations*, 103–104 (1981), 251–289.

46. Thus Fisher to J. S. Huxley, 6 May 1930: "... if I had had so large an aim as to write an important book on Evolution, I should have had to attempt an account of very much work about which I am not really qualified to give a useful opinion. As it is there is surprisingly little in the whole book that would not stand if the world had been created in 4004 B.C., and my primary job is to try to give an account of what Natural Selection *must* be doing, even if it had never done anything of much account until now." J. H. Bennett, ed., *Natural Selection, Heredity, and Eugenics. Including Selected Correspondence of R. A. Fisher with Leonard Darwin and Others* (Oxford: Clarendon Press, 1983), p. 222.

47. M. B. Williams, "Deducing the Consequences of Evolution: A Mathematical Model," *The Journal of Theoretical Biology*, 29 (1970), 343–385, and "The Logical Status of the Theory of Natural Selection and other Evolutionary Controversies," in M. Bunge, ed., *The Methodological Unity of Science* (Dordrecht: D. Reidel, 1973), pp. 84–101. Hull, in his *Philosophy of Biological Science* (note 25), brings out the advantages and limitations of Williams's approach. See also Sober, "Fact, Fiction and Fitness" (note 35).

48. M. Ruse, "The Structure of Evolutionary Theory," in his *Is Science Sexist?* (note 1).

49. S. J. Gould and R. C. Lewontin, "The Spandrels of San Marco and the Panglossian Paradigm: A Critique of the Adaptationist Programme," *Proceedings of the Royal Society of London, Series B*, 205 (1977), 581–598.

50. Fisher's views on drift and related issues are discussed authoritatively, in this volume, in J. R. G. Turner, "Random Genetic Drift, R. A. Fisher, and the Oxford School of Ecological Genetics." For Fisher on determinism, free will, and associated subjects, see the correspondence published in J. H. Bennett, *Natural Selection, Heredity, and Eugenics* (note 46), and two published papers: "Indeterminism and Natural Selection," *Philosophy of Science*, 1 (1934), 99–117; "The Renaissance of Darwinism," *The Listener*, 37 (1947), 1001. Both are in the *Collected papers of R. A. Fisher*, edited by J. H. Bennett (Adelaide: University of Adelaide, 1971–1974). See further J. R. G. Turner, "Fisher's Evolutionary Faith and the Challenge of Mimicry," *Oxford Surveys in Evolutionary Biology* 2 (1985): 159–196.

51. The classic sources are S. Wright, "Evolution in Mendelian Populations," *Genetics*, 16 (1931), 97–159, and "The Roles of Mutation, Inbreeding, Crossbreeding and Selection in Evolution," *Proceedings of the VIth Congress of Genetics*, 1 (1932), 356–366. Wright's metaphysical views are instructively related to his biology in his "Biology and the Philosophy of Science," *The Monist*, 48 (1964), 265–90. W. B. Provine has a biography of Wright in preparation.

52. Here, issues are relevant that go back to H. J. Muller and, beyond him, to Weismann, and are now pursued in, for example, J. Maynard Smith, *The Evolution of Sex* (Cambridge: Cambridge University Press, 1978), and G. C. Williams, *Sex and Evolution* (Princeton: Princeton University Press, 1975).

53. B. J. Norton, "The Biometric Defense of Darminism," *Journal of the History of Biology*, 6 (1973), 283–316.

54. (Added in proof) On causation, probability, and explanation, see now W. C. Salmon, *Scientific Explanation and the Causal Structure of the World* (Princeton: Princeton University Press, 1984). For all of these topics the reader should now see Sober's *The Nature of Selection* (note 1).

INDEX

Printed and bound by CPI Group (UK) Ltd, Croydon, CR0 4YY

21/10/2024

01777095-0017